Lecture Notes in Artificial Intelligence   8436

Subseries of Lecture Notes in Computer Science

LNAI Series Editors

Randy Goebel
University of Alberta, Edmonton, Canada
Yuzuru Tanaka
Hokkaido University, Sapporo, Japan
Wolfgang Wahlster
DFKI and Saarland University, Saarbrücken, Germany

LNAI Founding Series Editor

Joerg Siekmann
DFKI and Saarland University, Saarbrücken, Germany

Marina Sokolova   Peter van Beek (Eds.)

Advances in Artificial Intelligence

27th Canadian Conference
on Artificial Intelligence, Canadian AI 2014
Montréal, QC, Canada, May 6-9, 2014
Proceedings

 Springer

Volume Editors

Marina Sokolova
University of Ottawa
Faculty of Medicine and
School of Electrical Engineering and Computer Science
Dept. of Epidemiology & Community Medicine, Room 3105
451 Smyth Road
Ottawa, ON, Canada K1H 8M5
E-mail: sokolova@uottawa.ca

Peter van Beek
University of Waterloo
Cheriton School of Computer Science
200 University Avenue West
Waterloo, ON, Canada N2L 3G1
E-mail: vanbeek@cs.uwaterloo.ca

ISSN 0302-9743
ISBN 978-3-319-06482-6
DOI 10.1007/978-3-319-06483-3
Springer Cham Heidelberg New York Dordrecht London

e-ISSN 1611-3349
e-ISBN 978-3-319-06483-3

Library of Congress Control Number: 2014936254

LNCS Sublibrary: SL 7 – Artificial Intelligence

Typesetting: Camera-ready by author, data conversion by Scientific Publishing Services, Chennai, India

Printed on acid-free paper

Springer is part of Springer Science+Business Media (www.springer.com)

Preface

The 27th Canadian Conference on Artificial Intelligence (AI 2014) continued
the success of its predecessors. The conference was held in Montréal, Québec,
Canada, during May 6–9, and was collocated with the 40th Graphics Interface
Conference (GI 2014) and the 11th Conference on Computer and Robot Vision
(CRV 2014).

AI 2014 attracted 94 submissions from Canada and around the world. Each
submission was thoroughly reviewed by the Program Committee, and for the
final conference program and for inclusion in these proceedings, 22 regular papers
were accepted, for an acceptance rate of 23.4%. Regular papers were allocated
12 pages in the proceedings. In addition, 18 short papers were accepted. Short
papers were allocated six pages in the proceedings. The volume also presents
three papers from the Graduate Student Symposium.

We invited four distinguished researchers to provide their insights on the
state-of-art in Artificial Intelligence: Doina Precup, McGill University, and Yoshua
Bengio, Université de Montréal, gave tutorial talks and Stan Matwin, Dalhousie
University, and Khaled El Emam, University of Ottawa and Privacy Analytics
Inc., gave the conference talks. The AI 2014 program greatly benefited from
these exceptional contributions.

The organization of such a successful conference benefited from the collaboration of many individuals. We would like to thank the members of the Program
Committee and external reviewers for providing significant and timely reviews.
To manage the submission and reviewing process, we used the EasyChair Conference system. We thank Aminul Islam, Cristina Manfredotti, Shamima Mithun
for organizing and chairing the Graduate Student Symposium and its Program
Committee. We express our appreciation to Alain Désilets for organizing and
chairing the Industry Track session. We would like to take this opportunity to
thank Atefeh Farzindar, the AI/GI/CRV general chair, local arrangement chairs
Philippe Langlais and Guillaume-Alexandre Bilodeau, and Michel Gagnon, Local
AI contact for handling the local arrangements and registration.

AI 2014 was sponsored by the Canadian Artificial Intelligence Association
(CAIAC). We thank Nathalie Japkowicz and members of the CAIAC Executive
Committee for all their efforts in making AI 2014 a successful conference.

May 2014

Marina Sokolova
Peter van Beek

Organization

The 27th Canadian Conference on Artificial Intelligence (AI 2014) was sponsored by the Canadian Artificial Intelligence Association (CAIAC), and held in conjunction with the 40th Graphics Interface Conference (GI 2014) and the 11th Conference on Computer and Robot Vision (CRV 2014).

General Chair

Atefeh Farzindar   NLP Technologies Inc. & Université
                                      de Montréal, Canada

Local Arrangements Chair

Philippe Langlais   Université de Montréal, Canada
Guillaume-Alexandre Bilodeau   École Polytechnique de Montréal, Canada

Local AI Contact Person

Michel Gagnon   École Polytechnique de Montréal, Canada

Program Co-Chairs

Marina Sokolova   University of Ottawa & Institute for Big Data
                                      Analytics, Canada
Peter van Beek   University of Waterloo, Canada

Graduate Student Symposium Co-Chairs

Aminul Islam   Dalhousie University, Canada
Cristina Manfredotti   Université Pierre et Marie Curie, France
Shamima Mithun   Concordia University, Canada

AI Industry Track Chair

Alain Désilets   Alpaca Technologies, Canada

Program Committee

Esma Aimeur	Université de Montréal, Canada
Xiangdong An	York University, Canada
Dirk Arnold	Dalhousie University, Canada
Ebrahim Bagheri	Ryerson University, Canada
Andrew Baker	University of Ottawa, Canada
Gábor Bartók	ETH Zurich, Switzerland
Virendra Bhavsar	University of New Brunswick, Canada
Narjes Boufaden	KeaText Inc., Canada
Scott Buffett	National Research Council Canada, Canada
Cory Butz	University of Regina, Canada
Eric Charton	École Polytechnique de Montréal, Canada
Colin Cherry	National Research Council Canada, Canada
David Chiu	University of Guelph, Canada
Lyne Da Sylva	Université de Montréal, Canada
Joerg Denzinger	University of Calgary, Canada
Chrysanne Dimarco	University of Waterloo, Canada
Chris Drummond	National Research Council Canada, Canada
Ahmed A.A. Esmin	Federal University of Lavras, Brazil
Jocelyne Faddoul	Saint Francis Xavier University, Canada
Atefeh Farzindar	Université de Montréal, Canada
Paola Flocchini	University of Ottawa, Canada
Michel Gagnon	Polytechnique Montreal, Canada
Yong Gao	UBC Okanagan, Canada
Ali Ghorbani	University of New Brunswick, Canada
Cyril Goutte	National Research Council Canada, Canada
Kevin Grant	University of Lethbridge, Canada
Howard Hamilton	University of Regina, Canada
Holger H. Hoos	University of British Columbia, Canada
Jimmy Huang	York University, Canada
Frank Hutter	University of Freiburg, Germany
Diana Inkpen	University of Ottawa, Canada
Ilya Ioshikhes	University of Ottawa, Canada
Aminul Islam	Dalhousie University, Canada
Christian Jacob	University of Calgary, Canada
Nathalie Japkowicz	University of Ottawa, Canada
Igor Jurisica	University of Toronto, Canada
Youssef Kadri	Privacy Analytics, Canada
Vlado Kešelj	Dalhousie University, Canada
Fazel Keshtkar	Southeast Missouri State University, USA
Svetlana Kiritchenko	National Research Council Canada, Canada
Ziad Kobti	University of Windsor, Canada
Grzegorz Kondrak	University of Alberta, Canada
Leila Kosseim	Concordia University, Canada
Adam Krzyżak	Concordia University, Canada

Additional Reviewers

Afra Abnar	Lydia Odilinye
Nima Aghaeepour	Alexandre Patry
Falah Al-Akashi	Armin Sajadi
Bradley Ellert	Mohammad Salameh
Alexandre Fréchette	Martin Scaiano
Chuancong Gao	Rushdi Shams
Behrouz Haji Soleimani	Abhishek Srivastav
Fadi Hanna	Craig Thompson
Amir Hossein Razavi	Milan Tofiloski
Magdalena Jankowska	Amine Trabelsi
Levi Lelis	Mauro Vallati
Guohua Liu	Lei Yao
Xiangbo Mao	Kui Yu
Erico Neves	Bartosz Ziółko

Sponsoring Institutions and Companies

Canadian Artificial Intelligence Association (CAIAC)
https://www.caiac.ca/

GRAND (Graphics, Animation and New Media) research network
http://grand-nce.ca/

Nana Traiteur par l'Assommoir, Montréal
http://www.nanatraiteur.com/

Grévin, Montréal
http://www.grevin-montreal.com

Polytechnique Montréal
http://www.polymtl.ca/

NLP Technologies
http://www.nlptechnologies.ca

CRIM
http://www.crim.ca/en/

Invited Presentations

Deep Learning of Representations
(Invited Tutorial)

Yoshua Bengio

Université de Montréal
yoshua.bengio@umontreal.ca

Abstract. Deep learning methods have been extremely successful recently, in particular in the areas of speech recognition, object recognition and language modeling. Deep representations are representations at multiple levels of abstraction, of increasing non-linearity. The success of machine learning algorithms generally depends on data representation, and we hypothesize that this is because different representations can entangle and hide more or less the different explanatory factors of variation behind the data. Although specific domain knowledge can be used to help design representations, learning with generic priors can also be used, and the quest for AI is motivating the design of more powerful representation-learning algorithms implementing such priors. This talk introduces basic concepts in representation learning, distributed representations, and deep learning, both supervised and unsupervised.

Biography

Yoshua Bengio is a Full Professor in the Department of Computer Science and Operations Research at Université de Montréal, head of the Machine Learning Laboratory (LISA), CIFAR Fellow in the Neural Computation and Adaptive Perception program, Canada Research Chair in Statistical Learning Algorithms, and he also holds the NSERC-Ubisoft industrial chair. His main research ambition is to understand principles of learning that yield intelligence. He teaches a graduate course in Machine Learning and supervises a large group of graduate students and post-docs. His research is widely cited (over 10,300 citations found by Google Scholar in 2012, with an H-index of 45). Yoshua is currently action editor for the Journal of Machine Learning Research, editor for Foundations and Trends in Machine Learning, and has been associate editor for the Machine Learning Journal and the IEEE Transactions on Neural Networks. He has also been very active in conference organization including Program Chair for NIPS'2008, General Chair for NIPS'2009, co-organizer (with Yann Le Cun) of the Learning Workshop, and co-founder (also with Yann Le Cun) of the International Conference on Representation Learning.

Artificial Intelligence Applications in Clinical Monitoring (Invited Tutorial)

Doina Precup

McGill University
dprecup@cs.mcgill.ca

Abstract. Health care has traditionally been one of the early adopters of artificial intelligence and machine learning technologies. AI methods have been used extensively in patient diagnosis, medical image analysis, disease outbreak detection and assistive technologies. However, most of the clinical monitoring, especially in emergency situations, is performed by physicians and nurses. In the last 10-15 years, however, clinical monitoring has also started to embrace the use of AI techniques, due to the fact that the sensing capabilities have improved, providing a flood of different signals which are more difficult to analyze by people and increase the variability in clinical decisions. Moreover, hospital crowding makes it harder for nurses and doctors to monitor patients continuously. This tutorial will describe AI approaches for bed-side clinical monitoring, with examples from labor monitoring and monitoring of cardio-respiratory patterns. The goal in these applications is to accurately predict an unfavourable clinical situation early enough that a clinical intervention can be enacted. An emerging theme will be the interplay and complementarity of AI methods and state-of-art signal processing.

Biography

Dr. Doina Precup is an Associate Professor in the School of Computer Science of McGill University. She earned her B.Sc. degree from the Technical University Cluj-Napoca, Romania (1994) and her M.Sc. (1997) and Ph.D. (2000) degrees from the University of Massachusetts Amherst, where she was a Fulbright fellow. Her research interests lie in the area of machine learning, with an emphasis on reinforcement learning and time series data, as well as applications of machine learning and artificial intelligence to activity recognition, medicine, electronic commerce and robotics. She currently serves as chair of the NSERC Discovery Evaluation Group for Computer Science.

Deploying Secure Multi-party Computation in Practice: Examples in Health Settings (Invited Talk)

Khaled El Emam

Privacy Analytics & University of Ottawa
kelemam@privacyanalytics.ca

Abstract. There is significant pressure to link and share health data for research, public health, and commercial purposes. However, such data sharing must be done responsibly and in a manner that is respectful of patient privacy. Secure multi-party computation (SCM) methods present one way to facilitate many of these analytic purposes. In fact, in some instances SCM is the only known realistic way allow some of these data disclosures and analyses to happen (without having to change the law to selectively remove privacy protections). This talk will describe two recent real-world projects where SMC was applied to address such data sharing concerns. The first was to measure the prevalence of antimicrobial resistant organism (e.g., MRSA) infections across all long term care homes in Ontario. A SMC system was deployed to collect data from close to 600 long term care homes in the province and establish a colonization and infection rate baseline. The second project pertains to securely linking large databases to allow de-duplication and secure look-up operations without revealing the identity of patients. This system performs approximate matching while maintaining a constant growth in complexity. In both of these cases a number of theoretical and engineering challenges had to be overcome to scale SCM protocols to operate efficiently and to transition them from the laboratory into practice.

Biography

Khaled El Emam is founder and CEO of Privacy Analytics Inc., a software company which develops de-identification tools for hospitals and registries to manage the disclosure of health information to internal and external parties. Khaled is also an Associate Professor at the University of Ottawa, Faculty of Medicine and the School of Information Technology and Engineering, a senior investigator at the Children's Hospital of Eastern Ontario Research Institute, and a Canada Research Chair in Electronic Health Information at the University of Ottawa. His main area of research is developing techniques for health data anonymization and secure disease surveillance for public health purposes.

Text Mining: Where Do We Come From, Where Do We Go? (Invited Talk)

Stan Matwin

Dalhousie University
stan@cs.dal.ca

Abstract. In this talk we will review text mining, a "hot" area of current AI research. There is little doubt that text is an important kind of data, and that it calls for special analysis and information extraction techniques. It has been stated that text data constitutes 80% of the overall data volume. In the last fifteen years, Machine Learning and Applied Computational Linguistics have come together to significantly contribute to intelligent text mining. In this talk we will review the history of the field and some of its more recent accomplishments, e.g. selected modern text mining techniques such as Conditional Random Fields and Latent Dirichlet Allocation. We will speculate where the field may be going. In particular, we will ponder the Big Data argument, suggesting that the ability to process huge data (and text) repositories limits, or even eliminates, the need for knowledge-intensive methods.

Biography

Stan Matwin is Canada Research Chair (Tier 1) in Visual Text Analytics at Dalhousie University. His other honours include Fellow of the European Coordinating Committee on Artificial Intelligence, Fellow of the Canadian Artificial Intelligence Association, and Ontario Champion of Innovation. He has held academic positions at many universities in Canada, the U.S., Europe, and Latin America including the University of Guelph, Acadia University, and the University of Ottawa where in 2011 he was named a Distinguished University Professor (on leave). In 2013, he moved to Dalhousie University, where he is a CRC and Director of the Institute for Big Data Analytics. Stan is recognized internationally for his work in text mining, applications of machine learning, and data privacy. He is author or co-author of more than 250 research papers. He is a former president of the Canadian Artificial Intelligence Association (CAIAC) and of the IFIP Working Group 12.2 (Machine Learning). Stan has significant experience and interest in innovation and technology transfer and he is one of the founders of Distil Interactive Inc. and Devera Logic Inc.

Table of Contents

Long Papers

Short Papers

Contributions from Graduate Student Symposium

A Novel Particle Swarm-Based Approach for 3D Motif Matching and Protein Structure Classification

Hazem Radwan Ahmed and Janice Glasgow

School of Computing, Queen's University at Kingston,
K7L 3N6 Ontario, Canada
{hazem,janice}@cs.queensu.ca

Abstract. This paper investigates the applicability of Particle Swarm Optimization (PSO) to motif matching in protein structures, which can help in protein structure classification and function annotation. A 3D motif is a spatial, local pattern in a protein structure important for its function. In this study, the problem of 3D motif matching is formulated as an optimization task with an objective function of minimizing the least Root Mean Square Deviation (*l*RMSD) between the query motif and target structures. Evaluation results on two protein datasets demonstrate the ability of the proposed approach on locating the true query motif of *all* 66 target proteins almost always (9 and 8 times, respectively, on average, out of 10 trials per target). A large-scale application of motif matching is protein classification, where the proposed approach distinguished between the positive and negative examples by consistently ranking *all* positive examples at the very top of the search results.

Keywords: Particle Swarm Optimization (PSO), 3D Motif Matching, Protein Structure Classification, least Root Mean Square Deviation (*l*RMSD).

1 Introduction

Proteins are complex macromolecules that perform a number of essential biological functions for any living cell. Not only do they transport oxygen, hormones and catalyze almost all chemical reactions in the cell, but they also play an important role in protecting our bodies from foreign invaders, through the human immune system. Proteins are made of long sequences of amino acid residues that fold into energetically-favorable three-dimensional structures to achieve minimal energy conformations of its individual residues for the sake of stability.

In general, protein structures are 3-10 times more evolutionary conserved than protein sequences [1]. Thus, the analysis of protein structures can provide better insights into understanding evolutionary history and detecting distant homology that may not be detectable by sequence comparison alone, in addition to providing insights into predicting protein functional annotation and classification. Proteins can be classified into different classes, folds, super-families and families [2]. Each protein family generally shares one or more common sub-structural patterns or motif(s), which can be used as signature patterns to characterize the family [3]. That is the reason why

M. Sokolova and P. van Beek (Eds.): Canadian AI 2014, LNAI 8436, pp. 1–12, 2014.
© Springer International Publishing Switzerland 2014

proteomic pattern analysis helps in protein classification by identifying common structural core(s) or conserved motif(s) of family-related proteins [4, 5].

One general approach for protein fold classification is concerned with identifying common substructures in two proteins based on an overall structure similarity search, whereas another more-focused approach is concerned with finding specific 3D motifs or patterns, which could help in protein structure classification by distinguishing between proteins of the same overall fold, but different super-families or families [6]. Most programs that measure pairwise similarity in protein structures perform an overall similarity search to find any regions of common substructures between protein pairs [7]. However, structural classification experts are usually more interested in finding specific 3D motifs in protein structures than performing a general structure similarity search between protein pairs for a number of reasons [8-10]. Firstly, structural differences within the motif are often more important than structural differences in other regions of protein structures. Secondly, a motif-based search helps focus attention on the conserved motif regions of a protein fold instead of the other less-conserved regions in protein structures [6]. This motivates the main objective of this study of finding all target structures that share a query 3D motif of a particular protein classification with known function. The retrieved high-scoring target structures can then be assigned a similar structural classification or functional annotation of that query motif.

One important question here is how to score local structural similarity between the target structures and the query 3D motif. While there are numerous scoring techniques to quantify 3D structural similarity between two protein conformations [11], one of the most widely accepted score is the least Root Mean Square Deviation (lRMSD) [12], which is the average atomic displacement of equivalent α-carbon atoms after optimal rigid-body superimposition using least-squares minimization [13]. In particular, the RMSD is calculated as the square root of the average of the squared positional deviations between corresponding atoms of the 3D structure pair.

Another important question is concerned with the structure searching technique. Largely, most protein structure comparison and searching tools in the literature use heuristics [11], such as Monte Carlo simulated annealing [14], Genetic Algorithms [15], and Variable Neighborhood Search [16]. However, to the best of our knowledge, none of the available protein structural comparison tools employed Particle Swarm Optimization as a simple, efficient and robust heuristic technique to this problem.

The paper is organized as follows: Section 2 provides a brief background on Particle Swarm Optimization (PSO), optimal rigid-body superimposition using the Kabsch algorithm and zinc finger protein motifs. Section 3 describes the proposed RMSD-based Analysis using PSO (RA-PSO) algorithm for motif matching and protein structure classification. Section 4 presents the experimental results and observations on two datasets of different types of zinc finger proteins. Lastly, we conclude with a summary of the study key points and present possible directions for future research.

2 Background

2.1 Particle Swarm Optimization (PSO)

For more than a decade, PSO has had growing scientific attention as an efficient and robust optimization tool in various domains. PSO is a population-based computational technique that is essentially inspired by the intelligent social behavior of birds flocking. The emergence of flocking in swarms of birds has long intrigued a wide range of scientists from diverse disciplines: including physics, social psychology, social science, and computer science for many decades. Bird flocking can be defined as the collective motion behavior of a large number of interacting birds with a shared objective. The local interactions between birds (particles) generally determine the shared motion direction of the swarm. These interactions are based on the *"nearest neighbor"* principle where birds follow certain flocking rules to adjust their position and velocity based on their nearest flockmates, without any central coordination.

The secret of PSO success lies in the experience-sharing behavior, in which the experience of the best performing particles is continuously communicated to the entire swarm, leading the overall motion of the swarm towards the most promising areas detected so far in the search space [17]. Therefore, the moving particles, at each iteration, evaluate their current position with respect to the problem's objective function to be optimized, and compare the current fitness of themselves to their historically best positions, as well as to the positions of their best-performing particle. Then, each particle updates its private experience if the fitness of the current position is better than its historically best one, and adjusts its velocity to imitate the swarm's global best particle by moving closer towards it. Before the end of each iteration of PSO, the index of the swarm's global best particle is updated if the most-recent change of the position of any particle in the entire swarm happened to be better than the current position of the swarm's global best particle.

Several variants of the original PSO method have been proposed in the literature. This study adopts Clerc's constriction PSO [18], which is a popular PSO variant described by the following "velocity" and "position" update equations shown in (1) and (2). One advantage of the constriction PSO is that it defines a clear mathematical relationship between the constriction coefficient χ, and *cognitive* and *social* parameters (c_1 and c_2), as shown in (3).

$$v_{id}(t+1) = \chi[v_{id}(t) + c_1 r_{1d}(p_{id}(t) - x_{id}(t)) + c_2 r_{2d}(p_{gd}(t) - x_{id}(t))] \tag{1}$$

$$x_{id}(t+1) = x_{id}(t) + v_{id}(t+1) \tag{2}$$

$$\chi = \frac{2}{|2 - \Phi - \sqrt{\Phi^2 - 4\Phi}|}, \quad \Phi = c1 + c2 > 4 \tag{3}$$

Where: χ is the constriction coefficient typically set to 0.729 and Φ to 4.1 (i.e., $c_1 = c_2 = 2.05$) [19]. The first term, v_{id} in (1) represents the velocity of the ith particle in the dth dimension, and t denotes the iteration time step. The second term, ($p_{id}(t) - x_{id}(t)$), finds the difference between the current position of the ith particle in the dth dimension, x_{id}, and its historically best position, p_{id}, which implements a linear attraction towards the historically best position found so far by each particle [19]. This term represents the private-thinking, the self-learning or the "cognitive" component from each particle's flying experience, and is often referred to as "local memory", "self-knowledge", or "remembrance" [17]. The third term, $(p_{gd}(t) - x_{id}(t))$, finds the difference between the current position of the ith particle in the dth dimension, x_{id}, and the position of the swarm's global best particle, p_{gd}, which implements a linear attraction towards the global best position ever found by any particle [19]. This term represents the experience-sharing, the group-learning or the "social" component from the overall swarm's flying experience, and is often referred to as "cooperation", "group knowledge", or "shared information" [17]. r_{1d} and r_{2d} are two random numbers uniformly selected in the range of [0.0, 1.0] at each dimension d, which introduce useful randomness for the search strategy. Moreover, c_1 *and* c_2 are positive constant weighting parameters, called the *cognitive* and *social* parameters, respectively, which control the relative importance of particle's private experience versus swarm's social experience [19].

2.2   Least Root Mean Square Deviation (*l*RMSD)

As mentioned earlier, the Root Mean Square Deviation (RMSD) is one of the standard scoring techniques for protein structural similarity, which principally calculates the square root of the average of all square distances between two paired sets of atomic positions in the two structures, as shown in (4).

$$RMSD = \sqrt{\frac{\sum_{i=1}^{n} d_i^2}{n}} \qquad (4)$$

Where: n is the number of corresponding atoms, and d_i is the Euclidian distance between the pair of corresponding atoms i.

However, it is often the case that the atomic positions of different structural conformations do not follow the same coordinate reference system [12]. Therefore it is necessary to calculate RMSD after performing optimal rigid-body superimposition to find the least RMSD (*l*RMSD) over all possible relative positions and orientations of the pair of conformations under consideration. This is in order to accurately assess the structural similarity without artificially increasing the RMSD distance between two conformations just because their 3D coordinate information does not share the same origin [12].

One of the commonly-used techniques for the optimal rigid-body superimposition is the Kabsch algorithm [20], which calculates the optimal translation and rotation matrices of the two conformations using least-squares minimization. The algorithm starts by translating the centroid of both conformations to the origin of the coordinate system by subtracting their corresponding centroids from the 3D atomic positions of each structure. Then, it computes the optimal rotation matrix of both conformations using their covariance matrix and Lagrange multipliers [12]. The mathematical details of the Kabsch algorithm implemented in this study can be found here [20, 21].

2.3   Zinc Finger Protein Motif

As discussed in detail in the Results Section (Section 4.1), two datasets of different types of zinc finger proteins are used to evaluate the performance of the proposed PSO-based approach for 3D motif matching and motif-based structure classification. For classification purposes, the positive examples in the protein datasets are collected such that they contain the classic zinc finger motif (C2H2 type) in one dataset, and the CCHC-type zinc finger motif in the other dataset.

The term 'zinc finger' was first introduced in mid-1980s as a short chain of 20-30 amino acids that folds tightly and independently around a central zinc ion, which helps stabilizes its structure [22]. Zinc finger proteins are among the most abundant proteins in the human proteome, with a wide range of diverse functional roles. The ability to identify zinc-binding sites in zinc finger proteins with unknown functions is important for their function inference [23]. One of the most common zinc binding motifs is the classic C2H2 zinc finger motif, which is abundantly found in transcriptional regulatory proteins [24]. The structure of the classic zinc finger motifs contain a central zinc atom and a short anti-parallel beta sheet (β-hairpin), followed by a small loop and an α helix, as shown in Figure 1.

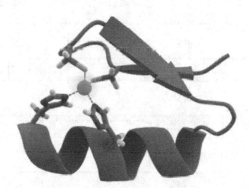

Fig. 1. Structural representation of the classic zinc-finger motif (Cys2His2 Type)

3   RMSD-Based Analysis Using PSO (RA-PSO)

Given a query motif, Q, of size L residues, the goal is to locate the best matching substructures similar to Q in a set of target proteins. All high-scoring target structures (with lowest RMSD) can then be assigned a similar structural classification and/or functional annotation to the query protein. The proposed RA-PSO method (RMSD-based Analysis using PSO) initializes each particle at a random residue of the target protein with a *3xL* window of coordinate positions, and calculate its fitness as the *least* RMSD between the *3xL* window and Q. The swarm objective is obviously to minimize the *l*RMSD scores and locate the best matching *3xL* window to the query motif, as shown in Figure 2.

For the sake of simplicity, the 'toy' example presented in Figure 2 only uses two particles, P_1 & P_2, over a set of 3 iterations to find the best matching sub-structure in the target protein with respect to a query motif of size 14 residues. Assuming the best matching *3x14* window in the target structure is located at the positions 4—17, and P_1 & P_2 are initially randomly placed at the positions 1 and 5, respectively. By the end of the first iteration, the global best particle is marked as P_2 (the particle with the best *I*RMSD or closest to the query motif), which will attract P_1 to move closer to in the next iteration. Thus, if P_1, at the second iteration, moved to position 4 (the global optimum position), it will be marked as the global best and will attract P_1 to its position. So eventually, or by the end of the third iteration in this example, the swarm will converge to the global optimum and will be able to accurately locate the best matching substructure to Q in the target protein, which is the main hypothesis of the study

It is worth mentioning that to prevent particles from going beyond the allowed search range, the restricted boundary condition [25] is enforced to reset the position of errant particles to the maximum possible position P_{max} and reverses their velocity direction towards the respective search space, where P_{max} = *Target Protein Length – Query Motif Size*.

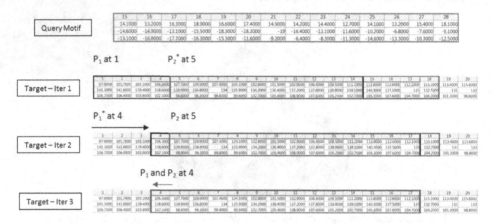

Fig. 2. A schematic illustration of the proposed methodology for motif matching using PSO

4   Results and Discussion

4.1   Protein Datasets

The first protein dataset used in this study consists of 68 proteins: 1 query protein with CCHC zinc finger motif (1mfsA: 15—28), 19 positive protein examples of the same protein family (Zinc finger, CCHC type) and 48 negative protein examples of different protein families.

The following is a list of all positive examples in the first protein dataset of small proteins with an average length of 40 residues, formatted as *AAAAC(n)*, where *AAAA* is the PDB ID, C the is Chain ID and n is the starting position of the zinc finger motif retrieved from the SITE record in the PDB file:

1altA(13), 1a6bB(24), 1aafA(13), 1bj6A(13), 1cl4A(51), 1dsqA(29), 1dsvA(56), 1eskA(13), 1f6uA(34), 1hvnE(1), 1hvoE(1), 1nc8A(7), 1ncpC(1), 1u6pA(24), 1wwdA(24), 1wweA(24), 1wwfA(24), 1wwgA(24), 2znfA(1).

The second protein dataset used in this study consists of 96 proteins: 1 query protein with classic zinc finger motif (1aijA: 137—157), 47 positive examples of the same protein family (Classic Zinc finger, C2H2 type), and similarly, the rest are negative protein examples of different protein families. Both protein datasets are generated from the PAST server [26]. The following is a list of all positive examples in the second protein dataset of medium-to-large proteins with an average length of 137 residues:

1alfA(137), 1algA(137), 1alhA(137), 1aliA(137), 1alkA(137), 1allA(137), 1aayA(137), 1ej6C(183), 1f2iG(1137), 1f2iH(2137), 1f2iI(3137), 1f2iK(5137), 1f2iL(6137), 1g2dC(137), 1g2fC(137), 1g2fF(237), 1jk1A(137), 1jk2A(137), 1llmD(206), 1meyC(7), 1meyF(35), 1meyG(63), 1p47A(137), 1sp1A(5), 1ubdC(327), 1yuiA(36), 1zaaC(37), 2drpD(113), 2givA(37), 2i13A(136), 2i13B(108), 2pq8A(210), 2prtA(385), 2rc4A(540), 2wbtA(103), 2y0mA(210), 3mjhD(43), 3qahA(210), 3tntA(210), 3tnhA(210), 1uk1C(471), 1uk1D(471), 4dlmA(210), 4f2jC(501), 4f6mA(496), 4f6nA(496), 4is1D(473).

4.2   Motif Matching Results

The first task in our experiments is to locate the true positions of the query motifs, displayed above in parentheses, for each target protein (or positive examples) in both datasets using the proposed RA-PSO method. Because PSO is a heuristic search, it is necessary to evaluate both the algorithm effectiveness and robustness in the search results by checking its ability on successfully and consistently locating the true positions of the query motif over multiple trials. Thus, the search process is repeated 10 times per each target protein and the success rate is measured as the number of successful runs over all number of 10 trials. For example, a success rate of 70% means the algorithm managed to successfully locate the true position of a query motif in 7 out of 10 runs, each with a maximum of 100 iterations.

Figure 3 and 4 show the success rates for *all* positive examples in both datasets, respectively, over 10 runs and varying swarm sizes from 10 to 50 particles. Not surprisingly, the performance on the first dataset (of small target proteins) is relatively better than the performance on the other dataset (of medium-to-large target proteins).

Moreover, as shown in Table 1, the performance on the first dataset appears less sensitive to the swarm size compared to the second dataset with a success rate ranging from 87% — 94% as opposed to 55% —82%. It is worth emphasizing that an average success rate of 92% does not mean that the proposed method managed to find the query motif in only 92% of the target proteins, but rather means it almost always managed to find the true positions of the query motif in *all* target proteins 92% of the time, or with an average of ~9 times out of 10 trials per each target.

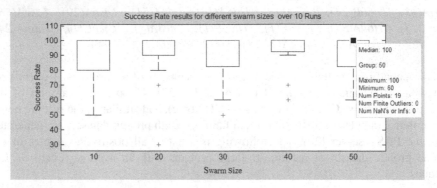

Fig. 3. Success rate results for the first dataset (CCHC-type Zinc Finger) of 19 small target proteins over 10 runs

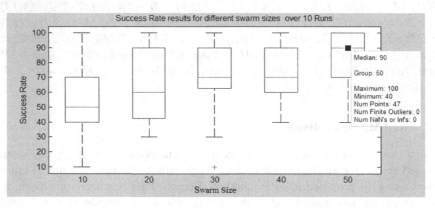

Fig. 4. Success rate results for the second dataset (Classic C2H2 Zinc Finger) of 47 medium-to-large target proteins over 10 runs

Table 1. Average success rate over 10 runs

Swarm Size	First Dataset (CCHC) 19 Targets	Second Dataset (C2H2) 47 Targets
10	87%	55%
20	91%	65%
30	91%	74%
40	94%	74%
50	92%	82%

4.3   Protein Classification Results

The second task in our experiments is to check the ability of the proposed method on distinguishing between positive and negative examples in each dataset. Thus, ideally all true positives in the first dataset that belong to the CCHC-Type Zinc protein family should be ranked at the top 19 results of lowest RMSD scores (out of 68 proteins), whereas all true positives in the second dataset that belong to the classic-C2H2 Zinc protein family should be ranked at the top 47 results of lowest RMSD scores (out of 96 proteins). To that end, each protein in both datasets is given a score according to its structural similarity to the respective query motif, and all proteins in each dataset are ordered ascendingly by their least RMSD scores.

Figures 5 and 6 confirm that all true positives of both datasets are indeed ranked at the very top of results with lowest RMSD scores. It is also worth mentioning that the query protein in each dataset was deliberately included among the target proteins to be compared with itself, which always resulted into the expected zero RMSD score over multiple trials.

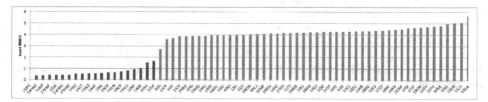

Fig. 5. First dataset of 68 proteins ordered by their RMSD scores. All true positives (of CCHC-Type Zinc protein family) are successfully ranked at the top 19 results with lowest scores.

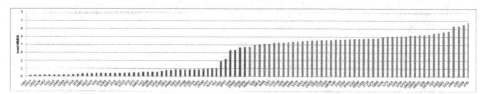

Fig. 6. Second dataset of 96 proteins ordered by their RMSD scores. All true positives (of classic C2H2-Zinc protein family) are successfully ranked at the top 47 results with lowest scores.

Table 2. Wilcoxon rank-sum significance test results

Significance	First Dataset (CCHC)	Second Dataset (C2H2)
P-Value	6.87E-18	7.07E-18
Z-Score	8.61709	8.61383
H	1	1

For validation purposes, the experiment is repeated 50 times, and the results in all trials are statistically compared using the Wilcoxon Rank-Sum test, which determines if the means of the RMSD scores for positive and negative examples are significantly different at 99% confidence level (i.e., P-Value <0.01 and H=1). Table 2 confirms the significance of the results and shows that the mean RMSD scores for positive and negative examples are indeed significantly different in both datasets.

Lastly, Figures 7 and 8 perform a boxplot analysis of the mean RMSD results over 50 trials for positive and negative examples in both datasets, respectively, which shows two important observations. Firstly, there is a distinguishable mean RMSD score between positive and negative examples, with a median of less than one RMSD over 50 trails for all positives examples in both datasets. Secondly, the small difference between the lower and upper quartiles in the boxplots indicates the robustness of the algorithm and the consistency of the results.

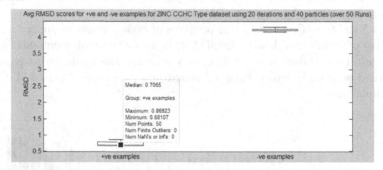

Fig. 7. The average RMSD scores over 50 runs for positive and negative examples in the first dataset of 68 small proteins

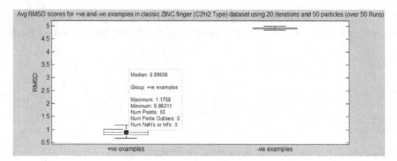

Fig. 8. The average RMSD scores over 50 runs for positive and negative examples in the second dataset of 96 medium-to-large proteins

5   Conclusions

This paper proposed a novel PSO-based method for motif matching and protein structure classification. To the best of our knowledge, this is the first study to investigate the applicability of PSO, as a robust and efficient optimization technique, to the aforementioned proteomic problems. The proposed method almost always managed to successfully locate the true motif positions of *all* 66 targets in two datasets of small and medium-to-large proteins 94% and 82% of the time, on average, for each dataset, respectively.

The successful detection of structural motifs in proteins can help predict their functions and families/super-families classification. Thus, one large-scale application of the motif matching problem is protein structure classification, which has been investigated in this study. The proposed approach correctly distinguished between the positive and negative examples in the two datasets of 68 and 96 proteins. The mean ranks of all positive examples over 50 trials are statistically validated to be significantly different than the mean ranks of all negative examples in both datasets at 99% confidence level using Wilcoxon Rank-Sum test. Moreover, the proposed method showed a highly robust performance with accurate results by consistently producing similar ranks for positive and negative examples over 50 runs.

Such encouraging results demonstrate that protein structure comparison/classification is yet another promising application of the fast-growing and far-reaching PSO algorithms, which was the main hypothesis of this study. Our plans for future research include comparing the effectiveness and efficiency of PSO with other population-based heuristics on solving the presented protein structure comparison and classification problems, as well as experimenting with different variants of PSO algorithms and more protein datasets.

References

1. Illergård, K., Ardell, D.H., Elofsson, A.: Structure is three to ten times more conserved than sequence—A study of structural response in protein cores. Proteins: Structure, Function, and Bioinformatics 77, 499–508 (2009)
2. Murzin, A.G., Brenner, S.E., Hubbard, T., Chothia, C.: SCOP: A structural classification of proteins database for the investigation of sequences and structures. Journal of Molecular Biology 247, 536–540 (1995)
3. Brejova, B., Vinar, T., Li, M.: Pattern discovery: Methods and software. Introduction to Bioinformatics, 491–522 (2003)
4. Kleywegt, G.J.: Recognition of spatial motifs in protein structures. Journal of Molecular Biology 285, 1887–1897 (1999)
5. Watson, J.D., Laskowski, R.A., Thornton, J.M.: Predicting protein function from sequence and structural data. Current Opinion in Structural Biology 15, 275–284 (2005)
6. Shapiro, J., Brutlag, D.: FoldMiner: Structural motif discovery using an improved superposition algorithm. Protein Science 13, 278–294 (2004)
7. Shi, S., Zhong, Y., Majumdar, I., Krishna, S.S., Grishin, N.V.: Searching for three-dimensional secondary structural patterns in proteins with ProSMoS. Bioinformatics 23, 1331–1338 (2007)
8. Andreeva, A., Howorth, D., Brenner, S.E., Hubbard, T.J., Chothia, C., Murzin, A.G.: SCOP database in 2004: Refinements integrate structure and sequence family data. Nucleic Acids Research 32, D226-D229 (2004)
9. Lesk, A.M.: Systematic representation of protein folding patterns. Journal of Molecular Graphics 13, 159–164 (1995)
10. Pearl, F.M.G., Bennett, C., Bray, J.E., Harrison, A.P., Martin, N., Shepherd, A., Sillitoe, I., Thornton, J., Orengo, C.A.: The CATH database: An extended protein family resource for structural and functional genomics. Nucleic Acids Research 31, 452–455 (2003)
11. Hasegawa, H., Holm, L.: Advances and pitfalls of protein structural alignment. Current Opinion in Structural Biology 9, 341–348 (2009)

12. Odegard, J.E.: Molecular Distance Measures (2007),
    http://cnx.org/content/m11608/1.23/
13. Hendrickson, W.A.: Transformations to optimize the superposition of similar structures.
    Acta Crystallographica Section A: Crystal Physics, Diffraction, Theoretical and General
    Crystallography 35, 158–163 (1979)
14. Sali, A., Blundell, T.L.: Definition of general topological equivalence in protein structures:
    A procedure involving comparison of properties and relationships through simulated an-
    nealing and dynamic programming. Journal of Molecular Biology 212, 403–428 (1990)
15. Szustakowski, J.D., Weng, Z.: Protein structure alignment using a genetic algorithm. Pro-
    teins: Structure, Function, and Bioinformatics 38, 428–440 (2000)
16. Pelta, D., González, J., Vega, M.M.: A simple and fast heuristic for protein structure com-
    parison. BMC Bioinformatics 9, 161 (2008)
17. Parsopoulos, K.E., Vrahatis, M.N.: Particle swarm optimization and intelligence: Ad-
    vances and applications. Information Science Reference Hershey (2010)
18. Clerc, M., Kennedy, J.: The particle swarm-explosion, stability, and convergence in a mul-
    tidimensional complex space. IEEE Transactions on Evolutionary Computation 6, 58–73
    (2002)
19. del Valle, Y., Venayagamoorthy, G.K., Mohagheghi, S., Hernandez, J.-C., Harley, R.G.:
    Particle swarm optimization: Basic concepts, variants and applications in power systems.
    IEEE Transactions on Evolutionary Computation 12, 171–195 (2008)
20. Kabsch, W.: A solution for the best rotation to relate two sets of vectors. Acta Crystallo-
    graphica Section A: Crystal Physics, Diffraction, Theoretical and General Crystallogra-
    phy 32, 922–923 (1976)
21. Kabsch, W.: A discussion of the solution for the best rotation to relate two sets of vectors.
    Acta Crystallographica Section A: Crystal Physics, Diffraction, Theoretical and General
    Crystallography 34, 827–828 (1978)
22. Laity, J.H., Lee, B.M., Wright, P.E.: Zinc finger proteins: New insights into structural and
    functional diversity. Current opinion in Structural Biology 11, 39–46 (2001)
23. Ebert, J.C., Altman, R.B.: Robust recognition of zinc binding sites in proteins. Protein
    Science 17, 54–65 (2008)
24. Simpson, R.J., Lee, S.H.Y., Bartle, N., Sum, E.Y., Visvader, J.E., Matthews, J.M.,
    Mackay, J.P., Crossley, M.: A classic zinc finger from friend of GATA mediates an inte-
    raction with the coiled-coil of transforming acidic coiled-coil 3. Journal of Biological
    Chemistry 279, 39789–39797 (2004)
25. Xu, S., Rahmat-Samii, Y.: Boundary conditions in particle swarm optimization revisited.
    IEEE Transactions on Antennas and Propagation 55, 760–765 (2007)
26. Täubig, H., Buchner, A., Griebsch, J.: PAST: Fast structure-based searching in the PDB.
    Nucleic Acids Research 34, W20–W23 (2006)

Rhetorical Figuration as a Metric in Text Summarization

Mohammed Alliheedi and Chrysanne Di Marco

Cheriton School of Computer Science
University of Waterloo
Waterloo, Ontario, Canada
{malliheedi,cdimarco}@uwaterloo.ca

Abstract. We show that surface-level markers of pragmatic intent can be used to recognize the important sentences in text and can thereby improve the performance of text summarization systems. In particular, we focus on using automated detection of rhetorical figures—characteristic syntactic patterns of persuasive language—to provide information for an additional metric to enhance the performance of the MEAD summarizer.

1 Introduction

Extractive summarization attempts to find the most "important" sentences in a document or document set for inclusion in a summary. One cue to importance which has hitherto received little attention is the writer's or speaker's use of rhetorical devices, especially those that involve repetition, to deliberately draw attention to important points. Over the past 20 years, research in automated text summarization has grown significantly in the field of Natural Language Processing. However, because the information available on the Web is ever-expanding, reading the sheer volume of information is a significant challenge. Although many automated text summarization systems have been proposed in the past twenty years, most of these systems have relied only on statistical approaches without incorporating much use of detailed linguistic knowledge. Our hypothesis is that rhetorical figuration, which involves the persuasive presentation of information in a text, generally at the sentence level, can provide such linguistic detail and can be detected computationally relatively easily. In particular, this research investigates the role of rhetorical figuration in determining the most important information to preserve in text summarization. Our experiments show that extractive summarization at the sentence level can be improved using the inclusion of metrics based on rhetorical figuration.

2 Related Work

Researchers have proposed various approaches in the area of text summarization. Luhn states that the frequency of certain words in the document reflects its relevance [14]. Subsequent researchers (e.g., [9], [13], [11], and [23]) have proposed

M. Sokolova and P. van Beek (Eds.): Canadian AI 2014, LNAI 8436, pp. 13–22, 2014.

various frequency measures. The position of sentences is another feature that has been proposed in different systems [2], [3], [13], and [16]. Lexical chains are also used in several summarization systems [1], [22], [19], and [25] as a representation of the text to produce a summary. Several researchers in text summarization used Machine Learning methods such as Naive Bayes in [13], Neural Networks in [24], and Hidden Markov Models in [6]. Maximal marginal relevance (MMR) is proposed in [7] which combines measurements for both query relevance and the novelty of information for a specific topic. Different systems have been proposed as well, such as SUMMON [17], MEAD [21], multi-lingual summarization system [10], Hub/Authority Framework [26], and a progressive summarization system [5].

3   MEAD

MEAD is an extractive multi-document summarizer that was developed by [20]. We built our approach on top of the MEAD summarizer [20] for the following reasons. Most significantly, MEAD is open source, can be downloaded online, and is capable of summarizing multiple natural language texts. MEAD is also domain-independent so different types of corpora can be used. In addition, MEAD implements various summarization algorithms such as position-based, term frequency, inverse document frequency, and others[20]. Moreover, MEAD has the capability to modify, add, include, and exclude new algorithms (features).

3.1   MEAD Components

MEAD uses centroid-based summarization to identify important words in a cluster of documents based on the cosine overlap between the words and the vector of that cluster. In other words, a centroid represents a group of words that are statistically significant in the cluster [20].

Various features are used to score a sentence to determine whether it should be retained in the summary: centroid value, positional value, and length feature.

The centroid value C_i is the sum of the centroid values of all words $C_{w,i}$ in a sentence S_i. For instance, assume that we have the centroid values of words (Obama = 44.68; Hillary = 32.50; Clinton = 29.40; morning = 12.25), and the sentence "President Obama meets with Secretary of State Hillary Rodham Clinton in the morning." In this case the total score of this sentence is 118.83 based on the following formula:

$$C_i = \sum_w C_{w,i} \tag{1}$$

The positional value for each sentence in a document is calculated as follows:

$$P_i = \frac{(n - i + 1)}{n} * C_{max} \tag{2}$$

While n, i are the total number of sentences in a cluster of documents and the order number of a sentence in the cluster respectively, C_{max} is the highest centroid value in the cluster. Note that the position value of the first sentence in the cluster is assigned C_{max} since the first sentence tends to possess more important information than other sentences.

The length feature is a cut-off feature that means every sentence with a length shorter than the threshold will receive a score of zero regardless of other feature scores. Thus, if a sentence has a length greater than the default threshold, which is nine words, it will receive a score that is the combination of other feature scores.

As no learning algorithm has been incorporated to predict the weight for each feature, we assign an equal weight for all features equal to one.

$$Score(S_i) = W_c\ C_i + W_p\ P_i\ ;\ \text{if Length}(S_i) > 9$$

$$Score(S_i) = 0;\ \text{if Length}(S_i) < 9$$

While the input is the cluster of d documents with n sentences, the output is the number of sentences from the cluster with the highest score multiplied by compression rate r.

4   What Is Rhetorical Figuration?

Rhetoric is the art of persuasive discourse, i.e., using language for the purpose of persuading, motivating, or changing the behaviour and attitudes of an audience. Corbett defines rhetorical figuration as "an artful deviation" from normal usage [8]. Rhetorical figures can be categorized into several types: schemes, tropes, and colours. Schemes are defined as "deviations from the ordinary arrangement of words" [4] and are generally characteristic syntactic patterns, such as repetition of the same word in successive sentences. We focus on figures of repetition because these are among the most commonly used and most effective rhetorically. Since figures of repetition often occur at the surface level, they can be detected and classified computationally with relative ease. Although there are over 300 rhetorical figures, we chose to focus on the following four figures of repetition in our experiments since they are among the most common and JANTOR demonstrated strong performance in detecting these figures:

Antimetabole
   A repetition of words in adjacent sentences or clauses but in reverse grammatical order[4]. For instance, *But I am here tonight and I am your candidate because the most important work of my life is to complete the mission we started in 1980. How do we complete it?* [27].

Isocolon
   A series of similar structured elements that have the same length, which is a kind of parallelism [4]. For example, in the sentence, *The ones who raise the*

family, pay the taxes, meet the mortgage [27], the phrases *raise the family, pay the taxes,* and *meet the mortgage* have a parallel structure.

Epanalepsis

The repetition at the end of a line, phrase, or clause of a word or words that occurred at the beginning of the same line, phrase, or clause [4]. For example, in the sentence, *I am a man who sees life in terms of missions - missions defined and missions completed* [27], the word *missions* appears three times.

Polyptoton

The repetition of a word, but in a different form, using a cognate of a given word in close proximity [4]. For example, *I may not be the most eloquent, but I learned early that eloquence won't draw oil from the ground* [27], the words *eloquent* and *eloquence* are derived from the same root, *loqui*[1], but appear in different forms.

5   Methodology

5.1   Basis of the Approach

Our proposed multi-document summarization system is divided into two components: (1) an annotator of rhetorical figures; and (2) the basic multi-document summarizer itself. For the first component, we make use of JANTOR [12], a computational annotation tool for detecting and classifying rhetorical figures. We called our system JANTOR-MEAD.

5.2   Rhetorical Figure Value

The rhetorical figure value RF_j is the new feature added to our system JANTOR-MEAD along with the basic MEAD features above. RF_j is the sum of occurrences of a specific rhetorical figure in a document divided by the total number of all figures in the document. It is calculated as follows:

$$RF_j = \frac{n}{N} \tag{3}$$

where j is a type of rhetorical figure such as polyptoton or isocolon, and n is the total number of i occurrences in a document. N is the total number of all figures, which includes antimetabole, epanalepsis, isocolon, and polyptoton, that occurred in the document. For example, suppose we have a document containing 149 rhetorical figures (12 antimetabole; 2 epanalepsis; 100 isocolon; 35 polyptoton) and we would like to have the value of each figure. We apply the rhetorical figure value as follows:

$RF_{antimetabole} = 12/149 = 0.081$
$RF_{epanalepsis} = 2/149 = 0.013$

---

[1] http://www.merriam-webster.com/dictionary/eloquent

$\mathrm{RF}_{isocolon} = 100/149 = 0.671$
$\mathrm{RF}_{polyptoton} = 35/149 = 0.235$

Since there are four rhtorical figures have been considered in this study, the score of each sentence will include the total value of all rhtorical figures that occurred in that sentence as follows:

$$RF_i = \sum_{j=0}^{4} RF_j \qquad (4)$$

Thus, the sentence score equation is modified to include rhetorical figure feature as follows:

$$\text{Score}(S_i) = \mathrm{W}_c\,\mathrm{C}_i + \mathrm{W}_p\,\mathrm{P}_i + \mathrm{W}_{rf}\,\mathrm{RF}_i$$

6 Data Set

We created a set of clusters for several U.S. presidents using presidential speeches from The American Presidency Project[1]. In our analysis we used four collections for the following presidents: Barack Obama, George W. Bush, Bill Clinton, and George Bush. Each collection contains three different types of speeches: State of the Union, campaign, and inaugural speeches, i.e., we have a cluster for each of the three types of speeches. Each cluster contains from 6 to 17 speeches, and these speeches contain from 90 to 700 sentences. These tested clusters were used for the evaluation process such that each cluster is summarized with ratio of 10% of actual texts.We choose presidential speeches because they tend to be rhetorical and persuasive texts. JANTOR has been tested previously by [12] using presidential speeches as data set and JANTOR was able to detect successfully many instances of different rhetorical figures include antimetabole, epanalepsis, polyptoton and isocolon.

7 Experiments with Rhetorical Figures

The following sections show the results of our experiments done for various rhetorical figures. Our tests involved four types of rhetorical figures: antimetabole, epanalepsis, isocolon, and polyptoton. JANTOR was used to detect these figures in all clusters of presidential speeches. However, the quality of performance varied for the different figures.

7.1 Antimetabole

Although antimetabole is defined as the repetition of words in adjacent clauses or phrases in reverse grammatical order, this definition does not specify whether

---

[1] http://www.presidency.ucsb.edu

or not different word forms and word types should be considered[12]. Gawryjolek [12] addressed this issue by considering all types of words, such as determiners and prepositions, and by looking for only the repetition of the exact words. Although it is obvious that including prepositions and determiners will definitely produce more antimetaboles, which are not salient in a text, it is important to include them in order not to decrease the recall value. Therefore, JANTOR identifies correct antimetaboles, but it may also identifies antimetaboles that are not rhetorically genuine.

7.2   Epanalepsis

The simple definition of epanalepsis—repetition of the same word or phrase in the beginning and the end of a sentence, clause, or line—allowed JANTOR to correctly detect most of the epanalepsis instances in our corpus.

7.3   Isocolon

Isocolon requires a set of similarly structured phrases or sentences that have the same length. JANTOR was able to successfully detect isocolon instances in our corpus. However, JANTOR sometimes encountered the same instance repeated and counted it as a new isocolon occurrence. This happens because JANTOR is not able to distinguish between a sentence and the same sentence with punctuation such as dash and semicolon. For example, "for all of our economic dominance" and "for all of our economic dominance -". In this example, JANTOR detects the same instance of isocolon twice, and is not able to identify them as one instance.

7.4   Polyptoton

JANTOR was able to correctly detect several instances of polyptoton in our test corpus. From our observation of the detection of polyptoton, however, we found that JANTOR has one problem with detecting this figure: it fails to detect some words that are in different stem forms.

8   Evaluation

Evaluation of summarization requires human judgment in several quality metrics such as coherence, grammaticality, and content. Although the quality metrics are significant for the evaluation of summarization systems, these metrics would require an enormous amount of human effort which is very expensive and difficult to manage [15]. Thus, over the past years, many researchers in the area of text summarization have attempted to develop an automatic evaluation metric for summarization tasks.

8.1 ROUGE

We used the intrinsic evaluation system ROUGE [15] to evaluate our summarization system. ROUGE metrics include ROUGE-N, ROUGE-L, ROUGE-S, ROUGE-SU. ROUGE-N is an n-gram recall measure that calculates the number of n-grams in common between the human summaries and system summaries. ROUGE-L is a measure based on Longest Common Subsequence (LCS) between reference summaries and model summaries. It uses the concept that the longer the subsequence of words that are in common between summaries, the greater the similarity will be. ROUGE-S is a skip-bigram co-occurrence statistical measure that counts any pair of words in their sentence order in a model summary that overlaps with a reference summary. ROUGE-S allows arbitrary gaps between bigrams. ROUGE-SU is an extension of ROUGE-S with a unigram as a counting unit.

9 Results

As we discussed in the data set section, the presidential speeches are annotated by the JANTOR tool for different rhetorical figures that include antimetabole, epanalepsis, isocolon, and polyptoton. We have two types of summaries for our data set: MEAD summaries, which are the baseline summaries, and our rhetorically based summaries. Two human assessors provided manual summaries for all data sets with a ratio of 10% of actual texts. These assessors are two graduate students from the English Language and Literature Department at the University of Waterloo. One is a PhD student and the other is a Master's student. The task of assessors was to review the presidential speeches and identify the most important sentences in each speech and manually extract 10% of the actual text. The assessors were able to create summaries for all presidential speeches in the data set. Thus, we have a total of 33 human-based summaries. Both summaries produced by the MEAD baseline and JANTOR-MEAD are compared with the human-based summaries, which are the gold standard, using the evaluation system ROUGE. We denote JMEAD in the tables for JANTOR-MEAD and MEAD for the baseline summarizer.

Table 1. Results of ROUGE metrics for both MEAD and JANTOR-MEAD

ROUGE	MEAD	JMEAD
R-1	0.690	0.715
R-L	0.679	0.702
R-S4	0.483	0.515
R-SU4	0.483	0.515

Table 1 shows the results of ROUGE metrics for both the baseline, which is the MEAD summarization system, and our rhetorically enhanced JANTOR-MEAD summarization system using all rhetorical figures includes antimetabole,

Table 2. Results of ROUGE metrics for both MEAD and JANTOR-MEAD with Porter Stemmer

ROUGE	MEAD	JMEAD
R-1	0.714	0.743
R-L	0.704	0.732
R-S4	0.521	0.560
R-SU4	0.521	0.560

epanalepsis, isocolon, and polyptoton. Table 1 demonstrates that JANTOR-MEAD summarization system slightly outperforms the MEAD system in every ROUGE metric. The correlation in R-SU4 and R-S4 between human-based summaries and JANTOR-MEAD is higher than MEAD which indicates that rhetorical figures tend to preserve salient information that can contribute to improve summaries.Table 2 show how the results changed when the Porter stemmer algorithm applied in ROUGE metrics. Once Porter stemmer is incorporated, it penalizes the polyptoton score and reduce it to minimal since polyptoton concerns the occurrences of words in different forms or part of speech and the stemmer counts words with same root as same words. Table 2, however, shows that JANTOR-MEAD still slightly outperforms MEAD.

In summary, we found that our system using all figures include antimetabole, epanalepsis, isocolon, and polytoton provides better results than MEAD in every ROUGE method. Surprisingly, Porter stemmer did imporve the performance of JANTOR-MEAD in Table 2. Thus, JANTOR-MEAD is better than MEAD in every ROUGE evalutation metric. Since only two human assessors were involved in this experiment (due to funding constraints) and there was no consideration of inter-agreement between assessors, we hypothesize that if there were more assessors participating and each assessor evaluates all documents in the data set, the results of JANTOR-MEAD would improve significantly.

10   Conclusion

We propose the use of rhetorical devices in the form of figuration patterns as additional metrics for improving the performance of text summarization. This hypothesis is based on the observation that rhetoric is persuasive discourse and so may be expected to highlight significant sentences in a text that should be preserved in a summary. Our experiment showed that rhetorical figures can provide additional information that in some cases allowed our summarizer to outperform the MEAD system. These results are promising and indicate that rhetoric, which provides additional linguistic information, may be a useful tool in natural language processing systems.

11   Future Work

Further investigation is planned to expand the work of JANTOR-MEAD and include new rhetorical figures such as Anaphora and Ploche as well as increasing the number of assessors to provide more accurate concrete gold standard summaries. We also plan to test our approach on other corpora. It is also feasible to test other corpora such as Document Understanding Conference (DUC) data sets since they are standard data sets and have been widely used by many summarization systems.

Acknowledgements. We would like to acknowledge and thank the first author's sponsor Al Baha University, Albaha, Saudi Arabia for funding this research. We would like also to thank Graeme Hirst for his valuable comments on earlier versions of the manuscript.

References

[1] Barzilay, R., Elhadad, M.: Using lexical chains for text summarization. In: Proceedings of the ACL Workshop on Intelligent Scalable Text Summarization, pp. 10–17 (1997)

[2] Baxendale, P.B.: Machine-made index for technical literature: An experiment. IBM Journal of Research and Development 2(4), 354–361 (1958)

[3] Brandow, R., Mitze, K., Rau, L.F.: Automatic condensation of electronic publications by sentence selection. Information Processing and Management 31(5), 675–685 (1995)

[4] Burton, G.: Silva rhetoricae: The forest of rhetoric (2007)

[5] Bysani, P.: Detecting novelty in the context of progressive summarization. In: Proceedings of the NAACL HLT 2010 Student Research Workshop, Los Angeles, California, pp. 13–18 (2010)

[6] Conroy, J., O'Leary, D.P.: Text summarization via Hidden Markov Models. In: Proceedings of the 24th Annual International ACM SIGIR Conference on Research and Development in Information Retrieval, New Orleans, Louisiana, pp. 406–407 (2001)

[7] Carbonell, J., Goldstein, J.: The use of MMR, diversity-based reranking for reordering documents and producing summaries. In: Proceedings of the 21st Annual International ACM SIGIR Conference on Research and Development in Information Retrieval, Los Angeles, California, pp. 335–336 (1998)

[8] Corbett, E.P.J.: Classical rhetoric for the modern student, 3rd edn. Oxford University Press, New York (1990)

[9] Edmundson, H.P.: New methods in automatic extracting. Journal of the ACM 16(2), 264–285 (1969)

[10] Evans, D.K., McKeown, K., Klavans, J.L.: Similarity-based multilingual multi-document summarization. IEEE Transactions on Information Theory, 49 (2005)

[11] Hovy, E., Lin, C.-Y.: Automated text summarization in SUMMARIST. Advances in automatic text summarization, pp. 81–94. MIT Press, Cambridge (1999)

[12] Gawryjolek, J.J.: Automated annotation and visualization of rhetorical figures. Master's Thesis, University of Waterloo (2009)

[13] Kupiec, J., Pedersen, J., Chen, F.: A trainable document summarizer. In: Proceedings of the 18th Annual International ACM SIGIR Conference on Research and Development in Information Retrieval, Seattle, Washington, pp. 68–73 (1995)

[14] Luhn, H.P.: The automatic creation of literature abstracts. IBM Journal of Research and Development 2(2), 159–165 (1958)

[15] Lin, C.-Y.: Rouge: A package for automatic evaluation of summaries. In: Proceedings of the Workshop on Text Summarization Branches Out (WAS 2004), vol. 16 (2004)

[16] Lin, C.-Y., Hovy, E.: Identifying topics by position. In: Proceedings of the Fifth Conference on Applied Natural Language Processing, Washington, DC, USA, pp. 283–290 (1997)

[17] McKeown, K., Radev, D.R.: Generating summaries of multiple news articles. In: Proceedings SIGIR 1995 Proceedings of the 18th Annual International ACM SIGIR Conference on Research and Development in Information Retrieval, Seattle, Washington, United States, pp. 74–82 (1995)

[18] McQuarrie, E.F., Mick, D.G.: Figures of rhetoric in advertising language. Journal of Consumer Research 22(4), 424–438 (1996)

[19] Nahnsen, T., Uzuner, O., Katz, B.: Lexical chains and sliding locality windows in content-based text similarity detection. In: Companion Volume to the Proceedings of Conference Including Posters/Demos and Tutorial Abstracts, pp. 150–154 (2005)

[20] Radev, D., Allison, T., Blair-Goldensohn, S., Blitzer, J., Celebi, A., Dimitrov, S., Drabek, E., Hakim, A., Lam, W., Liu, D.: MEAD-A platform for multidocument multilingual text summarization. In: Proceedings of the International Conference on Language Resources and Evaluation, vol. 2004, pp. 1–4 (2004)

[21] Radev, D.R., Jing, H., Budzikowska, M.: Centroid-based summarization of multiple documents: sentence extraction, utility-based evaluation, and user studies. In: Proceedings of the 2000 NAACL-ANLP Workshop on Automatic Summarization, Seattle, Washington, vol. 4 (2000)

[22] Silber, H.G., McCoy, K.F.: Efficiently computed lexical chains as an intermediate representation for automatic text summarization. Computational Linguistics - Summarization Archive 28(4), 487–496 (2002)

[23] Strzalkowski, T., Wang, J., Wise, B.: A robust practical text summarization. In: Proceedings of the AAAI Symposium on Intelligent Text Summarization, pp. 26–33 (1998)

[24] Svore, K., Vanderwende, L., Burges, C.: Enhancing single-document summarization by combining RankNet and third-party sources. In: Proceedings of the 2007 Joint Conference on Empirical Methods in Natural Language Processing and Computational Natural Language Learning (EMNLP-CoNLL), pp. 448–457 (2007)

[25] Verma, R., Chen, P., Lu, W.: A semantic free-text summarization using ontology knowledge. In: Proceedings of the Document Understanding Conference (2007)

[26] Zhang, J., Sun, L., Zhou, Q.: A cue-based hub-authority approach for multi-document text summarization. In: Proceedings of IEEE International Conference on Natural Language Processing and Knowledge Engineering, pp. 642–645 (2005)

[27] Bush, G.: Address Accepting the Presidential Nomination at the Republican National Convention. Online by Gerhard Peters and John T. Woolley. The American Presidency Project, New Orleans, Louisiana (1988)

Empirical Evaluation of Intelligent Mobile User Interfaces in Healthcare

Reem Alnanih[1,2], Olga Ormandjieva[1], and Thiruvengadam Radhakrishnan[1]

[1] Department of Computer Science and Software Engineering, Concordia University,
Montreal, Quebec, Canada
[2] Department of Computer Science, King Abdulaziz University, Jeddah, Saudi Arabia
{r_aln,ormandj,krishnan}@cse.concordia.ca

Abstract. The users of mobile healthcare applications are not all the same, and there may be considerable variation in their requirements. In order for the application to be attractive to potential adopters, the interface should be very convenient and straightforward to use, and easy to learn. One way to accomplish this is with an intelligent mobile user interface (IMUI), so that the application can be readily adapted to suit user preferences. This paper presents the results of adapting the IMUI for the various user stereotypes and contexts in the healthcare environment. We begin with a context model of the healthcare domain and an analysis of user needs, and then proceed to solution analysis, followed by product design, development, and user testing in a real environment. In terms of IMUI design, we focus on adapting the MUI features for users in the healthcare context, either at design time or at runtime.

Keywords: Intelligent user interface, Mobile user interface, Adaptation, User modeling, Context model, Rule-based knowledge, Controlled experiment, Quality measurement.

1   Introduction

Mobile computing has become one of the dominant computer usage paradigms, and there are a number of contrasting visions of how best to realize it. There has been a great deal of interest recently in the use of mobile applications to support healthcare [1]. As these applications become more sophisticated, a trend will inevitably develop towards providing comprehensive support for healthcare practitioners. It is expected that mobile healthcare applications will improve patients' quality of life, while at the same time reducing system costs, paper records, delays, and errors [2]. An intelligent environment would be an extremely useful way to help adapt applications to users with a wide range of needs [3].

The dynamic work habits of healthcare practitioners, along with the variety and complexity of the healthcare environments in which they operate, represent challenges for software designers, in terms of designing mobile medical applications that meet the needs of these users. Healthcare practitioners need interfaces tailored to their specific requirements to facilitate their work and to help them avoid the misunderstandings that can result in medical errors [4]. Designing adaptable mobile user

M. Sokolova and P. van Beek (Eds.): Canadian AI 2014, LNAI 8436, pp. 23–34, 2014.

interfaces (MUIs) coupled with artificial intelligence (AI) in healthcare applications will not only support healthcare practitioners with a wide range of needs, but also will result in the effective management of those applications.

In order to improve the user's experience with using UIs in mobile applications in healthcare environments, we investigated the incorporation of intelligent techniques into the mobile application through the user interface of mobile devices, specifically smartphones. An intelligent mobile user interface (IMUI) serves as the means of communication between a user seeking to execute a task and the user interface, in this case the IMUI, which is designed to enable the user to do so effectively. The IMUI could be used for different purposes. In our case, we apply it in an autonomous way by adapting the various UI features that already exist in smartphones, such as font size, voice data input, audio data output, etc., to the healthcare context.

Extending existing research, our main contribution is twofold. First, we develop an application providing safe information access (search, navigation, specific data points) and communication (texting), while the user is moving in a busy environment and performing various tasks by speaking or listening, such as entering a new order into the system (laboratory, radiology, or pharmacy) or reviewing patient results. Second, a controlled experiment in a real environment with real users using an adaptable IMUI was carried out to test and evaluate the usability of the IMUI from the end-user's point of view.

The significance of this research lies in the interdisciplinary nature of the method of IMUI operation, specifically the interaction and the perspectives involved: 1) the user perspective, based on user stereotype modeling and the set of facts extracted from the user, referred to as context descriptors; 2) the domain perspective, based on the context model; and 3) the designer perspective, based on an intelligent adaptation rule and the real, empirical evaluation of the design of the adaptation features, including voice, dialog, and audio features.

The rest of the paper is organized as follows: section 2 summarizes the literature on designing adaptable IMUIs and explains how our approach differs from those in existing published work. In section 3, a new direction for adapting IMUIs is presented. In section 4, a controlled experiment aimed at empirically validating our approach on an IMUI for healthcare application is presented. The results of the experiment are provided in section 5. We present our conclusion in section 6.

2   Motivation and Related Work

A great deal of work has been carried out recently to solve technical issues surrounding the design of an adaptable mobile application. Most users who work with these devices belong to different user stereotypes and have different backgrounds [5], and the application should designed so that it is convenient and easy for anyone to use, i.e. the system should be usable and adaptable [2]. Intelligent user interfaces (IUIs), in which AI concepts are applied to UIs, have already been devised to handle standard application requirements, such as helping users to manage complex systems [6]. Although IUIs are complex, researchers have been learning how to implement them efficiently for some time now, and various frameworks have been proposed for managing IUIs for mobile systems [6].

Incorporating adaptability and intelligence into an application provides users with significant opportunities, in terms of personalization and customization according to their preferences: "Adaptive architectures and Ubiquitous Computing are combined into a semantic interoperation-based approach to creating context-informed adaptive applications that make maximum use of rich content" [7]. However, in [7], the authors presented only the architecture phase of their approach. They give no clear indication as to how this approach would work in a real environment.

In [3], the authors focus on the eHealth environment and their approach differs from ours in some important respects. They aim to drive the interface by means of intelligent policies, which can be gradually refined using inputs from users through intelligent agents.

Sultan and Mohan (2012) [8] present an approach to adapting the user interface of a mobile health system with the goal of promoting the patient behaviors necessary for long-term use of the system. They focus on patients rather than doctors, and their type of adaptation can influence the sustainability of a mobile health initiative in the target patient market. In the next section, we present our direction for IMUI adaptation.

3   New Direction in Adaptive MUIs

Although a variety of research and development efforts have shown the potential of adaptive user interfaces, there remains considerable room for extending their flexibility and their interaction style to the healthcare domain specifically. Below, we describe our project, which is designed to explore new directions in the area of automated adaptive MUIs based on AI and considering user models.

Most experts are capable of expressing their knowledge in the form of rules for problem solving [9]. Our IMUI adaptation approach is based on a knowledge representation technique which is easy to use, and easy to modify and extend [8].

The term *knowledge-based system* is commonly used to describe a rule-based processing system. In this paper, rules are represented as recommendations and directives, and all the conditions presented in this work are based on our context model [10], and on the user model that has been extracted from experimental work and defined for healthcare applications [5]. Context-aware computing helps us to identify the criteria that an application might use as a basis for adapting its behavior [11]. However, there is as yet no agreement on the definition of context. In [10], we propose a context model for the design of an adaptable healthcare application from the MUI designer's perspective. The model covers general aspects related to the knowledge of the spatial context, such as location, time, and ambient conditions in terms of light and noise. The proposed model also identifies aspects that are highly dependent on the application domain, such as knowledge of the end-user through a user model [5], knowledge of the mobile device, and the nature of the task or activity in which an end-user is engaged in (see Fig. 1) [10]. Although the stereotypical model tailors the IMUI to the user's basic needs, preferences, and priorities, the context descriptors in fact allow for more adaptability and greater versatility in the form of real-life facts and individual user information.

The adaptation structure in our approach is based on the idea that designers solve problems by applying the knowledge (expressed as changes to the MUI parameters achieved intelligently using rule-based knowledge). The structure of IMUI adaptation consists from the following elements (see Fig. 1):

- The *knowledge base* contains domain knowledge which is useful for designers. In a rule-based approach, knowledge is represented by a set of rules, each rule specifying a relation, recommendation, or directive, and having the IF (condition) THEN (action) structure. When the condition of a rule is satisfied, the rule is said to "fire", and the action is executed. Decision table technique was used for supporting the representation process of that knowledge.
- The *context descriptor* includes a set of facts stored in a local persistent storage in the mobile device, and is used to match the IF (condition) parts of the rules stored in the knowledge base.
- The *inference engine* links the rules given in the knowledge base with the facts provided in the context descriptor.
- The IMUI, or adaptable MUI, is the means of communication between the user seeking a solution to the problem or performing a task intelligently and the application.

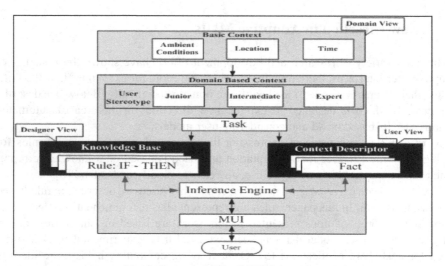

Fig. 1. Structure of the IMUI Adaptation

Below is an example of rule-based knowledge, illustrating the adaptation of IMUIs to the expert user stereotype:

- IF THE DOCTOR IS AN <u>EXPERT</u> AND IS IN THE <u>PATIENT'S ROOM</u> AT <u>10.00 AM</u>, THE <u>BRIGHTNESS</u> OF THE ROOM IS **HIGH**, AND THE <u>NOISE</u> LEVEL IN THE ROOM IS **HIGH**, THEN THE ACTIONS TO BE TAKEN ARE: CHANGE <u>FONT SIZE</u> FOR DATA DISPLAY TO **LARGE,** MOVE <u>BRIGHTNESS</u> LEVEL TO **INCREASE**, CHANGE <u>RING VOLUME</u> TO **USER DEFAULT**, CHANGE DATA OUTPUT MODE TO **HEADSET**, AND CHANGE DATA INPUT TO **MICROPHONE**.

In the following section, our IMUI adaptation approach is validated empirically through a carefully controlled experiment in a real environment.

4 Case Study

The case study involves a software application, the Phoenix Health Information System (PHIS), which is in use at King Abdulaziz University Hospital (KAUH). Our goal is to enhance communication between the user and IMUI features adapted to their needs. This application was developed to help doctors interact with patient information at any time through the performance of their daily tasks, such as reviewing patient results, ordering tests, and prescribing medications.

Identifying PHIS Users. PHIS is divided into three parts at KAUH: administrators, nurses, and doctors. Doctors obviously play a critical role in the health of the patient. The actions they take depend on the patient's test results displayed via the user interface and on the expertise of the doctor who interprets those results. It is vital, therefore, that patient information be displayed clearly and accurately. A cluttered view can cause confusion and frustration among users, which can affect their performance and even endanger a patient's life. Consequently, we have chosen to focus on the PHIS subsystem used by KAUH doctors and the tasks they perform on a daily basis. There are three categories of doctor at the KAUH: interns, residents, and consultants. These categories fit the Stereotype model for healthcare applications [5].

Doctors' Tasks. The daily tasks performed by the doctors registered in the PHIS system are the following:

1. Check patient profile;
2. Check patient laboratory, radiology, and pharmacology results;
3. Order a test, e.g. a lab test or an X-ray, or prescribe a medication;
4. View the Patient Summary.

New PHIS For Intelligent Adaptation (PHIS2-MA). After the PHIS was designed, PHIS2, the desktop version, was introduced and redesigned to take into account the principles of the HCI in terms of mental model, metaphor, visibility, affordability, and feedback [4]. A new, mobile-based version of PHIS2 (PHIS2-M) has since been introduced, in order to make the PHIS accessible from a mobile-based platform. A quality-in-use measurement model was developed specifically for the MUI to evaluate the quality of the mobile-based version of PHIS2 (PHIS2-M) [12].

The MUI features were adapted in an intelligent way using an AI rule-based technique to match the requirements of the mobile device to the user model. The device recognizes the context at runtime, and the designer adapts the interface widgets based on preprogramming requirements. The adaptable features were then implemented on an iPhone 4.1 loaded with PHIS2-M using Xcode 4 and SQLite, after a special framework for speech recognition (iSpeech iOS SDK) had been installed.

The result was a new version of PHIS2-M for mobile adaptation (MA), PHIS2-MA, on which the adaptable features of the MUI that match the various user stereotypes were evaluated. Below is a summary of the functionalities of the MUI features implemented on PHIS2-MA for different user types:

• Patient selection by voice;
• Main menu options, such as View Result, Add Order, View Profile or Summary by voice or recorded message;
• Output (e.g. patient results) displayed by audio or text;
• Adaptation of font size (small, medium, or large);
• Adaptation of brightness level (increase/ decrease) based on user type and time of day (morning shift, evening shift, night shift);
• Adaptation of audio features based on user type (on/off);
• Addition of a new option, Settings, to allow the user to change preferences;
• Adaptation of MUI settings at user sign-in;
• Saving of user preferences (which can only be accessed by the administrator).

5   Experimental Evaluation of Intelligent MUI Adaptation

The experimental evaluation presented in this section involves the usability testing of the PHIS2-M and PHIS2-MA applications on a mobile device used by a single participant in one environment.

Our participant samples were made up of doctors from different specialties in the Pediatric department during the fall of 2013 at KAUH. Forty-five doctors were selected at random for the study. In order to determine the effectiveness of adaptable MUI features, each participant was asked to perform the same set of tasks, once using PHIS2-M (with no adaptable features) and once using PHIS2-MA (with adaptable features). The two versions of the PHIS-based mobile application were loaded onto the same iPhone device. To ensure the accuracy of the experiment, half the participants started with PHIS2-M and the other half started with PHIS2-MA. The characteristics of the participants are presented in Table 1.

Table 1. Characteristics of the Participants

Doctors	Junior	Intermediate	Expert
Experience	Intern, 1st and 2nd year of residency	More than 2 years of residency and a fellow	Consultant
Total number of participants	20	15	12
Female: male ratio	13:7	7:8	4:8
Timeframe		8:00 am to 6:00 pm	
Test environment	Pediatric ward, ER, NICU, PICU, doctor's office, meeting room on the Pediatric floor, conference room in the Pediatric building		
Age group	<31	31- 45	>45
Domain experience	<6 years	6-10 years	>10 years
Smartphone experience	<1 year	1-3 years	>3 years

The test environment included the participants, the iPhone device, a timer installed in the application, and a tester. Prior to conducting the formal experimental evaluation, a list of materials to be used during testing was prepared, as suggested by Dumas and Redish [13], as explained below:

Task List. A list of three tasks corresponding to the functionalities most frequently used by the doctors on a daily basis, and to the ways in which the doctors interact with the application are: Search, Choose, Select, Read, and Write (Table 2). These tasks include various adapted MUI features that help researchers extract the effectiveness of MUI adaptation for different types of user. For example, the UI features that are adapted or modified temporarily are as follows: in Task 1, change to voice input while the doctor searches for the patient Medical Record Number (MRN); in Task 2, change to voice input while the doctor adds a new order; and in Task 3, change to audio output while the doctor reviews patient results.

Table 2. List of Tasks

Tasks	MUI Features	Task Functionality	Task Type
1. Select the Patient with the MRN (100)	Tap, text or speak	Select, search, or speak	Input-write
2. Order chest X-ray	Tap or speak	Choose, select, or speak	Input-write
3. Review the most recent complete blood count (CBC) results	Tap or sound	Choose, select, or speak and listen	Output-read

Objective Measures (Paper Log-in Form). Since the aim of testing the adapted MUI features was to make sure that they are easy to use and easy to learn for all the user stereotypes, compared with using the un-adapted MUI features, the new quality-in-use model for MUIs [12] was used as the testing method. The root of this new divided into objective and subjective factors. The objective factors are defined in terms of effectiveness, productivity, task efficiency, safety, and task navigation. To collect these measures, a paper log-in form was provided to each participant which includes criteria for the independent variables, such as the time taken to complete each task, along with the number of correct actions, the minimum number of correct actions, the number of incorrect actions, and the number of views.

Subjective Measures (User Satisfaction Questionnaire). A user satisfaction questionnaire was handed out at the end of the test to gather the impressions of the participants of the two applications, for subsequent evaluation. The questionnaire contained two types of questions: general questions, which focused on personal information and were applicable to any application; and specific questions, which only apply to the applications tested. We have to point out here that the responses to the questions relating to personal information, such as age group, gender, domain experience, and smartphone experience, were collected during the formal testing of PHIS2-MA when the participant signed in to create a user name before starting the test.

6   Results of the Experimental Evaluation

In order to empirically investigate the effectiveness of the MUI with and without adaptation for the same user belonging to a specific user stereotype in a real-world environment, two sets of hypotheses were formulated, one relating to the objective factors of the new quality-in-use model [12] and the other to its subjective factors. Sample hypotheses for both the subjective and objective factors are listed below:

HYP0: There is no significant difference between the *effectiveness* of PHIS2-M and the *effectiveness* of PHIS2-MA.

HYP1: There is a significant difference between the *effectiveness* of PHIS2-M and the *effectiveness* of PHIS2-MA.

HYPA0: There is no significant difference between the *user satisfaction* of the participant using PHIS2-M and using PHIS2-MA, in terms of the median of each feature for mobile users, in the case where it is the same for mobile adaptation users.

HYPA1: There is a significant difference between the *user satisfaction* of the participant using PHIS2-M and using PHIS2-MA, in terms of the median of each feature for mobile users, in the case where it is not the same for mobile adaptation users.

6.1   Objective Factor Results

The objective factors of the new quality-in-use model provide measures of the effectiveness, productivity, task efficiency, safety, and task navigation relating to 45 doctors belonging to 3 user stereotypes performing 3 tasks on PHIS2-M and PHIS2-MA.

User Stereotype Results. Since our data are normal, and we have two conditions (using PHIS2-M and using PHIS2-MA) for the same participants, we have paired them. Consequently, for statistical validation, the hypotheses are verified for each factor, based on the t-Test value and the P-values for the data analysis for all the participants in each stereotype (Table 3).

Junior stereotype: with the critical value approach of the t-Test at 19 degrees of freedom, $\alpha =0.025$ for the two-tailed test and the critical level of $t = \pm 2.09$. Our decision rule is to reject the null hypothesis if the computed t statistic is less than -2.09 or more than 2.09.

Intermediate stereotype: with the critical value approach of the t-Test at 12 degrees of freedom, $\alpha =0.025$ for the two-tailed test and the critical level of $t = \pm 2.17$. Our decision rule is to reject the null hypothesis if the computed t statistic is less than -2.17 or more than 2.17.

Experienced stereotype: with the critical value approach of the t-Test at 11 degrees of freedom, $\alpha =0.025$ for the two-tailed test and the critical level of $t = \pm 2.20$. Our decision rule is to reject the null hypothesis if the computed t statistic is less than -2.20 or more than 2.20.

The results of the hypotheses for each factor, based on the t-Test value and the P-value approach, are presented in Table 3. The symbol ↑↓ means that we can't reject the null hypothesis, HYP0, so there is no significant difference between using the mobile without adaptation and using the mobile with adaptation for the objective factor. The symbols ↑ and ↓ mean that we reject the null hypothesis, HYP0. So, there is a significant difference between using mobile with and without adaptation, and, based on the value of the t test that falls either in the positive or the negative region for each factor, we accept the alternative hypothesis, HYP1. If the symbol is ↑, one application type (either adaptation or non adaptation) is better than the other, and if the symbol is ↓, one application type is not better than the other.

Table 3. Results of the Hypothesis for the Objective Factors for User Stereotypes

New Quality-in-Use Objective Factors	Application Type	User Stereotype		
		Junior	Intermediate	Experienced
Effectiveness	Mobile	↓	↓	↑↓
	Mobile Adaptation	↑	↑	↑↓
Productivity	Mobile	↑	↑	↑
	Mobile Adaptation	↓	↓	↓
Task Efficiency	Mobile	↑↓	↑↓	↑↓
	Mobile Adaptation	↑↓	↑↓	↑↓
Safety	Mobile	↑↓	↑↓	↓
	Mobile Adaptation	↑↓	↑↓	↑
Task Navigation	Mobile	↑↓	↑↓	↑↓
	Mobile Adaptation	↑↓	↑↓	↑↓

Legend	↑	↓	↑↓
	Better than	Not better than	No difference
	HYP1	HYP1	HYP0

6.2 Subjective Factor Results

The subjective factors of the new quality-in-use model measure the user's level of enjoyment as a result of interacting with the IMUI application in a specified context of use in terms of the eight features listed below:

F1. Learning how to use the PHIS
F2. Using the PHIS
F3. Understanding how to navigate the PHIS
F4. Recovering from an error
F5. Performing the task anywhere and at any time
F6. Satisfaction with the whole application
F7. Satisfaction with performing a particular task
F8. Satisfaction with finding the desired features on the screen

The results of the questionnaire reveal that all the participants gave all the features of both types of mobile application ratings of 'easy' and 'very easy'.

Results for All Participants. For statistical validation, HYPA0 and HYPA1 are verified for all the features, based on the Wilcoxon signed ranks method for all the participants in each stereotype. The statistical significance was set at P <0.05. All the statistical analyses were performed with the SPSS 18 software (IBM, New York, USA). The Wilcoxon signed ranks method tests whether or not two related medians are equal. It is based on the absolute value of the difference between the two test variables.

- Junior stereotype group. The Wilcoxon test provides sufficient evidence against the null hypothesis for the difference between the medians of features 1 and 4 for mobile users and mobile adaptation users (the P-value is less than 0.05). Consequently, we can conclude that learning and recovery from errors is easier with PHIS2-M than with PHIS2-MA. For the other features, there is no significant difference between the medians for mobile users and mobile adaptation users (Table 4). However, this result is expected from this group. Since the junior users are young doctors and they don't want to spend time learning how the adaptation features work, or waiting until the speech recognition feature recognizes their voice input. They speak quickly, which causes errors in speech recognition, and so they repeat the input many times until the correct view is displayed. In this respect, the habits of these users don't match the framework that we used in the application.

- Intermediate stereotype group. The Wilcoxon test provides no evidence against the null hypothesis of the difference between the medians of all the features for mobile users and mobile adaptation users. Consequently, we can conclude that the new application results in no significant difference for all 8 features (Table 4). We can interpret this as meaning that the habits of these users do not match our framework, but, at the same time they don't have any problems with learning or using the application. It may be that by increasing the sample size of this group in the future, the results for using the adaptation application will be significant.

- Expert stereotype. The Wilcoxon test gives sufficient evidence against the null hypothesis for the difference between the median of feature 5 for mobile users and mobile adaptation users (P-value < 0.05). Consequently, we can conclude that PHIS2-MA is easier to use anywhere and at any time than PHIS2-M. For the rest of the features, there is no significant difference between the medians for mobile users and mobile adaptation users (Table 4). This result is expected from this group, since these users are mature individuals who move and talk more slowly than those in the other groups. For them, the adaptation application works very well, as their habits match our framework.

Table 4. Results of the Hypothesis for the Subjective Factors for User Stereotypes

New Quality-in-Use Subjective Factors	Application Type	User Stereotype		
		Junior	Intermediate	Experienced
F1: Learning the application	Mobile	↑	↑↓	↑↓
	Mobile Adaptation	↓	↑↓	↑↓
F4: Recovery from error	Mobile	↑	↑↓	↑↓
	Mobile Adaptation	↓	↑↓	↑↓
F5: Performance of the task anywhere and at any time	Mobile	↑↓	↑↓	↓
	Mobile Adaptation	↑↓	↑↓	↑

The participants were asked to evaluate each of the new features implemented in the mobile-based adaptation, such as data input by voice, audio data output, brightness level adjustment, font size change, and user preference customization under Settings, following the Dumas and Redish [13] classification. The performance levels for the tasks were the following [13]:

- Excellent: The system is very easy to use.
- Acceptable: Its users are satisfied with the system on this task.
- Unacceptable: Its users are not satisfied with the system on this task, because they have problems using the product of the task.

Table 5 presents the results for the adaptable features of PHIS-MA for all the user stereotypes, and, although the results were not statistically significant, the trend is toward the superiority of the adaptable version in terms of the adaptability of the latest iPhone features (note the higher percentages in the Excellent column).

Table 5. Adaptation Feature Results

IMUI Features	Junior doctors			Intermediate doctors			Expert doctors		
	Exc.	Acc.	Unacc.	Exc.	Acc.	Unacc.	Exc.	Acc.	Unacc.
Voice	60%	40%	-	62%	38%	-	83%	17%	-
Audio	85%	15%	-	77%	23%	-	83%	17%	-
Brightness	90%	10%	-	77%	23%	-	83%	17%	-
Font size	85%	15%	-	92%	8%	-	100%	-	-
Customization	75%	25%	-	85%	15%	-	92%	8%	-

7   Conclusion

The goal of this paper was to propose an adaptable IMUI design approach which has been evaluated empirically on a real-world case study from the healthcare domain. The results of our research show that the adoption of intelligent techniques may well offer a practical approach to the effective realization of the MUI features that exist in the latest smartphones. Adaptation has the following benefits: i) it allows considerable changes to be made in applications in terms of personal preferences and customization, as defined by the content being adapted; and ii) it allows information to be presented that is tailored to the user.

From this case study, we conclude the following: First, the proposed methodology for designing adaptable MUIs is sound and useful. Second, in this particular case study, where users are categorized into junior, intermediate, and expert doctors, we found that it is much easier for the experts to use the adapted version of the system in terms of effectiveness and safety, since they have patience and take the time to learn about its features. However, the other two groups (junior and intermediate doctors) either tend not to spend their time adapting or do not need the adapted features. Our future research in other contexts will include more experimentation to study the needs of the other user stereotypes in depth.

It is interesting to compare the results of this work with those of our previous experimental work carried out for the junior user stereotype [14]. In [14], ten junior doctors with previous experience using the PHIS application at KAUH participated in a controlled experiment conducted in a simulated environment in a university lab, which was different from a real hospital environment. The connection to the network was a fast one, the environment was not crowded as in a hospital, and the doctors were not under the usual pressures of work. The results of that experiment support PHIS2-M for productivity and task efficiency, while for the remaining factor there is no significant difference between PHIS2-M and PHIS2-MA. However, when the sample of junior users was increased to 20 in the experiment reported in this paper, which was conducted in a hospital, PHIS2-MA scored higher in terms of the effectiveness factor, but not the productivity or task efficiency factors. A possible explanation for this might be the time available: entering input by voice using the voice recognition functionality on a mobile platform is slower than typing or tapping, and most of the tasks evaluated in the experiment for the adaptable version depend on voice input. Since productivity depends on time, PHIS2-M scores better on this feature than PHIS2-MA, because of a slower Internet connection in the hospital.

To solve this problem, we plan to use a different speech recognition modality which does not require a network connection. We will then be able to examine the effect of using the new framework with junior users. In our future work, we will also consider the effectiveness of the adaptable mobile healthcare application for use by doctors with a speech impediment.

References

1. Orwat, C., Graefe, A., Faulwasser, T.: Towards pervasive computing in health care–A literature review. BMC Medical Informatics and Decision Making 8(1), 26 (2008)
2. Archer, N.: Mobile E-Health: Making the Case (2007)
3. Tavasoli, A., Archer, N.: A proposed intelligent policy-based interface for a mobile eHealth environment. Springer (2009)
4. Al-Nanih, R., Al-Nuaim, H., Ormandjieva, O.: New health information systems (HIS) quality-in-use model based on the GQM approach and HCI principles. In: Jacko, J.A. (ed.) HCI International 2009, Part IV. LNCS, vol. 5613, pp. 429–438. Springer, Heidelberg (2009)
5. Alnanih, R., Ormandjieva, O., Radhakrishnan, T.: Context-based User Stereotype Model for Mobile User Interfaces in Health Care Applications. Procedia Computer Science 19, 1020–1027 (2013)
6. O'Grady, M.J., O'Hare, G.M.: Intelligent user interfaces for mobile computing. In: Lumsden, J. (ed.) Handbook of Research on User Interface Design and Evaluation for Mobile Technology (2008)
7. O'Connor, A., Wade, V.: Informing context to support adaptive services. In: Wade, V.P., Ashman, H., Smyth, B. (eds.) AH 2006. LNCS, vol. 4018, pp. 366–369. Springer, Heidelberg (2006)
8. Sultan, S., Mohan, P.: Transforming usage data into a sustainable mobile health solution. Electronic Markets 23(1), 63–72 (2012)
9. Grosan, C., Abraham, A.: Rule-Based Expert Systems. In: Grosan, C., Abraham, A. (eds.) Intelligent Systems. ISRL, vol. 17, pp. 149–185. Springer, Heidelberg (2011)
10. Alnanih, R., Radhakrishnan, T., Ormandjieva, O.: Characterising Context for Mobile User Interfaces in Health Care Applications. Procedia Computer Science 10, 1086–1093 (2012)
11. Schmidt, A., Beigl, M., Gellersen, H.-W.: There is more to context than location. Computers & Graphics 23(6), 893–901 (1999)
12. Alnanih, R., Ormandjieva, O., Radhakrishnan, T.: A New Quality-in-Use Model for Mobile User Interfaces. In: Proceedings of the 23rd International Workshop on Software Measurement, IWSM-MENSURA (2013)
13. Dumas, J., Redish, J.: A Practical Guide to Usability Testing, 3rd edn. Intellect, Wiltshire (1999)
14. Alnanih, R., Ormandjieva, O., Radhakrishnan, T.: Context-based and Rule-based Adaptation of Mobile User Interfaces in mHealth. Procedia Computer Science 21, 390–397 (2013)

Learning to Measure Influence in a Scientific Social Network

Shane Bergsma, Regan L. Mandryk, and Gordon McCalla

University of Saskatchewan, Saskatoon, Saskatchewan, Canada S7N 5C9
shane.bergsma@usask.ca, {regan,mccalla}@cs.usask.ca

Abstract. In research, *influence* is often synonymous with *importance*; the researcher that is judged to be influential is often chosen for the grants, distinctions and promotions that serve as fuel for research programs. The influence of a researcher is often measured by how often he or she is cited, yet as a measure of influence, we show that citation frequency is only weakly correlated with influence ratings collected from peers. In this paper, we use machine learning to enable a new system that provides a better measure of researcher influence. This system predicts the influence of one researcher on another via a range of novel social, linguistic, psychological, and bibliometric features. To collect data for training and testing this approach, we conducted a survey of 74 researchers in the field of computational linguistics, and collected thousands of influence ratings. Our results on this data show that our approach significantly outperforms measures based on citations alone, improving prediction accuracy by 56%. We also perform a detailed analysis of the key features in our model, and make some important observations about the scientific and non-scientific factors that most predict researcher influence.

1 Introduction

This paper is concerned with understanding and quantifying researcher influence. We define influence as the capacity of a researcher to have an effect on another researcher's opinions, ideas, experimental approach, or choice of research topics.

Studying influence is important. Knowing about influential individuals can helps us understand how behaviours spread [1], and inform technology such as paper recommendation systems [2]. Also, since measures of influence are used to evaluate research, better measures of influence could help us make better strategic decisions, improving processes ranging from hiring, funding, promotion, and award-giving, to the assessment of organizations and funding programs.

Both quantitative and qualitative factors are commonly used to assess researcher influence. Quantitative measures based on citation counts, such as the h-index [3], provide a convenient and objective indicator of influence. But are these measures reliable? While citations "are correlated with other assessments of scientists' impact or influence, such as awards, honors, and Nobel laureateships," [4] citation-based measures are less useful in fields with "less heavy citation traffic" and are susceptible to manipulation [5]. Alternatively, qualitative assessments

M. Sokolova and P. van Beek (Eds.): Canadian AI 2014, LNAI 8436, pp. 35–46, 2014.

(e.g. recommendation letters) can be more reliable, but are time-consuming and subjective. Our aim is to develop an improved measure of researcher influence that can account for both quantitative and qualitative information, based on both scientific and non-scientific (social, psychological) factors.

Our main contribution is a learned model that predicts, for a pair of researchers, the influence that one researcher has on another. Formally, this model treats influence as function $Infl(x, y) \to [0, \infty)$: the influence of researcher x on researcher y, expressed as a non-negative real number. It is possible to aggregate these predictions in order to develop measures of total influence, but in this work we focus on pairwise, directed influence as a first step toward global models.

We train and test our approach within the scientific domain of computational linguistics. We first extracted a dataset of researchers from the information available through the ACL Anthology Network (Sec. 2) (derived from publications of the Association for Computational Linguistics). We then used the ACL data to build individual *scientific social networks* for the ACL authors; we elaborate on this concept in Sec. 3. Next, we solicited researchers to perform an online survey where they can select those members of their scientific networks that most influence them; respondents provided high-quality ratings (Sec. 4). We then used these ratings to train the influence model using techniques from supervised machine learning (Sec. 5). The heart of our model is a creative set of features (Sec. 6) that enables dramatic improvements in the prediction of influence, substantially improving accuracy over baselines based on citation counts alone (Sec. 8). In our results, we also provide an instructive analysis of what factors most affect influence. Overall, this paper provides important new tools, ideas, and directions for work at the intersection of machine learning and bibliometrics.

2   The ACL Anthology Network

To compute our function $Infl(x, y)$, we require information about researchers x and y and their relationship to one another. In this paper, we extract this information for researchers in the field of computational linguistics. We selected this field partly because of the availability of the ACL Anthology Network or AAN [6]. The AAN comprises a majority of worldwide papers in computational linguistics since the 1960s. The 2012 release of the AAN provides 20K full-text papers and 95K paper citations. Note, however, that these citations only include citations to other papers in the AAN. Each author is associated with a unique ID, from which we can trace citations and co-authorships within the AAN.

3   Building Scientific Social Networks

While the domain of $Infl(x, y)$ could comprise all pairs of researchers, if we define $Infl(x, y)$ in terms of how often y cites x then, by definition, y is only influenced by the set of researchers $\{x\}$ whom y has cited. Rather than restrict our analysis to the set of citing researchers, we expand the set of researchers that might influence y to a group we call y's *scientific social network*.

A researcher x is part of y's scientific social network if and only if:

1. y cites a paper authored by x (y's citation network)
2. x is a co-author of y (y's degree-1 network)
3. x and y share a co-author (y's degree-2 network)
4. x is among y's most-similar authors by paper content (y's topic network)

We calculate the author similarity using a vector-space approach [7]. We first build *tf-idf* vectors for each author from the combined text of all their papers. We took steps to exclude names, affiliations, references, stopwords, and infrequent terms. We compute the similarity between two authors by computing the cosine similarity between their *tf-idf* vectors. y's topic network comprises y's 50 most similar authors, provided the cosine values exceed a minimum threshold (.06).

The average network size is 90.3 people, which includes overlapping contributions of, on average, 6 people via the degree-1 network, 52 people via degree-2, 32 via citations and 47 via topic similarity.

4   A Researcher-Specific Survey of Scientific Influence

To train and test our influence model, we require a gold standard set of influence ratings. The gold standard in bibliometrics has long been direct peer assessment [8,5]. We used a survey to obtain peer assessments of the researchers within our data. We took steps to safeguard the confidentiality of the responses and our methodology received ethical approval from our institution.

Respondents (users) clicked on a hyperlink that took them to an online form (Fig. 1). The screens were customized for each user, with 11 clickable names randomly drawn from that user's scientific social network. Eleven names were chosen as a reasonably large number that could still fit on one screen without the need to scroll. Users were instructed to "click on the researcher... who has most affected your personal opinions, ideas, experimental approach or choice of research topics." The task of selecting a single name, as opposed to ranking or ordering the names, was chosen both to make the task easy for the users, and because such choices can still imply a global influence ranking (Sec. 5). When a name is clicked, the selected name and the 10 unselected ones are recorded, and a new set of names is randomly chosen. While users have the option of skipping screens, one user reported that making one choice "sent me into a deep introspective philosophical debate that I will likely be weeks recovering from."

114 researchers were contacted and 74 responded, resulting in a response rate of 65.5%. Users completed 40 screens on average (2957 in total). 86% of users were male, while 76% were from North America. The users ranged from new researchers to those who have been publishing for several decades, and included 10 students, 9 postdocs, 34 professors/government scientists, and 21 in industry.

Survey Reliability. One measure of response quality is how often users contradict themselves, i.e., how often they rate researcher A higher than researcher B on one screen and then B higher than A on another. Because the average user completed 40 screens and had 90 people in their network, there were many chances to do

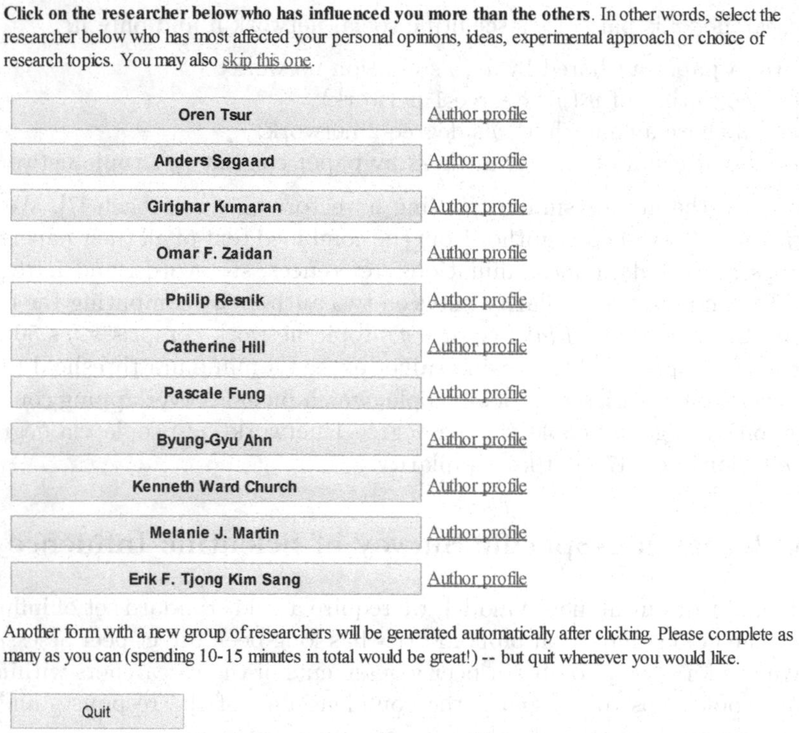

Click on the researcher below who has influenced you more than the others. In other words, select the researcher below who has most affected your personal opinions, ideas, experimental approach or choice of research topics. You may also skip this one.

Oren Tsur	Author profile
Anders Søgaard	Author profile
Girighar Kumaran	Author profile
Omar F. Zaidan	Author profile
Philip Resnik	Author profile
Catherine Hill	Author profile
Pascale Fung	Author profile
Byung-Gyu Ahn	Author profile
Kenneth Ward Church	Author profile
Melanie J. Martin	Author profile
Erik F. Tjong Kim Sang	Author profile

Another form with a new group of researchers will be generated automatically after clicking. Please complete as many as you can (spending 10-15 minutes in total would be great!) -- but quit whenever you would like.

Quit

Fig. 1. Screenshot of online survey form Clicking on the 'Author profile' links will open the selected author's profile on the AAN website, providing information about the author's publications, affiliations, collaborators, etc

this. However, we found that of all the thousands of ratings, only one user did this three times, while five other users did this only once.

Another indicator of response quality is the pattern of how frequently each name in the form is clicked, by position on the screen. Because the names are ordered randomly, the null hypothesis is that the clicks would be evenly distributed, with roughly the same number of clicks on names presented in the first position as those in the second, third, etc. The alternative hypothesis is that users click names in the first half significantly more than those in the bottom, since they read from top to bottom and may not bother to read all the names. While the clicks were fairly evenly distributed (Fig. 2), a one-tailed binomial test shows that users click on names in the top half significantly more ($p=0.024$). But since the names are randomly drawn, this behaviour results in random noise. Nevertheless, we are considering ways to account for and model this behaviour as part of our training algorithm. We also investigated if users were more likely to click on a name close to where they clicked on the previous screen. Again, any effect here would be noise rather than systematic bias. However, the observed next-click distribution closely tracked the expected distribution (Fig. 3).

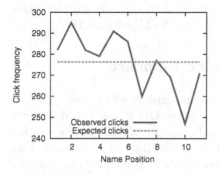 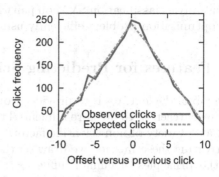

Fig. 2. Distribution of responses by position of researcher name on screen

Fig. 3. Distribution of responses by distance from previous click on screen

Altogether, these analyses show the users did an excellent job on the surveys. However, there is some noise in the ratings, and while our training algorithm (described in the following section) is robust to such effects, there is clearly an upper bound to the accuracy achievable on our test data.

5 An SVM Ranking Approach to Influence Prediction

While peer assessment ratings have typically been used to *evaluate* citation-based metrics [8,5], here we use our peer assessments to *train* a model of influence.

Formally, this model expresses the influence of researcher x on researcher y as a function $Infl(x, y) \rightarrow [0, \infty)$. We model $Infl(x, y)$ using a linear function, $Infl(x, y) = \mathbf{w} \cdot \mathbf{f}(x, y)$, where $\mathbf{w} = (w_1, w_2 \ldots w_N)$ are the parameters of the model: the N weights chosen by the machine learning algorithm during training. The function $\mathbf{f}(x, y) = (f_1(x, y), f_2(x, y) \ldots f_N(x, y))$ is an N-dimensional feature function that describes the relationship between researcher x and y. The particular binary and real-valued features that we use are described in Sec. 6.

Our survey was designed to provide us with *relative* influence ratings for learning the weights of the $Infl(x, y)$ model. When a user y selects the kth researcher, x_k, from the set of 11 researchers, $\{x_i\}$, the user is indicating that this kth researcher was more influential than the others presented in that set. This implies relative preference rankings, $Infl(x_k, y) > Infl(x_i, y), \forall i \neq k$. Ten such preferences are implied for each screen, and with roughly 3000 screens completed by our survey respondents, we obtain close to 30,000 implied preferences.

Our training algorithm has the basic goal of finding the set of weights, w, that can satisfy as many of these preferences as possible. Joachims [9] describes the similar problem of optimizing a ranking function for search engines, via the preferences implied by click-through data. We follow Joachims in using a support vector machine (SVM) solution. The SVM objective is to find the set of weights that results in the maximum separation between $Infl(x_k, y)$ and $Infl(x_i, y)$ (with slack variables and regularization). With this objective, Joachims shows how the constraints can be reduced to the form $\mathbf{w} \cdot (\mathbf{f}(x_k, y) - \mathbf{f}(x_i, y)) > 1$, which is essentially a classification SVM on pairwise difference vectors. We can thus solve the optimization problem efficiently using standard SVM software (Sec. 7).

6   Features for Predicting Scientific Influence

The job of the features is to provide information to explain why a user, y, deems one researcher, x, to be more influential than another. One of our key hypotheses is that a variety of non-scientific factors can help explain scientific influence. For motivating these factors, we draw on the ideas of Cialdini [10], a popular work on the principles of human influence. Of course, these principles may operate differently here, and the effectiveness of our features may be the result of other not-yet-understood processes. However, our aim is to both improve our system, and gain insights about what sorts of information are most useful.

The particular features were selected based on development experiments.

Basic Network. These features encode the information that we originally used to create the scientific social networks (Sec. 3). Features indicate the (a) no. of papers where y cites x, (b) no. of times y and x are co-authors, (c) no. of times x and y share a common co-author, and (d) cosine similarity of x and y's papers.

ReverseCite. Cialdini [10]'s first principle of influence is *reciprocity*: we are influenced by people who do something for *us*. As a potential instance of this in science, we include a feature for the number of times x cites y, i.e., the opposite of what we typically measure. Note we are not assessing here whether pure citation reciprocity is at play; we are assessing whether the fact someone cites you makes you more likely to rate them as genuinely *influencing* you.

Authority. We are also influenced by people in positions of authority. This is best exemplified by Milgram's famous study where participants performed acts against their personal conscience when instructed to do so by an authority figure [11]. To encode the authority of x, we include features for x's total number in the data of (a) citations and (b) co-authors; these measures are known to correlate with authoritative positions such as program committee membership [12]).

Similarity. It has also been established that people are most influenced by *similar* people [10]. We measure similarity in three ways:

(1) Affiliation: Since our data includes affiliation information for each researcher, we include features for whether x and y have the same affiliation and whether the final token in their affiliations match (indicating they are in the same country, state, or province). We also labeled each affiliation with whether it represents an academic, government, or industrial organization type; we then include a feature for whether x and y have the same organization type.

(2) Name: Two researchers may also share a similar ethnicity. We capture this in our system by including a feature for the semantic similarity of x and y's first names. We compute this similarity using recent data from Bergsma et al. [13],

who built clusters of names based on communication patterns on Twitter. The name clusters were shown to be useful for predicting a user's ethnicity, race, and native language. Our feature is the cosine similarity between x and y's cluster-membership-vector (the vector of similarities to each cluster centroid).

(3) Gender: People of one gender may be more influenced by people of the same gender. For each x and y, we compute the most likely gender of their names via the gender data of Bergsma and Lin [14], and include a feature for which of the four x/y gender configurations (*M/M, M/F, F/M, F/F*) is under consideration.

SocialProof. A final, widely-exploited form of influence is commonly called *social proof*: we often adopt the behaviour of our peers as a default habit [10]. In our case, if y's academic peers are influenced by researcher x, then y might be too. An example of this kind of default thinking in research is the hundreds of citations to a paper by Salton, where the paper does not actually exist [15]; many researchers were simply copying the citations of their peers without looking up the original paper. To capture this kind of influence, we add a feature to our classifier for the number of times that y's co-authors cite x.

TimeInactive. In development experiments, we discovered one other feature that was a good *negative* predictor of influence: time since the researcher's last publication, in years. After researchers become inactive, their influence decreases.

All. When we use all the above features together, we refer to it as the *All* system. When using *All*, we also incorporate a new technique for improving the feature types that rely on citations. Rather than just counting the number of papers where y cites x, we count how often the *surname* of x is mentioned in the papers of y. While more noisy than citations, there are two primary reasons for using surname counts: (1) Zhu et al. [16] recently showed that citations mentioned more than once in a paper are more likely to be rated by the authors as being "influential", and surname counts indirectly capture citation counts, and (2) since the AAN data is a closed set (Sec. 2), we can indirectly capture citations to papers by the researchers that occur in non-AAN venues (e.g., papers by researchers who regularly publish in both machine learning and NLP).

7   Experiments

Our experiments address two main questions: (1) How much can our approach improve over the standard way of measuring influence via citations? and (2) How important are each of the scientific and non-scientific factors in the prediction of influence? For the latter question, we test the value of each feature type by seeing how much accuracy drops when features are removed from the system.

For evaluation, we divide the gold ratings by *user*: we take the ratings from 54 users for training and from 20 other users for final test data. We train our models using SVM-Rank [17]. Since we do not have a surfeit of data, we performed development experiments and tuned the SVM regularization parameter by performing 54 rounds of leave-one-out evaluation on the training data. For the final results, we train on all 54 users and test on the held-out test data.

Table 1. Main results: Baselines, *All* system, and with features removed. *Signif.* gives p-value (McNemar's) for whether system's Top-1 Acc. is significantly worse than *All*.

Type	System	Top-1 Acc. (%)	MRR (%)	Signif.
Baseline	Random	9.1	27.5	*p<0.001*
	Most-cited	29.6	51.4	*p<0.001*
RankSVM	Features: *ResearcherID*	27.6	47.2	*p<0.001*
RankSVM	Features: *Basic Network*	37.4	56.0	*p<0.001*
	Features: *All*	**46.1**	**63.4**	-
	Features: *All - ReverseCite*	45.2	62.9	*Not signif.*
	Features: *All - Authority*	45.0	62.9	*p<0.2*
	Features: *All - Similarity*	44.4	62.2	*p<0.1*
	Features: *All - SocialProof*	45.2	62.8	*Not signif.*
	Features: *All - TimeInactive*	44.2	62.4	*p<0.05*

Table 2. Most-highly-ranked researchers: Ranked in descending order of the learned weights of the *ResearcherID* model (trained on ratings of 54 survey respondents)

Fernando Pereira, Dekang Lin, Chris Manning, Aravind Joshi, Ken Church, Mark Steedman, Hal Daume, Graeme Hirst, Jason Eisner, Chris Quirk, Ray Mooney, Philip Resnik

Our unit of evaluation is a screen completed by a user, y: one selected researcher x and 10 alternatives. Our evaluation metrics are: (1) Top-1 Acc.: the proportion of screens for which we perfectly predict the user-selected researcher from the 11 choices, and (2) Mean reciprocal rank: (MRR) of the correct researcher, a standard IR metric for evaluating rankings (closer to 1 is better).

We compare to three baseline systems:

1. Random: Select a researcher, x, randomly from the 11 choices
2. Most-cited: Select the researcher, x, that y cites the most
3. *ResearcherID*: A learned classifier with one feature: the ID of researcher x

ResearcherID ignores y and thus the relationship between x and y, and thus lets us test whether a global ranking of researchers can generate accurate predictions.

8   Results

Table 1 presents the main results of our study. *All* is substantially and significantly better than the baselines. In fact, the *Basic Network* system is also significantly better than *Most-cited* (p<0.01, McNemar's test), showing the power of combining even simple features via machine learning. The feature ablation shows that *Similarity* and *TimeInactive* are the most important new features, but Top-1 Acc. drops by a percentage or two when any of the feature classes are removed. The overall strong performance of *All* is thus due to the collective contributions of all feature types, even though these contributions may not individually be statistically significant. So, even when we consider how many papers x and y have co-authored, and how often y has cited x, etc., there is still valuable information in more subtle clues, such as whether x and y are in the same country, whether x has cited y, whether y's colleagues have cited x, etc.

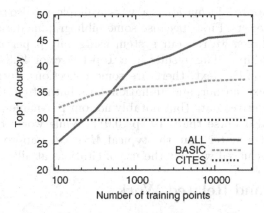

Fig. 4. Learning curve of different systems (x-axis is log-scale): *All* is better than the alternatives after 50 training screens, and continues to improve across all training sizes

Table 3. Features and their learned weights in the *All* system

Degree-1: Number of co-authored papers	0.82
Similarity of first names	0.75
Affiliation perfect match	0.56
Topic-Sim: Cosine similarity of publication term vectors	0.49
Cites: Number of times *you* mention the *researcher's* last name	0.42
Degree-2: Number of shared co-authors	0.26
ReverseCite: Number of times the *researcher* mentions *your* last name	0.19
Affiliation match of country/state	0.18
Affiliation match of type: academic, government, or industrial	0.18
Authority: Number of overall citations researcher has	0.18
SocialProof: Number of times your *co-authors* cite the researcher	0.12
Cites: Number of times *you* cite the researcher	0.11
TimeInactive: Number of years since researcher's last published paper	-0.27

The *ResearcherID* system performs worse than *Most-cited*, illustrating the importance of modeling the relationship between an influencer and a user, rather than just relying on a global ranking of influencers (further experiments also showed adding these features impairs the *All* system). It is nevertheless interesting to consider the resulting ranking of researchers using *ResearcherID* (Table 2). These researchers are certainly some of the leaders of the ACL community.

Fig. 4 shows how system performance depends on the amount of training data. Even with very little training data (50 or so completed screens), the *All* system exceeds both the *Most-cited* baseline and the *Basic Network* system performance. At the same time, *All* continues to improve up to the full 3K training instances, and so collecting further data may well increase accuracy further.

Table 3 provides the feature weights of *All* and thus a picture of how *All* computes its scores. We regard these numbers cautiously as the feature *values* themselves have different dynamic ranges; also when there are two similar features, the SVM divides credit between them (e.g. the two *Cites* features).

Since not all the above information is always available, we also ran experiments with reduced feature sets. First, because some bibliographic databases may not provide full paper texts, we ran our system using only paper *meta*-data (e.g. citations, co-authorships). This results in a Top-1 Acc. of 42.8% (significantly worse than *All*, p<0.01). Next, there are some collections, such as arxiv.org, which provide full-text, author, and affiliation data, but not citations. When excluding all citation-derived data (but notably not our last-name-count features), we obtain 42.6% (also worse than *All*, p<0.01). While worse than *All*, these systems are still much better than the typical *Most-cited* approach, and in the latter case, can be achieved without the use of citations at all!

9   Discussion and Related Work

Why are citations alone not the best predictor of influence? First of all, studies have documented that there are many reasons to cite a paper aside from acknowledging its influence [4]. You may simply be paying homage to an early pioneer, or perhaps criticizing another person's work. Secondly, our work here hints at influence beyond the medium of publication. For example, two people at the same institution are more likely to have a general influence on one another, even when such sharing is not manifested in citations or even co-authorship.

Recognizing that not all citations imply influence, Zhu et al. [16] used machine learning to predict the most important citations in a paper. Like us, they used peer assessment to gather training data. Unlike our approach, their method does not capture the valuable and effective social factors that we consider above.

A few recent papers have made use of the AAN in order to investigate the spread of scientific ideas [18,19]. Radev et al. [20] considered citation networks and collaboration networks separately, but did not integrate these into a joint model of influence, as we do. Johri et al. [21] used topic models to identify different types of collaboration in ACL articles (e.g. apprenticeship, synergistic, etc.); we could potentially exploit this data as an additional information source in order to refine the co-authorship features in our system.

There are ties between research in bibliometrics and research in broader social networks. For example, studies of influence have been performed within online social networks [1]. Academic social networks have also been used as case-study social-networks in many publications, including the foundational paper of the important *link prediction problem* in social networks [22].

Our models of influence have immediate application in the real world. For example, Fig. 5 shows the predicted influence on various researchers of the first author of this paper versus whether they responded to his request to complete the survey. When the predicted influence was high, people were more likely to respond to the request. Since it takes effort to contact people and configure their surveys, we could save time by using the predicted influence as a guide for whom to contact. As another example, note that our results suggest that the more people you cite, the more influence you will have. Unlike other conferences, AI'2014 does not allow authors an extra page for citations. This might be limiting the potential influence of work published at this venue!

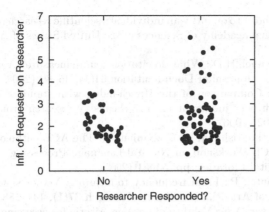

Fig. 5. The predicted influence on researchers (via *All*) of the person requesting participation in the survey, and whether the researchers complied with the request

10   Conclusion and Future Work

We have proposed a novel approach to the prediction of scientific influence. We extracted a dataset of researchers from the field of computational linguistics, and used readily-available, publication-derived data to create *scientific social networks* for each researcher. To generate training data for our approach, we asked a subset of these researchers to perform an online survey to identify the other researchers who most influenced them. Our survey enjoyed a 66% response rate, and users provided very high-quality ratings, rarely contradicting themselves and clicking across the whole range of options as expected. Our system, trained on these ratings, strongly outperformed the baselines and standard citation-based approach. The overall large gains in accuracy were attributed to the combination of small contributions from each of a variety of novel and interesting features.

The next step for this line of research is to apply our approach to other domains. Do we need peer assessments to train the system in each domain? Could we calibrate our system with only a few ratings? Or are the correlations we observed in our domain actually universal across science? We are also interested in aggregating our scores, both to compare the overall influence of researchers, as well as to answer questions such as, "how much influence does researcher x have in a particular geographic region or sub-community?" We are also interested in studying the dynamics of influence as people move into and out of the field.

References

1. Aral, S., Walker, D.: Identifying influential and susceptible members of social networks. Science 337(6092), 337–341 (2012)
2. Tang, T.Y., Winoto, P., McCalla, G.I.: Further thoughts on context aware paper recommendations for education. In: Manouselis, N., et al. (eds.) Recommender Systems for Technology Enhanced Learning, 16 p. Springer (in press, 2013)

3. Hirsch, J.E.: An index to quantify an individual's scientific research output. Proceedings of the National academy of Sciences of the United States of America 102(46), 16569–16572 (2005)
4. Bornmann, L., Daniel, H.D.: What do citation counts measure? A review of studies on citing behavior. Journal of Documentation 64(1), 45–80 (2008)
5. Van Raan, A.F.: Comparison of the Hirsch-index with standard bibliometric indicators and with peer judgment for 147 chemistry research groups. Scientometrics 67(3), 491–502 (2006)
6. Radev, D.R., Muthukrishnan, P., Qazvinian, V.: The ACL anthology network corpus. In: Proc. ACL Workshop on Natural Language Processing and Information Retrieval for Digital Libraries, pp. 54–61 (2009)
7. Turney, P.D., Pantel, P.: From frequency to meaning: Vector space models of semantics. Journal of Artificial Intelligence Research 37(1), 141–188 (2010)
8. Lawani, S.M., Bayer, A.E.: Validity of citation criteria for assessing the influence of scientific publications: New evidence with peer assessment. Journal of the American Society for Information Science 34(1), 59–66 (1983)
9. Joachims, T.: Optimizing search engines using clickthrough data. In: Proc. KDD, pp. 133–142 (2002)
10. Cialdini, R.B.: Influence: The Psychology of Persuasion. HarperCollins (2007)
11. Milgram, S.: Behavioral study of obedience. The Journal of Abnormal and Social Psychology 67(4), 371–378 (1963)
12. Liu, X., Bollen, J., Nelson, M.L., Van de Sompel, H.: Co-authorship networks in the digital library research community. Information Processing & Management 41(6), 1462–1480 (2005)
13. Bergsma, S., Dredze, M., Van Durme, B., Wilson, T., Yarowsky, D.: Broadly improving user classification via communication-based name and location clustering on twitter. In: Proc. NAACL-HLT, pp. 1010–1019 (2013)
14. Bergsma, S., Lin, D.: Bootstrapping path-based pronoun resolution. In: Proc. Coling-ACL, pp. 33–40 (2006)
15. Dubin, D.: The most influential paper Gerard Salton never wrote. Library Trends 52(4), 748–764 (2004)
16. Zhu, X., Turney, P., Lemire, D., Vellino, A.: Measuring academic influence: Not all citations are equal. Journal of the American Society for Information Science and Technology, JASIST (to appear, 2013)
17. Joachims, T.: Training linear SVMs in linear time. In: Proc. KDD (2006)
18. Hall, D., Jurafsky, D., Manning, C.D.: Studying the history of ideas using topic models. In: Proc. EMNLP, pp. 363–371 (2008)
19. Gerrish, S., Blei, D.M.: A language-based approach to measuring scholarly impact. In: Proc. ICML, pp. 375–382 (2010)
20. Radev, D.R., Joseph, M.T., Gibson, B., Muthukrishnan, P.: A bibliometric and network analysis of the field of computational linguistics. Journal of the American Society for Information Science and Technology, JASIST (2009)
21. Johri, N., Ramage, D., McFarland, D., Jurafsky, D.: A study of academic collaborations in computational linguistics using a latent mixture of authors model. In: Proc. 5th ACL-HLT Workshop on Language Technology for Cultural Heritage, Social Sciences, and Humanities, pp. 124–132 (2011)
22. Liben-Nowell, D., Kleinberg, J.: The link prediction problem for social networks. In: Proc. CIKM, pp. 556–559 (2003)

Filtering Personal Queries
from Mixed-Use Query Logs

Ary Fagundes Bressane Neto, Philippe Desaulniers,
Pablo Ariel Duboue, and Alexis Smirnov*

Radialpoint SafeCare Inc.
2050 Bleury Street,
Suite 300 Montreal,
Quebec H3A 2J5 Canada
{firstname.lastname}@radialpoint.com

Abstract. Queries performed against the open Web during working hours reveal missing content in the internal documentation within an organization. Mining such queries is thus advantageous but it must strictly adhere to privacy policy and meet privacy expectations of the employees. Particularly, we need to filter queries related to non-work activities. We show that, in the case of technical support agents, 78.7% of personal queries can be filtered using a words-as-features Maximum Entropy approach, while losing only 9.3% of the business related queries. Further improvements can be expected when running a data mining algorithm on the queries and when filtering private information from its output.

1 Introduction

Mining query logs [24] has the promise of unveiling structural information needs within an organization. This approach must strictly adhere to privacy policy and meet privacy expectations of the members of the organization. Situations that require extreme privacy guarantees have received plenty of attention of late [2,3], but they result in a loss of utility for the gained knowledge [27]. Moreover, techniques based in query distortion are still not well understood and might leak private information nevertheless [22]. In this work, we focus on a simpler case, where there is an employer-employee relation between the people generating the query logs and the people interested in the mining results.

We present a working system (Section 4) that classifies queries sent to a commercial search engine into "personal" and work-related ("business") queries. This system is a component of Radialpoint Reveal[TM], a product[1] that gathers work-related queries to allow query log analysis [24,26] and the identification of knowledge gaps in the technical support centre's internal documentation. According to a survey of several technical support centres we have performed, a

---

* To whom correspondence should be addressed.
[1] http://reveal.radialpoint.com

M. Sokolova and P. van Beek (Eds.): Canadian AI 2014, LNAI 8436, pp. 47–58, 2014.

large percentage of support sessions requires technical support agents[2] to perform a Web search in order to access the knowledge necessary to resolve the support request. Having the information available in their internal documentation repository ("knowledge base," in technical support *lingo*) reduces its access time. A system that results in reduction of handling time would be highly valuable for technical support call centres.

Our system is trained on a hand annotated sample of close to 9,000 real queries (more than 5,000 unique). It achieves a filtering accuracy of 78.7%, while losing only 9.3% of the business queries when evaluated on queries from different technical support agents than the ones used to train the system. Further improvements can be expected after applying a data mining algorithm, using *post hoc* privacy detection techniques [8]. We believe this is the first corpus of such queries that have been gathered and we are making it available for research purposes.[3]

This paper is organized as follows: we will discuss our problem and data in detail next. In Section 4, we present our system and in Section 5, our results. Related work and discussions for future work conclude this paper.

2   Scenario

We will now discuss Radialpoint Reveal[TM] scenario in detail, stressing the real-life constraints of our problem. Our working dataset is discussed in the next section. When technical support agents perform a Web search, the Radialpoint Reveal[TM] browser extension automatically highlights the most useful technical solutions as recommended and shared by the entire technical support team.

The query logs needed for our analysis are queries that technical support agents perform when solving issues for customers. However, in some circumstances, technical support agents may be using their work PCs for personal reasons, such as searching for information related to personal needs or interests during breaks. These logs of personal queries are not useful for the analysis performed by Radialpoint Reveal[TM], and the very fact that they are being processed and later be shown to supervisors may impede the adoption of the tool in certain call centres.

Therefore, we need to remove the personal queries from the query log data. Various approaches have been considered to achieve this.

One possible approach is to simply ask the technical support agents to use a different search engine[4] or Web browser when they are performing personal queries than when they are searching for technical support related information.

---

[2] While the term 'agent' is a technical term in Artificial Intelligence literature, in this paper we refer to humans working in a technical support centre. To avoid any confusion, we fully qualify the term as "technical support agent."

[3] The queries are available for non-commercial research purposes. Contact the corresponding author for details.

[4] Radialpoint Reveal[TM] currently only instruments searches against the Google commercial search engine.

We expect this approach would not be feasible, as it complicates the technical support agent workflow and is error prone. We also expect that privacy sensitive technical support agents would simply stop using the Web browser with Radialpoint Reveal™ installed, even for technical support queries, in order to simplify their workflows.

Another approach is to obtain timing information of technical support agents work breaks (when they would presumably be using the Web for personal purposes), and compare it to the query logs timestamps. Although they typically follow relatively well-organized shift schedules, and that technical support agents use of time is often (but not always) very precisely measured using various tools, it is impractical to attempt to use this data to isolate personal queries, because this would require software integration with a wide variety of time tracking tools, as well as resolving synchronization issues between these systems.

The approach we selected (Figure 1) was to collect enough raw data from a willing group of technical support agents, in order to train a classifier that would determine whether each query is a personal or technical support related query.

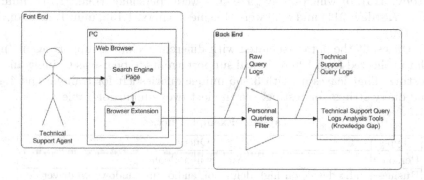

Fig. 1. Scenario

3   Data

We had access to a group of 145 technical support agents that willingly participated in this experiment. We collected four months of queries via Radialpoint Reveal™ Web browser extension. This resulted in 8,819 searches over 5,219 unique search queries.

One of the authors familiar with the type of customer issues being addressed in that technical support centre annotated the queries as being "personal" or "business." The criteria employed was

> *Would this query be issued by a technical support agent in the process of answering a customer call? If unsure, annotate it as "personal."*

Some example entries from the query log are shown in Table 2. Using the above criteria, we found that of the 8,819 queries, 3,336 are "personal" queries and 5,483 are "business" queries, see the first two columns of Table 1.

Table 1. Annotation numbers, including training and test set numbers

Class	Queries #	Queries %	Unique queries #	Unique queries %	Training Unique #	Training Unique %	Test Unique #	Test Unique %
personal	3336	37.82%	1622	31.10%	1378	35.4%	267	18.8%
business	5483	62.18%	3597	68.90%	2517	64.6%	1150	81.2%
Total	8819		5219		3895		1417	

We took a sample of 100 queries and annotated them by a second author that was not involved in the original annotation process (using the criteria stated above). We measured a Cohen's κ of 0.8797, which speaks of the reproducibility of the results and a well defined task [6].

Because we expect that the topics that constitute a "personal" query might vary among the technical support agents, we decided to split the data by technical support agent. When analyzing the data by the technical support agent, we see that each agent issued on average 103 queries (max: 3,097, min: 1, median: 18, stdev: 313), of which on average 40.7 were "personal" (max: 2,141, min: 1, median: 7, stdev: 211) and 82.3 were "business" (max: 1,132, min: 1, median: 17, stdev: 190).

We thus split the data associated with different technical support agents into the data collected from 100 technical support agents to be use as training and 45 for testing. That leaves us with 3,895 unique queries left for training and 1,417 unique query strings for testing, see the last two columns of Table 1.[5]

Table 2. Example Queries

Type	Query
Personal	Amoxicillin alcohol
Business	audio device on high definition audio bus windows xp driver dell
Business	Avaya Voip
Personal	Best poutine store in Montreal

4   Methods

In an earlier version of our system, we experimented with a number of machine learning techniques (particularly, Logistic Regression [19] and Support Vector Machines [14]) and data manipulation approaches (some basic data classes, TF*IDF weighting [13] and Singular Value Decomposition [4]). They all resulted in comparable results[6] and we thus prefer to report on a simpler system that is easier to explain and understand.

---

[5] Given the level of variability between the number of private queries issued by each technical support agent, it comes as no surprise the priors for each class are different when splitting by the technical support agent.

[6] These experiments used an earlier version of the dataset and splitting by queries rather than by technical support agents. There is no value reporting the actual numbers as they are not comparable to our current results.

We filter queries with a classifier with a words-as-features approach [25] using the MALLET Java-based library for machine learning applications in statistical natural language processing [18]. We use a Serial Pipe to process the text using a regular expression that discards non-alphabetic terms. The system uses a Maximum Entropy classifier described next.[7]

Maximum Entropy (MaxEnt) is a machine learning algorithm used in supervised learning to classify data based on a training set. As a supervised learning technique, the algorithm is divided into a training and a test phase. In the training phase a dataset with two fields is provided: (1) the data (queries in the context of this work) and (2) the class ("business" or "personal") associated to each query. In the test phase, the algorithm receives a list of queries and assigns a class to each one.

To train the model, MaxEnt extracts features from the dataset. In the case of text input, the algorithm can extract words, bigrams, phrases, other elements or a combination of them as features. After the extraction the algorithm creates a vector to each class where the content of the cells are the features extracted. Each cell is assigned a weight that indicates the influence of that feature in that class. In our case, we are using sequences of alphabetic characters (without discarding capitalization, but discarding all numbers) as features.

To find the weights, the model uses the maximum entropy principal which in the context of a classification problem means that in a system without any previous assumption the distribution must be the most uniform as possible [20]. The vector of weights is estimated by finding the distribution with the maximum entropy according the conditional probability function that satisfies the given constraints:

$$P_{ME}(c|q, w) = \frac{exp[\sum_i w_i f_i(c, q)]}{\sum_C exp[\sum_i w_i f_i(c, q)]}$$

Where c is the class, C is the set of classes, q is the query and w is the weight vector.

We decided to use a filtering solution rather than a distortion solution as distortion techniques (see Section 6.1) are still not properly understood in terms of query logs [23] and a filtering technique provides clear warranties in terms of privacy preservation. Other solutions involving searching cached sites might be explored in future work [12]. We also chose to filter the data at the backend rather than sanitizing at the acquisition site because that reduces the value of extracted data [27] and we have a small set of queries compared to commercial search engines.

5   Results

Using the split by technical support agents as described in Section 3, we trained a maximum entropy classifier on the 3,895 unique queries (1,378 of which were

---

[7] A MaxEnt classifier with two classes is equivalent to Logistic Regression classifier, as mentioned by an anonymous reviewer.

"personal") and test them on the 1,417 unique query strings, (267 of which were "personal"). The trained classifier identified correctly as "personal" 210 out of 267 (Recall of 78.7%), identifying 317 queries as "personal" (Precision of 66.2%).

These results are more conservative than when doing a split at the query level. Earlier experiments using such approach resulted in a filtering accuracy (recall) higher than 90%.

When evaluating on queries counting duplicates, our system correctly identifies as "personal" 440 out of 590 queries (Recall of 76.5%) and it labels as "personal" a total of 575 queries (Precision of 69.8%).

5.1 Learning Curve

We computed precision and recall on a system trained on 10% of our training data, 20% of our training data and so forth, tested against the full test set. For each cut point, we took a random sample and repeated the sampling and evaluation 5 times. Figure 2 shows the minimum, maximum and median for the different cut points. We can see that for the current approach, adding more data after 1,500 queries has no major impact in terms of precision and recall. At very small amounts of data, recall in the worse cases goes as low as the prior but the median picks up very quickly. A possible explanation for this behavior lies in the simplicity of the words-as-features approach: the machine learning cannot profit from more available training data. More elaborated techniques might fare better.

5.2 Qualitative Error Analysis

We took the 164 prediction errors (57 of which were wrongly predicted as "business" queries) and analyzed them by hand. Almost 90% of the errors in misclassifying "personal" queries could be attributed to two types of errors (Table 3). We also found three types of errors that account for 60% of the (less important for our application) cases of "business" queries misclassified as "personal" queries. We will describe now these types of errors:

Queries containing stop-words (personal). It seems long queries containing many stop-words (prepositions, etc) confuse the current classifier. We might want to train a separate model for this type of queries at a later time.

Personal queries with business terms (personal). Queries looking for help regarding technical issues unrelated to the problems of users (for example, *home banking*). These queries are very challenging for the approach, and most likely will not be addressed at this level of filtering. However, they pose very little privacy threat as they usually deal with fairly standard topics.

Queries with only one term (business). Most likely this *hapax* term is unseen in the training. Given the nature of our task, it makes sense for the system to treat these terms in a more conservative manner and consider them asprivate

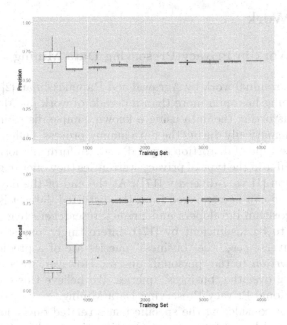

Fig. 2. Learning Curves. Precision and Recall are measured against the retrieval task for the 'personal' class. The ticks indicate maximum and minimum values in the sample run. The curve is through the median values.

matter. These errors would not affect our long term application, as technical support agents tend to extend their queries with extra information over time.
Only numbers (business). A very similar case to the previous type, although as explained in Section 4, we completely omit numbers at this time. Here we should try to distinguish regular numbers from phone numbers and IP addresses. Still, errors of this type are not large enough to have an impact.
Error in gold standard (business). It is very reassuring that there are very few errors of this type.

Table 3. Qualitative Error Analysis

Type	Erroneously Predicted As	Count	Percentage
Queries containing stop-words	Business	31	18.9%
Personal queries with business terms	Business	20	12.2%
Queries with only one term	Personal	54	32.9%
Only numbers	Personal	7	4.3%
Error in gold standard	Personal	6	3.6%
Other errors	Personal	46	28.0%

6   Related Work

6.1   Related Work in Privacy Preserving Data Mining

Starting with the seminal work by Agrawal and Ramakrishnan [2], Privacy Preserving Data Mining has spun more than a decade of work [3]. Most techniques use the idea of distorting the data using a known, simple distribution and then removing it mathematically during the data mining process. Different algorithms exist based on the type of distortion applied, e.g., uniform vs. normal vs. Laplacian. They also differ in the type of privacy warranties they offer (e.g., the concept of k-anonymization [1] vs. l-diversity [17]). At the end of the day, whether the private information can be recovered from the distorted dataset is just an arms race between algorithm developers and privacy researchers (e.g., k-anonymity has been shown to be insufficient by [17]). Interestingly, our situation can be considered the inverse case: the "business" queries are of a predictable nature, applied as a *distortion* to the "personal" queries. However, we are interested in doing data mining over the "business" queries. We believe this set our problem apart from traditional Privacy Preserving Data Mining.

Moreover, when considering the specific issues related query logs, things are not as clear as record-based data mining. Our situation can be considered similar to publishing query logs [11] or systems put forward by privacy-conscious users of commercial search engines to data mining attempts [21]. These techniques do not seem to work well in practice, though [22]. Our filtering approach, compared to privacy preserving data mining offers clear and tangible output that does not need to define the type of attacks it is resilient against. Once a private query has been identified, it is removed from further processing.

6.2   Related Work in Text Classification

We compare related work in classifying short text. In [14], the author classifies articles from the Reuters dataset into classes using Support Vector Machines. One example of classification using short text is classifying scientific publications using the abstract [16]. In this case the shorter text provides less information which increase the complexity of the task, similar to our situation with query logs.

The classification task in social networks has new challenges comparing with the other applications because the text is not written in formal language, there is no revision process and the text is usually very short (in the case of Twitter, the message has at most 140 characters). In [13] the authors created a system that classify messages into different newsgroups using relevance feedback algorithm and term frequency analysis. Examples of more complex applications involving social network data are ongoing research on how to analyze tweets to extract information such as sentiments [9,5] (using Multinomial Naive Bayes, Stochastic Gradient Descent, Hoeffding Tree and Distant Supervision) and political affiliation [23] using Gradient Boosted Decision Trees.

6.3   Other Related Work

Our problem is also similar to collaborative techniques for spam detection [15], although again, we are interested in the background class (that is, using the terms used in spam detection, we are interested in filtering out the *ham* and keeping the *spam*). Our problem is also similar to existing techniques to filter sensitive (adult-themed) queries in multimedia search [7]. While our methods are also similar, our goal is to keep the cohesive class, rather than removing it.

7   Conclusions

We have presented the component of Radialpoint Reveal™ that filters "personal" (non-work related) queries from mixed-use logs captured during work hours at a place of employment. We have shown that a system using a words-as-features approach can accomplish filtering 78.7% of the "personal" queries with a relatively modest amount of training data (5,000+ queries) and only a 9.3% reduction of the available business queries.

Two questions remain at this point. First, there is the issue of whether 78.7% is good enough for maintaining the expectation of privacy of the employees. For an application showing individual queries, that would not be enough. But such application will require a high level of privacy, only Privacy Preserving Data Mining techniques (as discussed in Section 6.1) would be of use. Our long term application (see future work below) is to cluster the queries. We believe this clustering layer (where we also plan to filter "personal" queries a second time, this time at the cluster level [8]) will provide the ultimate warranty in terms of privacy.

As our work is preliminary, we were surprised by the 78.7% accuracy In a sense, "personal" queries are in topics as varied as the queries issued to regular commercial search engines. In the opposite problem to spam detection, we believe our system works not because it is able to identify "personal" queries, but because the "business" queries belong to a much restricted sublanguage that can be easily predicted.

However, there is a threat to the validity of our results that will require a deeper, multi-year study: the sublanguage of technical support queries is time-dependent. As new devices are introduced into the market, new terms will start moving from the "personal" category to the "business" category (imagine, for example, what would happen if Apple introduces a new party invitation system called "Cake"). We hope that a system that is continuously retrained will learn over time to adapt to these changes, something we will discuss in the next section.

Given the sensitivity of this data and its commercial value, there is few available data of this type for research purposes. As an extra contribution of this work, we are making these queries and labels available to other researchers in hope of establishing a benchmark dataset for the task.

To conclude, we have anecdotal information that shows that, even with an explicit feedback mechanism in place for the quality of the information available for technical support agents, they find it ineffective, because when they provide the feedback, it might take several days for the new information to appear in their internal documentation system. As such, the perceived utility of that approach is very low. Our system based on implicit feedback should have a much higher perceived utility, which in turn justifies the use of data mining techniques in the eyes of its users [29].

Future Work

Radialpoint Reveal™ final goal is to cluster the filtered queries to identify gaps in the existing internal documentation available to the technical support agents. We believe the "personal" queries that were not eliminated by our filtering approach will behave as outliers that will not be picked as top level clusters, but this still needs to be experimentally demonstrated. Nonetheless, we also plan to add a final filtering, similar to [8].

Regarding improving the classifier itself, we can improve it by using better features. As our error analysis in Section 5.2 showed, we should have a smart way of dealing with numbers and classes of numeric terms (such as telephone numbers). Moreover, there seem to be a need to divide the classification problem between queries with full phrases and simpler queries.

Moving the classifier to an ensemble of classifiers might also allow us to use temporal information, that is, the predicted class of the query issued by a technical support agent immediately before and after the current query. Such cascades of classifiers have proved useful for high-performance classification [10].

In terms of maintenance for our solution, an intriguing possibility we are considering is to identify weak labels for queries looking at whether there is any known technical support Web site in the first ten results as returned by a commercial search engine. We can compare this technique to an alternative were we evolve the classifier by continuously retraining the classifier using newly classified data by the existing model as gold standard to train a new model.

Finally, we are also interested in alternatives to our current server-based filtering solution. We plan to investigate solutions involving searching cached sites [12] or moving our trained classifier to the client-side, using client-side machine learning algorithms implemented in JavaScript [28].

References

1. Aggarwal, C.C., Philip, S.Y.: A general survey of privacy-preserving data mining models and algorithms. Springer (2008)
2. Agrawal, R., Srikant, R.: Privacy-preserving data mining. ACM Sigmod Record 29(2), 439–450 (2000)
3. Anitha, J., Rangarajan, R.: Researches on privacy-preserving data mining-A comprehensive review. International Journal of Computational Intelligence Research 6(4) (2010)
4. Berry, M.W., Dumais, S.T., O'Brien, G.W.: Using linear algebra for intelligent information retrieval. SIAM Review 37(4), 573–595 (1995)

5. Bifet, A., Frank, E.: Sentiment knowledge discovery in Twitter streaming data. In: Pfahringer, B., Holmes, G., Hoffmann, A. (eds.) DS 2010. LNCS, vol. 6332, pp. 1–15. Springer, Heidelberg (2010)
6. Carletta, J.: Assessing agreement on classification tasks: The kappa statistic. Computational Linguistics 22(2), 249–254 (1996)
7. Chuang, S.-L., Chien, L.-F., Pu, H.-T.: Automatic subject categorization of query terms for filtering sensitive queries in multimedia search. In: Shum, H.-Y., Liao, M., Chang, S.-F. (eds.) PCM 2001. LNCS, vol. 2195, pp. 825–830. Springer, Heidelberg (2001)
8. Fule, P., Roddick, J.F.: Detecting privacy and ethical sensitivity in data mining results. In: Proceedings of the 27th Australasian Conference on Computer Science, vol. 26, pp. 159–166. Australian Computer Society, Inc. (2004)
9. Go, A., Bhayani, R., Huang, L.: Twitter sentiment classification using distant supervision. Processing, 1–6 (2009)
10. Gondek, D.C., Lally, A., Kalyanpur, A., Murdock, W., Duboue, P.A., Zhang, L., Pan, Y., Qiu, Z., Welty, C.: A framework for merging and ranking of answers in DeepQA. IBM Journal of Research and Development 56(3/4), 14:1–14:12 (2012)
11. Gotz, M., Machanavajjhala, A., Wang, G., Xiao, X., Gehrke, J.: Publishing search logs-A comparative study of privacy guarantees. IEEE Transactions on Knowledge and Data Engineering 24(3), 520–532 (2012)
12. Jakobsson, M., Juels, A., Ratkiewicz, J.: Privacy-preserving history mining for web browsers. In: Web 2.0 Security and Privacy (2008)
13. Joachims, T.: A probabilistic analysis of the Rocchio algorithm with TF*IDF for text categorization. Technical report, DTIC Document (1996)
14. Joachims, T.: Text categorization with support vector machines: Learning with many relevant features. In: Nédellec, C., Rouveirol, C. (eds.) ECML 1998. LNCS, vol. 1398, pp. 137–142. Springer, Heidelberg (1998)
15. Li, K., Zhong, Z., Ramaswamy, L.: Privacy-aware collaborative spam filtering. IEEE Transactions on Parallel and Distributed Systems 20(5), 725–739 (2009)
16. Liu, Y., Wu, F., Liu, M., Liu, B.: Abstract sentence classification for scientific papers based on transductive SVM. Computer and Information Science 6(4), 125–131 (2013)
17. Machanavajjhala, A., Kifer, D., Gehrke, J., Venkitasubramaniam, M.: l-diversity: Privacy beyond k-anonymity. ACM Transactions on Knowledge Discovery from Data (TKDD) 1(1), 3 (2007)
18. McCallum, A.K.: Mallet: A machine learning for language toolkit (2002), http://mallet.cs.umass.edu
19. Menard, S.: Applied logistic regression analysis, vol. 106. Sage (2002)
20. Nigam, K., Lafferty, J., McCallum, A.: Using maximum entropy for text classification. In: IJCAI 1999 Workshop on Machine Learning for Information Filtering, pp. 61–67 (1999)
21. Oh, Y., Kim, H., Obi, T.: Privacy-enhancing queries in personalized search with untrusted service providers. IEICE Transaction on Information and Systems 95(1), 143–151 (2012)
22. Peddinti, S.T., Saxena, N.: On the privacy of web search based on query obfuscation: a case study of trackmenot. In: Atallah, M.J., Hopper, N.J. (eds.) PETS 2010. LNCS, vol. 6205, pp. 19–37. Springer, Heidelberg (2010)
23. Pennacchiotti, M., Popescu, A.-M.: Democrats, republicans and starbucks afficionados: User classification in Twitter. In: Proceedings of the 17th ACM SIGKDD International Conference on Knowledge Discovery and Data Mining, pp. 430–438. ACM (2011)

24. Richardson, M.: Learning about the world through long-term query logs. ACM Transactions on the Web (TWEB) 2(4), 21 (2008)
25. Schutze, H.: Dimensions of meaning. In: Proceedings of Supercomputing 1992, pp. 787–796. IEEE (1992)
26. Silvestri, F.: Mining query logs: Turning search usage data into knowledge. Foundations and Trends in Information Retrieval 4(1-2), 1–174 (2010)
27. Sramka, M., Safavi-Naini, R., Denzinger, J., Askari, M., Gao, J.: Utility of knowledge extracted from unsanitized data when applied to sanitized data. In: Sixth Annual Conference on Privacy, Security and Trust, PST 2008, pp. 227–231. IEEE (2008)
28. Umbel, C.: Apparatus, a collection of low-level machine learning algorithms for node.js (January 2012), https://github.com/NaturalNode/apparatus
29. Wahlstrom, K., Roddick, J.F., Sarre, R., Estivill-Castro, V., deVries, D.: On the ethical and legal implications of data mining. Technical Report SIE-06-001, School of Informatics and Engineering, Flinders University, Adelaide, Australia (2006)

Analysis of Feature Maps Selection in Supervised Learning Using Convolutional Neural Networks

Joseph Lin Chu and Adam Krzyżak

Department of Computer Science and Software Engineering, Concordia University,
Montreal, Quebec, Canada
jo_chu@encs.concordia.ca, krzyzak@cs.concordia.ca

Abstract. Artificial neural networks have been widely used for machine learning tasks such as object recognition. Recent developments have made use of biologically inspired architectures, such as the Convolutional Neural Network. The nature of the Convolutional Neural Network is that each convolutional layer of the network contains a certain number of feature maps or kernels. The number of these used has historically been determined on an ad-hoc basis. We propose a theoretical method for determining the optimal number of feature maps using the dimensions of the feature map or convolutional kernel. We find that the empirical data suggests that our theoretical method works for extremely small receptive fields, but doesn't generalize as clearly to all receptive field sizes. Furthermore, we note that architectures that are pyramidal rather than equally balanced tend to make better use of computational resources.

1 Introduction

The earliest of the hierarchical Artificial Neural Networks (ANNs) based on the visual cortex's architecture was the Neocognitron, first proposed by Fukushima & Miyake [7]. This network was based on the work of neuroscientists Hubel & Wiesel [8], who showed the existence of Simple and Complex Cells in the visual cortex. Subsequently, LeCun et al [10] developed the Convolutional Neural Network (CNN), which made use of multiple Convolutional and Subsampling layers, while also using stochastic gradient descent and backpropagation to create a feed-forward network that performed exceptionally well on image recognition tasks. The Convolutional Layer of the CNN is equivalent to the Simple Cell Layer of the Neocognitron, while the Subsampling Layer of the CNN is equivalent to the Complex Cell Layer of the Neocognitron. Essentially they delocalize features from the visual receptive field, allowing such features to be identified with a degree of shift invariance. This unique structure allows the CNN to have two important advantages over a fully-connected ANN. First, is the use of the local receptive field, and second is weight-sharing. Both of these advantages have the effect of decreasing the number of weight parameters in the network, thereby making computation of these networks easier.

M. Sokolova and P. van Beek (Eds.): Canadian AI 2014, LNAI 8436, pp. 59–70, 2014.

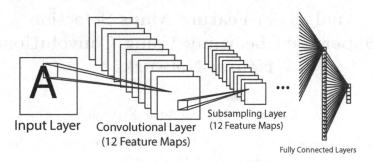

Input Layer   Convolutional Layer   Subsampling Layer
               (12 Feature Maps)   (12 Feature Maps)

Fully Connected Layers

Fig. 1. The basic architecture of the CNN

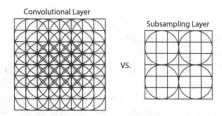

Fig. 2. A comparison between the Convolutional layer and the Subsampling layer. Circles represent the receptive fields of the cells of the layer subsequent to the one represented by the square lattice. On the left, an 8 x 8 input layer feeds into a 6 x 6 convolutional layer using receptive fields of size 3 x 3 with an offset of 1 cell. On the right, a 6 x 6 input layer feeds into a 2 x 2 subsampling layer using receptive fields of size 3 x 3 with an offset of 3 cells.

2   Theoretical Analysis

CNNs have particularly many hyper-parameters due to the structure of the network. Determining the optimal hyper-parameters can appear to be a bit of an art. In particular, the number of feature maps for a given convolutional layer tends to be chosen based on empirical performance rather than on theoretical justifications [14]. Numbers in the first convolutional layer range from very small (3-6) [12], [10], to very large (96-1600) [9], [3], [4], [5].

One wonders then, if there is some sort of theoretical rationale that can be used to determine the optimal number of feature maps, given other hyper-parameters. In particular, one would expect that the dimensions of the receptive field ought to have some influence on this optimum.

A receptive field of width r consists of r^2 elements or nodes. If we have feature maps m then, the maximum number of possible feature maps before duplication, given an 8-bit grey scale image, is 256^{r^2}. Since the difference between a grey level of say, 100 and 101, is roughly negligible, we simplify and reduce the number of bins in the histogram so to speak from 256 to 2. Looking at binary features as a way of simplifying the problem is not unheard of [2]. So, given a binary image the number of possible binary feature maps before duplication is

$$\Omega = 2^{r^2},\qquad(1)$$

which is still a rapidly increasing number.

Let's look at some very simple receptive fields. Take one of size 1x1. How many feature maps would it take before additional maps become completely redundant? Ω would suggest two. For receptive field sizes 2x2, 3x3, and 4x4, we get 16, 512, and 65536, respectively. These values represent an upper bound, beyond which additional feature maps should not improve performance significantly. But clearly, not even all of these feature maps would be all that useful. If we look again at a 2x2 receptive field, regarding those 16 non-redundant feature maps, shown in Figure 3, are all of them necessary? Assuming a higher layer that combines features, many of these are actually redundant in practice.

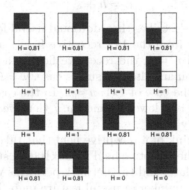

Fig. 3. The 16 possible binary feature maps of a 2x2 receptive field, with their respective entropy values

So how do we determine which ones are useful? Borrowing from Information Theory, we can look at how much information each map encodes. Consider Shannon entropy or the amount of information given

$$H(X) = -\sum_{i=1}^{n} p(x_i) log_2 p(x_i).\qquad(2)$$

We can calculate the Shannon entropy of each feature map, again, shown in Figure 3. What's interesting here is that we can group the Shannon entropy values together. In the 2x2 case, there are six patterns equal to an entropy of 1, while there are eight patterns with an entropy of around 0.81, and two patterns with an entropy of 0. Thus we have three bins of entropy values so to speak.

Thus, we hypothesize a very simple theoretical method, one that admittedly simplifies a very complex problem. Shannon entropy has the obvious disadvantage that it does not tell us about the spatial relationship between neighbouring pixels. And again, we assume binary feature maps. Nevertheless, we propose this as an initial attempt to approximate the underlying reality.

The number of different possible entropy values for the total binary feature map set of a particular receptive field size is determined by considering the number of elements in a receptive field r^2. The number of unique ways you can fill the binary histogram of the possible feature maps then is $r^2 + 1$. But roughly half the patterns are inverses of each other. So the actual number of unique entropy values is $(r^2 + 1)/2$ if r is odd. Or $(r^2)/2 + 1$ if r is even.

Given the number of different entropy values

$$h(r) = \begin{cases} \frac{r^2+1}{2} & if\ r\ is\ odd \\ \frac{r^2}{2} + 1 & if\ r\ is\ even \end{cases} \tag{3}$$

the number of useful feature maps is

$$u = h + s, \tag{4}$$

where s is a term that describes the additional useful feature maps above this minimum of h. We know from the 1x1 receptive field feature map set that s is at least 1, because when $r = 1$, the receptive field is a single pixel filter, and optimally functions as a binary filter. In such a case, $u = \Omega = 2$. In the minimum case that $s = 1$, then in the case of 1x1, $u = 2$. In the case of 2x2, $u = 4$. In the case of 3x3, $u = 6$. This is a lower bound on u that we shall use until we can determine what s actually is.

To understand this, think of a receptive field that takes up the entire image. So for a 100x100 image, the receptive field is 100x100. In such a case, each feature map is in essence a template, and the network performs what is in essence template matching. Thus the number of useful feature maps is the number of useful templates. With enough templates, you can approximate the data set, any more would be unnecessary. To determine how many such templates would be useful, consider the number of different Shannon entropies in the feature set. While it is not guaranteed that two templates with the same entropy would be identical, two templates with different entropies are certainly different. Also consider the difference between that 100x100 receptive field, and a 99x99 receptive field. The differences between the two in terms of number of useful feature maps intuitively seems negligible. This suggests that s is either a constant, or at most a linearly increasing term. Also, as h increases, the distance between various entropies decreases, to the point where many of the values start to become nearly the same. Thus, one can expect that for very high values of r, u will be too high.

Any less than u and the theory predicts a drop in performance. Above this number the theory is agnostic about one of three possible directions. Either the additional feature maps won't affect the predictive power of the model, so we could see a plateau, or the Curse of Dimensionality will cause the predictive power of the model to begin to drop, or as seen in other papers such as [4], the predictive power of the model will continue to increase, albeit at a slower rate.

Thus far we have taken care of the first convolutional layer. For the second convolutional layer and beyond, the question arises of whether or not to stick to this formula for u, or whether it makes more sense to increase the number of feature maps in some proportion to the number in the previous convolutional layer. Upper convolutional layers are not simply taking the pixel intensities, but instead, combining the feature maps of the lower layer. In which case, it makes sense to change the formula for u for upper layers to:

$$u_l = vu_{l-1}, \tag{5}$$

where v is some multiplier value and l is the layer. Candidates for this value range from u_{l-1} itself, to some constant such as 2 or 4.

There is no substitute for empirical evidence, so we test the theory by running experiments to exhaustively search for the hypothetical, optimal number of feature maps.

3   Methodology

To speed up and simplify the experiments, we devised, using the Caltech-101 dataset [6], a specialized dataset, which we shall refer to as the Caltech-20. The Caltech-20 consists of 20 object categories with 50 images per category total, divided into a training set of 40 images per category, and a test set of 10 images per category. The 20 categories were selected by finding the 20 image categories with the most square average dimensions that also had at least 50 example images. The images were also resized to 100 x 100 pixels, with margins created by irregular image dimensions zero-padded (essentially blacked out). To simplify the task so as to have one channel rather than three, the images were also converted to greyscale. The training set totalled 800 images while the test set consisted of 200 images. Some example images are shown in figure 4.

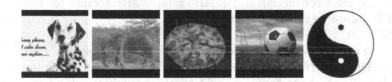

Fig. 4. Images from the Caltech-20 data set

CNNs tend to require a fairly significant amount of time to train. One way to improve temporal performance is to implement these ANNs such that they are able to use the Graphical Processing Unit (GPU) of certain video cards rather than merely the CPU of a given machine [15], [16], [13], [9]. NVIDIA video cards in particular have a parallel computing platform called Compute Unified Device Architecture (CUDA) that can take full advantage of the many cores on a typical video card to greatly accelerate parallel computing tasks. ANNs are quite parallel in nature, and thus quite amenable to this. Thus, for our implementation of the CNN, we turned to the Python-based Theano library (http://deeplearning.net/software/theano/) [1]. We were able to find appropriate Deep Learning Python scripts for the CNN. Our tests suggest that the speed of the CNN using the GPU improved by a factor of eight, as compared to just using the CPU.

CNNs require special consideration when implementing their architecture. A method was devised to calculate a usable set of architecture parameters. The relationship between layers can be described as follows. To calculate the reasonable dimensions of a square layer from either its previous layer (or next layer) in the hierarchy requires at least some of the following variables to be assigned. Let x be the width of the previous (or current) square layer. Let y be the width of the current (or next) square layer. Let r be the width of the square receptive field of nodes in the previous (or current) layer to each current (or next) layer node, and f be the offset distance between the receptive fields of adjacent nodes in the current (or next) layer. The relationship between these variables is best described by the equation below.

$$y = \frac{x - (r - f)}{f}, \tag{6}$$

where, $x \geq y$, $x \geq r \geq f$, and $f > 0$.
For convolutional layers this generalizes because f = 1, to:

$$y = x - r + 1. \tag{7}$$

For subsampling layers, this generalizes because r = f, to:

$$y = \frac{x}{f}. \tag{8}$$

From this we can determine the dimensions of each layer. To describe a CNN, we adopt a similar convention to [3]. An example architecture for the CNN on the NORB dataset [11] can be written out as:

$96 \times 96 \rightarrow 8C5 \rightarrow S4 \rightarrow 24C6 \rightarrow S3 \rightarrow 24C6 \rightarrow 100N \rightarrow 5N$,

where the number before C is the number of feature maps in a convolutional layer, the number after C is the receptive field width in a convolutional layer, the number after S is the receptive field width of a subsampling layer, and the number before N is the number of nodes in a fully connected layer.

For us to effectively test a single convolutional layer, we use a series of architectures, where v is a variable number of feature maps:

$100 \times 100 \rightarrow vC1 \rightarrow S2 \rightarrow 500N \rightarrow 20N$
$100 \times 100 \rightarrow vC2 \rightarrow S3 \rightarrow 500N \rightarrow 20N$
$100 \times 100 \rightarrow vC3 \rightarrow S2 \rightarrow 500N \rightarrow 20N$
$100 \times 100 \rightarrow vC5 \rightarrow S3 \rightarrow 500N \rightarrow 20N$
$100 \times 100 \rightarrow vC99 \rightarrow S1 \rightarrow 500N \rightarrow 20N$

The reason why we sometimes use 3x3 subsampling receptive fields is that the size of the convolved feature maps are divisible by 3 but not 2. Otherwise we choose to use 2x2 subsampling receptive fields where possible. We find that, with the exception of the unique 99x99 receptive field, not using subsampling produces too many features and parameters and causes the network to have difficulty learning. The method of subsampling we use is max-pooling, which involves taking the maximum value seen in the receptive field of the subsampling layer.

For testing multiple convolutional layers, we use the following architecture, where v_i is a variable number of feature maps for each layer:
$$100 \times 100 \rightarrow v_1C2 \rightarrow S3 \rightarrow v_2C2 \rightarrow S2 \rightarrow v_3C2 \rightarrow S3 \rightarrow v_4C2 \rightarrow S2 \rightarrow v_5C2 \rightarrow S1 \rightarrow 500N \rightarrow 20N$$

All our networks use the same basic classifier, which is a multi-layer Perceptron with 500 hidden nodes and 20 output nodes. Various other parameters for the CNN were also experimented with to determine the optimal parameters to use in our experiments. We eventually settled on 100 epochs of training. The CNN learning rate and learning rate decrement parameters were determined by trial and error. The learning rate was initially set to 0.1, and gradually decremented to approximately 0.001.

4   Analysis and Results

The following figures are intentionally fitted with a trend line that attempts to test the hypothesis that a cubic function approximates the data. It should not be construed to suggest that this is in fact the underlying function.

Figure 5 shows the results of experimenting with different numbers of feature maps on the accuracy of the CNN trained on the Caltech-20, and using a 1x1 receptive field.

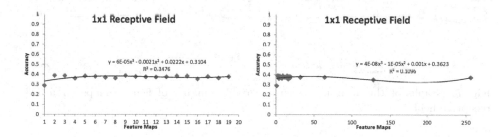

Fig. 5. Graphs of the accuracy given a variable number of feature maps for a 1x1 receptive field

As can be seen, the accuracy quickly increases between 1 and 2 feature maps, and then levels off for more than 2 feature maps. This is consistent with the theory, albeit, the plateau beyond seems to be neither increasing nor decreasing, which suggests some kind of saturation point around 2.

Figure 6 shows the results of experimenting with different numbers of feature maps on the accuracy of the CNN trained on the Caltech-20, and using a 2x2 receptive field.

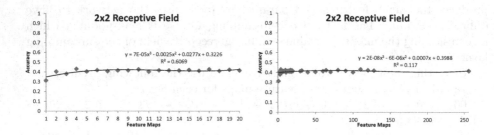

Fig. 6. Graphs of the accuracy given a variable number of feature maps for a 2x2 receptive field

As can be seen, the accuracy quickly increases between 1 and 4 feature maps, and then levels off for more than 4 feature maps. This is consistent with the theory where $s = 1$. The plateau beyond seems to be neither increasing nor decreasing, which suggests some kind of saturation point around 4.

Figure 7 shows the results of experimenting with different numbers of feature maps on the accuracy of the CNN trained on the Caltech-20, and using a 3x3 receptive field.

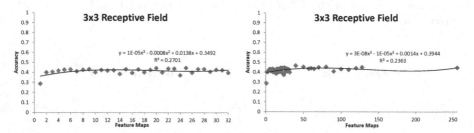

Fig. 7. Graphs of the accuracy given a variable number of feature maps for a 3x3 receptive field

As can be seen, the accuracy increases between 1 and 6 feature maps, and then proceeds to plateau somewhat erratically. Unlike the previous receptive field sizes however, there are accuracies greater than that found at u. The plateau also appears less stable.

Figure 8 shows the results of experimenting with different numbers of feature maps on the accuracy of the CNN trained on the Caltech-20, and using a 5x5 receptive field.

As can be seen, the accuracy quickly increases between 1 and 4 feature maps, and then plateaus for a while before spreading out rather chaotically. This may be explained by some combination of overfitting, the curse of dimensionality, or too high a learning rate causing failure to converge.

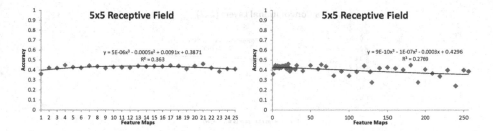

Fig. 8. Graphs of the accuracy given a variable number of feature maps for a 5x5 receptive field

Figure 9 show the results of experimenting with different numbers of feature maps on the accuracy of the CNN trained on the Caltech-20, and using a 99x99 receptive field.

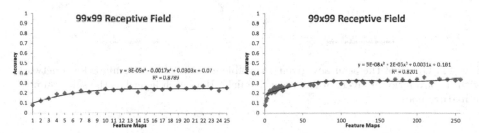

Fig. 9. Graphs of the accuracy given a variable number of feature maps for a 99x99 receptive field

As can be seen, the accuracy steadily increases between 1 and 210 feature maps, and then begins to plateau. As expected, our u value of 4901 is much too high for the very large value of $r = 99$.

Lastly, we look at the effect of multiple convolutional layers. Figure 10 shows what happens when the first layer of a 12 layer CNN with 5 convolutional layers is held constant at 4 feature maps, and the upper layers are multiplied by the number in the layer before it. So for a multiple v of 2, the feature maps for each layer would be 4, 8, 16, 32, 64, respectively. We refer to this as the pyramidal structure.

Clearly, increasing the number of feature maps in the upper layers has a significant impact. Perhaps not coincidentally, the best performing architecture we have encountered so far was this architecture with 4, 20, 100, 500, and 2500 feature maps in each respective layer. It achieved an accuracy of 54.5% on our Caltech-20 data set. This can be contrasted with the effect of having the same number of feature maps in each layer as shown in figure 11. We refer to this as the equal structure.

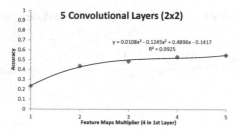

Fig. 10. Graph of the accuracy given a variable number of feature maps for a network with 5 convolutional layers of 2x2 receptive field. Here the higher layers are a multiple of the lower layers.

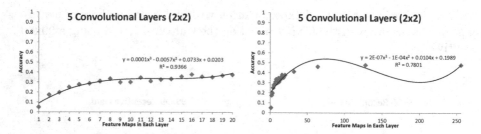

Fig. 11. Graph of the accuracy given a variable number of feature maps for a network with 5 convolutional layers of 2x2 receptive field. Here each layer has the same number of feature maps.

While the accuracy rises with the number of feature maps as well, it should be noted that for the computational cost, the pyramidal structure appears to be a better use of resources than the equal structure.

5   Discussion

It appears that the theoretical method seems to hold well for receptive fields of size 1x1, and 2x2. For larger sizes, the data is not as clear. The data from the 3x3 and 5x5 receptive field experiments suggests that there can be complicating factors involved that cause the data to spread. Such factors could include, the curse of dimensionality, and also, technical issues such as failure to converge due to too high a learning rate, or overfitting. As our experimental set up is intentionally very simple, we lack many of the normalizing methods that might otherwise improve performance. The data from the 99x99 receptive field experiment is interesting because it starts to plateau much sooner than predicted by the equation for u. However, we mentioned before that this would probably happen with the current version of u. The number of different entropies at $r = 99$ are probably very close together and an improved u equation should take this into account.

It should also be emphasized that our results could be particular to our choice of hyper-parameters such as learning rate and our choice of a very small dataset.

Nevertheless, what we do not find, is the clear and simple monotonically increasing function seen in [4], and [5]. Rather, the data shows that after an initial rise, the function seems to plateau and it is uncertain whether it can be construed to be rising or falling or stable.

This is not the case with highly layered networks however, which do appear to show a monotonically increasing function in terms of increasing the number of feature maps. However, this could well be due to the optimal number of feature maps in the last layer being exceedingly high due to multiplier effects.

One thing that could considerably improve our work would be finding some kind of measure of spatial entropy rather than relying on Shannon entropy. The problem with Shannon entropy is of course, that it does not consider the potential information that comes from the arrangement of neighbouring pixels. We might very well improve our estimates of u by taking into consideration the spatial entropy in h, rather than relying on the s term.

Future work should likely include looking at what the optimal receptive field size is. Our experiments hint at this value as being greater than 3x3 and [4] suggests that it is less than 8x8, but performing the exhaustive search without knowing the optimal number of feature maps for each receptive field size is a computationally complex task.

As with [9], we find that more convolutional layers seems to improve performance. The optimal number of such layers is something else that should be looked at in the future.

6   Conclusions

Our experiments provided some additional data to consider for anyone interested in optimizing a CNN. Though the theoretical method is not clear beyond certain extremely small or extremely large receptive fields, it does suggest that there is some relationship between the receptive field size and the number of useful feature maps in a given convolutional layer of a CNN. It nevertheless may prove to be a useful approximation.

Our experiments also suggest that when comparing architectures with equal numbers of feature maps in each layer with architectures that have pyramidal schemes where the number of feature maps increase by some multiple, that the pyramidal methods are a more effective use of computing resources.

References

1. Bergstra, J., Breuleux, O., Bastien, F., Lamblin, P., Pascanu, R., Desjardins, G., Turian, J., Warde-Farley, D., Bengio, Y.: Theano: A CPU and GPU math expression compiler. In: Proceedings of the Python for Scientific Computing Conference, June 2010, SciPy (2010) oral Presentation

2. Boureau, Y.L., Ponce, J., LeCun, Y.: A theoretical analysis of feature pooling in visual recognition. In: Proceedings of the 27th International Conference on Machine Learning (ICML 2010), pp. 111–118 (2010)
3. Ciresan, D., Meier, U., Schmidhuber, J.: Multi-column deep neural networks for image classification. In: 2012 IEEE Conference on Computer Vision and Pattern Recognition (CVPR), pp. 3642–3649. IEEE (2012)
4. Coates, A., Ng, A.Y., Lee, H.: An analysis of single-layer networks in unsupervised feature learning. In: International Conference on Artificial Intelligence and Statistics, pp. 215–223 (2011)
5. Eigen, D., Rolfe, J., Fergus, R., LeCun, Y.: Understanding Deep Architectures using a Recursive Convolutional Network. ArXiv e-prints (December 2013)
6. Fei-Fei, L., Fergus, R., Perona, P.: Learning generative visual models from few training examples: An incremental bayesian approach tested on 101 object categories. In: IEEE CVPR 2004, Workshop on Generative-Model Based Vision (2004)
7. Fukushima, K., Miyake, S.: Neocognitron: A new algorithm for pattern recognition tolerant of deformations and shifts in position. Pattern Recognition 15(6), 455–469 (1982)
8. Hubel, D.H., Wiesel, T.N.: Receptive fields, binocular interaction and functional architecture in a cats visual cortex. Journal of Physiology (London) 160, 106–154 (1962)
9. Krizhevsky, A., Sutskever, I., Hinton, G.: Imagenet classification with deep convolutional neural networks. Advances in Neural Information Processing Systems 25, 1106–1114 (2012)
10. LeCun, Y., Bottou, L., Bengio, Y.: Gradient-based learning applied to document recognition. Proceedings of the IEEE 86(11), 2278–2324 (1998)
11. LeCun, Y., Huang, F., Bottou, L.: Learning methods for generic object recognition with invariance to pose and lighting. In: IEEE Computer Society Conference on Computer Vision and Pattern Recognition (CVPR), vol. 2, pp. 97–104 (2004)
12. Nguyen, G.H., Phung, S.L., Bouzerdoum, A.: Reduced training of convolutional neural networks for pedestrian detection. In: International Conference on Information Technology and Applications (2009)
13. Scherer, D., Schulz, H., Behnke, S.: Accelerating large-scale convolutional neural networks with parallel graphics multiprocessors. In: Diamantaras, K., Duch, W., Iliadis, L.S. (eds.) ICANN 2010, Part III. LNCS, vol. 6354, pp. 82–91. Springer, Heidelberg (2010)
14. Simard, P., Steinkraus, D., Platt, J.C.: Best practices for convolutional neural networks applied to visual document analysis. In: ICDAR, vol. 3, pp. 958–962 (2003)
15. Uetz, R., Behnke, S.: Large-scale object recognition with cuda-accelerated hierarchical neural networks. In: IEEE International Conference on Intelligent Computing and Intelligent Systems, ICIS 2009, vol. 1, pp. 536–541. IEEE (2009)
16. Uetz, R., Behnke, S.: Locally-connected hierarchical neural networks for gpu-accelerated object recognition. In: NIPS 2009 Workshop on Large-Scale Machine Learning: Parallelism and Massive Datasets (2009)

Inconsistency versus Accuracy of Heuristics

Hang Dinh*

Department of Computer & Information Sciences
Indiana University South Bend
htdinh@iusb.edu

Abstract. Many studies in heuristic search suggest that the accuracy of the heuristic used has a positive impact on improving the performance of the search. In another direction, historical research perceives that the performance of heuristic search algorithms, such as A* and IDA*, can be improved by requiring the heuristics to be consistent – a property satisfied by any perfect heuristic. However, a few recent studies show that inconsistent heuristics can also be used to achieve a large improvement in these heuristic search algorithms. These results raise a natural question: *which property of heuristics, accuracy or consistency/inconsistency, should we focus on when building heuristics?*

In this article, we investigate the relationship between the inconsistency and the accuracy of heuristics with A* search. Our analytical result reveals a correlation between these two properties. We then run experiments on the domain for the Knapsack problem with a family of practical heuristics. Our empirical results show that in many cases, the more accurate heuristics also have higher level of inconsistency and result in fewer node expansions by A*.

1 Introduction

Heuristic search has been playing a practical role in solving hard problems. One of the most popular heuristic algorithms is A* search [6], which is essentially best-first search with an additive evaluation $f(x) = g(x) + h(x)$, where $g(x)$ is the cost of the current path from the start node to node x, and $h(x)$ is an estimation of the least cost $h^*(x)$ from x to a goal node. The function h is called a *heuristic function*, or *heuristic* for short. An important property of A* search is its admissibility: A* will always return an optimal solution if the heuristic h it uses is *admissible*, meaning $h(x)$ never exceeds $h^*(x)$.

Research on A* and other similar heuristic search algorithms, such as IDA* [11], has focused on understanding the impact of properties of the heuristic function on the quality of the search. A well-studied subclass of admissible heuristics is the one with the *consistency* property. Heuristic h is called *consistent* if $h(x) \leq c^*(x, x') + h(x')$ for all pairs of nodes (x, x'), where $c^*(x, x')$ is the least cost from x to x'. The definition of consistency can be simplified due to its equivalence to *monotonicity*.

---

* This work was supported by IU South Bend Faculty Research Grant.

M. Sokolova and P. van Beek (Eds.): Canadian AI 2014, LNAI 8436, pp. 71–82, 2014.

Consistency was introduced in the original A* paper [6] and later became a desirable property of admissible heuristics for two perceptions. First, since the perfect heuristic h^* is consistent, it is expected that a good heuristic should also be consistent. As indicated in [14, p. 82], consistency is believed to enable A* to forego reopening nodes, and thus can reduce the number of node expansions. Second, inconsistent admissible heuristics seem rare. In fact, it is assumed by many researchers [12] that "almost all admissible heuristics are consistent."

The portrait of inconsistent heuristics was usually painted negatively until a discovery by Zahavi et al. [19] that inconsistency is actually not that bad. They demonstrated by empirical results that in many cases, inconsistency can be used to achieve large performance improvements of IDA*. They then promoted the use of inconsistent heuristics and showed how to turn a consistent heuristic into an inconsistent heuristic using the *bidirectional pathmax* (BPMX) method of Felner et al. [3]. Follow-up studies [4, 21] have also provided positive results of inconsistent heuristics with A* search and encouraged researchers to explore inconsistency as a means to further improve the performance of A*.

In another line of research on heuristics, there have been extensive investigations on the impact of the heuristic accuracy on the performance of A* (and IDA*). Most studies [16, 5, 13, 17, 1, 2] in this line support the intuition that in many search spaces, improving accuracy of the heuristic can improve the efficiency of A*. There are, however, a few opposite results [9, 10, 7] which show that improving heuristic accuracy does not lead to any significant improvement in the search performance. Some of the opposite results [9, 10] on the benefit of heuristic accuracy were actually obtained under the assumption that the heuristic is consistent. Other opposite results only apply to optimal planning domains [7], or contrived search spaces with an overwhelming number of goal nodes [2].

In light of the newly discovered benefit of inconsistent heuristics and the well-established positive results on the accuracy of heuristics, it is natural to ask *which property, consistency/inconsistency or accuracy, of heuristics would have stronger impact on the performance of A*?. Is there any relationship between these properties of heuristics?* The goal of study is to partially address these questions.

In this work, we first analyze a correlation between inconsistency and accuracy of heuristics. Our analytical result reveals that the level of inconsistency of a heuristic can serve as an upper bound on the level of accuracy of the heuristic (see Theorem 1 for details.) We then investigate the relationship between the inconsistency and accuracy of heuristics, as well as their impacts on the performance of A*, by running experiments on a practical domain for the Knapsack problem taken from [2].

Our study differs from the previous works [4, 21] on inconsistent heuristics with A* in both the search space used and the construction of heuristics. While Felner et al. [4] and Zhang et al. [21] use undirected graphs and focus on the reduction in node re-expansions as a benefit of inconsistency, our experiments are done on a directed acyclic graph on which A* will never reopen nodes, regardless of the heuristic used. For this search graph, we use a family of heuristics that arise in practice, which allow us to compare the inconsistency level and the accuracy level of many heuristics within this family. Recall that Felner et al. [4] and Zhang et al. [21] incorporated BPMX into A* and compared the performance of A* with other less well-known heuristic algorithms (B, B', C).

Zahavi et al. [20] also introduced a method for predicting the performance of IDA* with inconsistent heuristics. While this prediction can be accurate in many domains, it only allows us to predict the performance of the search, i.e., the number of node expansions, rather than the accuracy of the heuristics, i.e., how close a heuristic value is to the actual value. Since a good (or bad) performance of A* search does not always imply that the heuristic used is accurate (or inaccurate), the relationship between inconsistency and accuracy of the heuristics remains open from the study of Zahavi et al. [20].

In summary, our work shall focus on the relationship between inconsistency and accuracy of heuristics, while previous works focus on the relationship between either inconsistency or accuracy of heuristics and the performance of the search. We are not attempting to predict the performance of A* based on the heuristic's inconsistency and/or accuracy, but rather use the search's performance as a factor to investigate the relationship between these two properties of heuristics.

2   Preliminaries

Firstly, we would like to review basic background on A* search and introduce our notation.

A typical search problem for A* is defined by an edge-weighted search graph G with a start node and a set of goal nodes. For each graph G, we will use $V(G)$ and $E(G)$ to denote the set of vertices and the set of edges of G. We will denote a general search space for A* as (G, c, x_0, S), where G is a directed graph, $c : E(G) \to \mathbb{R}^+$ is a function assigning a positive cost to each edge, $x_0 \in V(G)$ is the start node, and $S \subset V(G)$ is the set of goal nodes. When x_0 and S are not important in the current context, we may only write (G, c). Given a search space (G, c), for each node $x \in V(G)$, let $h^*(x)$ denote the cost of a least-cost path from x to a goal node.

A heuristic function on a search space (G, c) is a function $h : V(G) \to \mathbb{R}^+$, where $h(v)$ is an estimation of $h^*(x)$, for each $x \in V(G)$. Since $h^*(s) = 0$ for every goal node s, we will assume that a heuristic function must have value zero at every goal node. We will write A*(h) to refer to the A* search using heuristic h. Recall that A*(h) is a specialized best-first search algorithm with the evaluation function $f(x) = g(x) + h(x)$, where $g(x)$ is the cost of the *current* path from the start node to node x. Details of the A*(h) search on search space (G, c, x_0, S) can be found in [14, p. 64]. The efficiency of A* is usually measured by the number of node expansions.

Consistency and monotonicity. Consistency is in fact equivalent to *monotonicity* [14, Thm. 8, p. 83]. Precisely, heuristic h on a search space (G, c) is *consistent* if and only if $h(x) \leq c(x, x') + h(x')$ for all edges $(x, x') \in E(G)$.

3   Inconsistency and Accuracy

We now analyze the relationship between inconsistency and accuracy of heuristics. We begin with introducing metrics characterizing the inconsistency of a heuristic.

To characterize the level of inconsistency of a heuristic h, Zahavi et al. [19] defined the following two terms:

- *Inconsistency rate of an edge* (IRE): For each edge $e = (u, v)$, let $IRE(h, e) = |h(u) - h(v)|$. The IRE of h is the average $IRE(h, e)$ over all edges e of the search space.
- *Inconsistency rate of a node* (IRN): For each node v, let $IRN(h, v)$ be the maximal value of $|h(u) - h(v)|$ for any node u adjacent to v. The IRN of h is the average $IRN(h, v)$ over all nodes v of the search space.

Note that neither IRN nor IRE defined above takes into account the edge costs. If the search space has uniform edge cost, we can say that a consistent heuristic has IRN or IRE at most 1. But if the search space has nonuniform edge costs, we are unable to determine if a heuristic is consistent by just looking up its IRN or IRE. Additionally, the metrics IRN and IRE of Zahavi et al. [19] were defined for undirected graphs, which are not suitable for the case of search graphs considered in this paper. Therefore, we define other metrics for the inconsistency to overcome these shortcomings.

Definition 1. *Let h be a heuristic on a search space (G, c). The weighted inconsistency rate of h at edge $e = (x, x')$ is*

$$\mathsf{WIRE}(h, e) \stackrel{\text{def}}{=} \frac{h(x) - h(x')}{c(x, x')}.$$

The weighted inconsistency rate of h, denoted $\mathsf{WIRE}(h)$, is defined as the average of $\mathsf{WIRE}(h, e)$ over all edges $e = (x, x') \in E(G)$ where x is a non-goal node.

The notion of WIRE can be seen as a weighted analog of IRE with two minor caveats. First, we use $(h(x) - h(x'))$ instead of the absolute value $|h(x) - h(x')|$, since the graphs we consider are directed. Second, when computing $\mathsf{WIRE}(h)$, we do not count $\mathsf{WIRE}(h, e)$ for edges e from a goal node, because there will be no node expansion made from a goal node. More precisely, if $e = (x, x')$ and x is a goal node, then $\mathsf{WIRE}(h, e)$ has no impact on the search quality.

Clearly, if h is consistent, then $\mathsf{WIRE}(h) \leq 1$. The converse, however, is not necessarily true. Thus, we define the following metric that can be used to determine if a heuristic is consistent or inconsistent.

Definition 2. *Let h be a heuristic on a search space (G, c). We say that h is inconsistent at node x if $h(x) > c(x, x') + h(x')$ for some direct successor x' of x, i.e., $(x, x') \in E(G)$. The inconsistent node rate of h, denoted $\mathsf{INR}(h)$, is the ratio of the number of non-goal nodes at which h is inconsistent over the number of all non-goal nodes.*

In other words, $\mathsf{INR}(h)$ is the probability that h is inconsistent at a random non-goal node. Intuitively, the larger $\mathsf{INR}(h)$, the more inconsistent the heuristic h is. Note that since the heuristic value of any goal node is zero, a heuristic is never inconsistent at a goal node. Hence, we have the following fact:

Fact 1 *Let h be any heuristic. Then h is inconsistent if and only if $\mathsf{INR}(h) > 0$.*

For the accuracy metrics of heuristics, we will adopt the accuracy notion that measures the distance between the heuristic value and the actual value by a multiplicative factor, which has also been adopted in many previous works [5, 13, 17, 1, 2]

Definition 3. *Let h be a heuristic function on a search space (G, c, x_0, S). For any non-goal node x, we define the accuracy rate of h at node x to be*

$$\mathsf{ARN}(h, x) \stackrel{\text{def}}{=} \frac{h(x)}{h^*(x)}.$$

The accuracy rate of h, denoted $\mathsf{ARN}(h)$, is the average of $\mathsf{ARN}(h, x)$ for all non-goal nodes $x \in V(G) \setminus S$. The accuracy rate of h at the start node x_0 will be denoted $\mathsf{ARS}(h)$. That is,

$$\mathsf{ARS}(h) \stackrel{\text{def}}{=} \frac{h(x_0)}{h^*(x_0)}.$$

This notion of accuracy rate is particularly meaningful for admissible heuristics. Intuitively, if h is admissible, then the larger $\mathsf{ARN}(h, x)$, the more accurate the heuristic is at node x. The accuracy rate is in fact related to the informedness of admissible heuristics: for any two admissible heuristics h_1 and h_2 on the same search space, h_1 is more informed than h_2 iff $\mathsf{ARN}(h_1, x) > \mathsf{ARN}(h_2, x)$ for all non-goal node x.

We will now prove a basic relationship between weighted inconsistency rate and accuracy rate.

Theorem 1. *Let h be a heuristic on a search space (G, c) and $x \in V(G)$. If $\mathsf{WIRE}(h, e) \le \omega$ for all edges e along a least-cost path from x to a goal node, then $\mathsf{ARN}(h, x) \le \omega$.*

Proof. Let $(x_1, \ldots, x_\ell)$ be a least-cost path from x to a goal node, where $x_1 = x$ and x_ℓ is a goal node, and assume $\mathsf{WIRE}(h, e) \le \omega$ for all edges along this path. Then

$$h(x_i) - h(x_{i+1}) \le \omega \cdot c(x_i, x_{i+1}) \quad \forall i = 1, \ldots, \ell - 1.$$

On the other hand, $h^*(x) = \sum_{i=1}^{\ell-1} c(x_i, x_{i+1})$. It follows that

$$h(x_1) - h(x_\ell) = \sum_{i=1}^{\ell-1} (h(x_i) - h(x_{i+1})) \le \sum_{i=1}^{\ell-1} \omega \cdot c(x_i, x_{i+1}) = \omega h^*(x).$$

Since x_ℓ is a goal node, $h(x_\ell) = 0$. Thus, $h(x) = h(x) - h(x_\ell) \le \omega h^*(x)$.   □

Corollary 1. *For any heuristic h, if $\mathsf{WIRE}(h, e) \leq \omega$ for all edges e then we have $\mathsf{ARN}(h, x) \leq \omega$ for all nodes x.*

This means that an upper bound on the weighted inconsistency rates of a heuristic h is also an upper bound on the accuracy rates of h. In particular, if the heuristic h is consistent, then the less $\mathsf{WIRE}(h)$, the less accurate h can be. This suggests that imposing consistency on the heuristic can prevent improving the heuristic accuracy.

4   Experiments with Knapsack Problem

We will experimentally investigate the relationship between inconsistency and accuracy of heuristics on a practical domain namely the Knapsack problem. This problem is NP-complete and has applications in many fields, from business to cryptography. Our heuristics will also be built in a practical way, based on an approximation algorithm [18, p. 70] for the Knapsack problem.

4.1   Search Model for Knapsack

A Knapsack instance is denoted by a tuple $\langle X, p, w, C \rangle$, where X is a finite set of items, $p : X \to \mathbb{Z}^+$ is a function assigning profit to each item, $w : X \to \mathbb{Z}^+$ is a function assigning weight to each item, and $C > 0$ is the capacity of the knapsack. Recall that the Knapsack problem is to find a subset $X^* \subseteq X$ of items whose total weight does not exceed capacity C and whose total profit is maximal. We will write $p(X)$ and $w(X)$ to denote the total profit and the total weight, respectively, of all items in X, i.e., $w(X) = \sum_{i \in X} w(i)$ and $p(X) = \sum_{i \in X} p(i)$. For each positive integer n, let $[n] = \{1, 2, \ldots, n\}$, and we may simply write $[n]$ to represent a set of n items.

Here we will adopt the search model for the Knapsack problem that has been employed in [2]. In particular, consider the Knapsack instance $\langle [n], p, w, C \rangle$. The search graph for this instance is a directed graph, in which each node (or state) is a nonempty subset $X \subseteq [n]$ and each edge (X, X') corresponds to the removal of an item $i \in X$ so that $X \setminus \{i\} = X'$. The cost of such an edge (X, X') is the profit of the removed item i. See Figure 1 for an example of edges from a node $X = \{1, 2, 3, 4\}$. The start node is the set $[n]$. A node X is designated as a goal node if $w(X) \leq C$.

An important property of this search space is that every path from node X to node X' has the same total cost, which equals the total profit of items in $X \setminus X'$. Thanks to this property, A* will avoid reopening nodes. Thus, consistent heuristics are not needed in this case.

4.2   Heuristic Construction

Consider the search space for a Knapsack instance $\langle [n], p, w, C \rangle$. We construct efficient admissible heuristics on this search space in a similar way to the construction of Dinh et al. [2], but without constraints to obtain an accuracy guarantee,

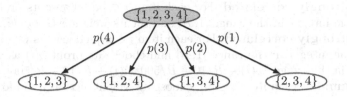

Fig. 1. Edges from a node in a Knapsack search space

which is a lower bound on the minimal accurate rate. The main ingredient of this construction is an FPTAS (*Fully Polynomial Time Approximation Scheme*) due to Ibarra and Kim [8], which is described in [18, p. 70]. This FPTAS is an algorithm, denoted $\mathcal{A}$, that returns a solution with total profit at least $(1-\epsilon)\mathrm{Opt}(X)$ to each Knapsack instance $\langle X, p, w, c\rangle$ and runs in time $O\left(|X|^3/\epsilon\right)$ [18, p. 70], for any given $\epsilon \in (0,1)$, where $\mathrm{Opt}(X)$ is the total profit of an optimal solution to the Knapsack instance $\langle X, p, w, c\rangle$. For each subset $X \subseteq [n]$, let $\mathcal{A}_\epsilon(X)$ denote the total profit of the solution returned by algorithm $\mathcal{A}$ with error parameter ϵ to the Knapsack instance $\langle X, p, w, c\rangle$. Then for any $\epsilon \in (0,1)$,

$$(1 - \epsilon)\mathrm{Opt}(X) \leq \mathcal{A}_\epsilon(X) \leq \mathrm{Opt}(X).$$

Since $h^*(X) = p(X) - \mathrm{Opt}(X)$, it follows that

$$p(X) - \frac{\mathcal{A}_\epsilon(X)}{1 - \epsilon} \leq h^*(X) \leq p(X) - \mathcal{A}_\epsilon(X). \tag{1}$$

Note that the lower bound $p(X) - \frac{\mathcal{A}_\epsilon(X)}{1-\epsilon}$ can fall below zero, especially for large ϵ. Hence, for each parameter $\epsilon \in (0,1)$, we define the following heuristic h_ϵ whose admissibility is guaranteed: for any non-goal node X,

$$h_\epsilon(X) \stackrel{\mathrm{def}}{=} \max\left\{p(X) - \frac{\mathcal{A}_\epsilon(X)}{1 - \epsilon}, 0\right\}.$$

Since the running time to compute $\mathcal{A}_\epsilon(X)$ is $O\left(|X|^3\epsilon^{-1}\right)$, the running time to compute $h_\epsilon(X)$ is also $O\left(|X|^3\epsilon^{-1}\right)$, which is polynomial in both n and ϵ^{-1}.

While there is no accuracy guarantee on h_ϵ, it is intuitive to expect the growth in the accuracy of h_ϵ by reducing the FPTAS error parameter ϵ. It then remains to find if the inconsistency of h_ϵ will also grow as ϵ decreases.

4.3 Experiments

For our experiments, we generate hard Knapsack instances $\langle [n], p, w, C\rangle$ from the following Knapsack instance distributions, or "types," which are identified by Pisinger [15] as difficult instances for best-known exact algorithms:

Strongly correlated: For each item $i \in [n]$, choose its weight $w(i)$ as a random integer in the range $[1, R]$ and set its profit $p(i) = w(i) + R/10$. This correlation between weights and profits reflects real-life situations where the profit of an item is proportional to its weight plus some fixed charge.

Inverse strongly correlated: For each item $i \in [n]$, choose its profit $p(i)$ as a random integer in the range $[1, R]$ and set its weight $w(i) = p(i) + R/10$.

Almost strongly correlated: For each item $i \in [n]$, choose its weight $w(i)$ as a random integer in the range $[1, R]$ and choose its profit $p(i)$ as a random integer in the range $[w(i) + R/10 - R/500, w(i) + R/10 + R/500]$.

Subset sum: For each item $i \in [n]$, choose its weight $w(i)$ as a random integer in the range $[1, R]$ and set its profit $p(i) = w(i)$. Knapsack instances of this type are instances of the subset sum problem.

Uncorrelated with similar weight: For each item $i \in [n]$, choose its weight $w(i)$ as a random integer in the range $[100000, 100100]$ and choose its profit $p(i)$ as a random integer in $[1, R]$.

Profit ceiling: For each item $i \in [n]$, choose its weight $w(i)$ as a random integer in the range $[1, R]$ and set its profit $p(i) = 3 \lceil w(i)/3 \rceil$. This family of instances is denoted pceil(3), which resulted in sufficiently difficult instances for experiments of Pisinger [15].

Here we set the data range parameter $R := 1000$. The knapsack capacity is chosen as $C = (t/101)w([n])$, where t is a random integer in the range $[30, 70]$.

In our experiments, we generate one random Knapsack instance $\langle [n], p, w, C \rangle$ of each type above. For each Knapsack instance generated, we run a series of A*(h_ϵ) with different values of ϵ, as well as breath-first search. We chose the sample points for ϵ with two consecutive points differed by a factor of 2, so as to clearly see the change in the number of node expansions made by A*(h_ϵ).

The main challenge of these experiments is to compute ARN(h_ϵ). It is typically too expensive to compute ARN of a heuristic because it requires computing $h^*(x)$ exactly for all non-goal nodes x. For the Knapsack search space, we can also rely on the given FPTAS $\mathcal{A}$ to compute $h^*(X)$ for each node $X \subseteq [n]$. Our computation is based on the following proposition:

Proposition 1. *For any* $0 < \gamma < 1/\mathrm{Opt}(X)$,

$$h^*(X) = \lfloor p(X) - \mathcal{A}_\gamma(X) \rfloor . \tag{2}$$

Proof. Since $\mathcal{A}_\gamma(X) \geq (1 - \gamma)\mathrm{Opt}(X)$, we have

$$p(X) - \mathcal{A}_\gamma(X) \leq p(X) - (1 - \gamma)\mathrm{Opt}(X) = h^*(X) + \gamma\mathrm{Opt}(X) < h^*(X) + 1 .$$

On the other hand, from Equation (1), we have $h^*(X) \leq p(X) - \mathcal{A}_\gamma(X)$. The proof is completed by noting that $h^*(X)$ is an integer.   □

Since $\mathrm{Opt}(X) < \min\{p(X), \mathrm{Opt}([n]) + 1\}$ for all non-goal node $X \subseteq [n]$, we compute $h^*(X)$ as in Equation (2) with

$$\gamma = \frac{1}{\min\{p(X), \mathrm{Opt}([n]) + 1\}} . \tag{3}$$

The value of $\mathrm{Opt}([n])$ is obtained after running A*(h_ϵ), which returns the optimal solution cost $h^*([n]) = p([n]) - \mathrm{Opt}([n])$.

While using the FPTAS could save us a considerable amount of time, computing $\mathcal{A}_\gamma(X)$ with γ specified in (3) is still time-consuming – it actually has pseudo-polynomial time complexity. As such, we limit our experiments to Knapsack instances of relatively small size ($n = 20$), for which each $\text{ARN}(h_\epsilon)$ can be computed within 10 hours. A faster method would be to compute $\text{ARN}(h_\epsilon, x)$ for only a few non-goal nodes x chosen randomly and take the average of $\text{ARN}(h_\epsilon, x)$ at those sample nodes as an approximation for $\text{ARN}(h_\epsilon)$. However, this statistical method is meaningful only when the distribution of $\text{ARN}(h_\epsilon, x)$ over the set of all non-goal nodes is close to uniform. Such an ideal condition is rarely achieved in practical domains.

Detailed results of our experiments are shown in Table 1, in which the columns with ϵ give the values of the FPTAS error parameter ϵ, the rows with "BFS" indicate the breath-first search, the columns with "Node Exps" contain the number of node expansions made by each search. Other columns show data for $\text{ARS}(h_\epsilon)$, $\text{ARN}(h_\epsilon)$, $\text{INR}(h_\epsilon)$, and $\text{WIRE}(h_\epsilon)$. Figure 2 shows the trend of these measures averaged over all Knapsack instances in our experiments. Recall that all these Knapsack instances have the same number of items, $n = 20$, thus have the same search graph.

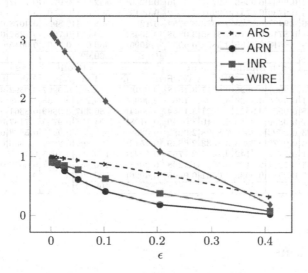

Fig. 2. Results averaged over all Knapsack instances from Table 1. The x-axis label is the FPTAS error parameter ϵ.

Our data show that when the accuracy metrics $\text{ARN}(h_\epsilon)$ and $\text{ARS}(h_\epsilon)$ grow, then so do the inconsistency metrics $\text{INR}(h_\epsilon)$ and $\text{WIRE}(h_\epsilon)$. Loosely speaking, the accuracy and the inconsistency level of heuristics h_ϵ are somewhat correlated. This could explain why the inconsistent admissible heuristics can improve the efficiency of A*. We observe, in addition, that for small ϵ (< 0.02), the values of ARN and INR are close to each other in many instances, such as Strongly

Correlated, Almost Strongly Correlated, and Subset Sum. Regarding the performance of A*, our data also show a significant reduction in the number of node expansions when the heuristic is more accurate, and thus more inconsistent.

Table 1. Results of randomly generated Knapsack instances (of size $n = 20$)

Strongly Correlated

ϵ	Node Exps	ARS	ARN	INR	WIRE
0.0016	332	0.9986	0.9918	0.9362	2.9378
0.0032	532	0.9973	0.9834	0.9363	2.9238
0.0064	627	0.9945	0.9667	0.9355	2.8956
0.0128	767	0.9889	0.9328	0.9328	2.8381
0.0256	9,921	0.9773	0.8644	0.9211	2.7249
0.0512	73,212	0.9538	0.7292	0.8709	2.4891
0.1024	446,925	0.9023	0.5104	0.7501	2.0087
0.2048	609,517	0.7782	0.2158	0.4711	1.1008
0.4096	609,535	0.4159	0.0054	0.0294	0.0519
BFS	609,501				

Inverse Strongly Correlated

ϵ	Node Exps	ARS	ARN	INR	WIRE
0.0016	4,095	0.9975	0.9651	0.7857	1.6031
0.0032	14,831	0.9950	0.9303	0.7759	1.5802
0.0064	19,576	0.9898	0.8586	0.7767	1.5349
0.0128	33,852	0.9794	0.7427	0.7552	1.4490
0.0256	99,985	0.9587	0.5826	0.6893	1.2883
0.0512	179,001	0.9149	0.3664	0.5455	0.9967
0.1024	190,346	0.8176	0.1467	0.3201	0.5434
0.2048	190,357	0.5886	0.0140	0.0573	0.0859
0.4096	190,352	0	0	0	0
BFS	190,352				

Almost Strongly Correlated

ϵ	Node Exps	ARS	ARN	INR	WIRE
0.0016	511	0.9992	0.9963	0.9782	4.7133
0.0032	766	0.9983	0.9925	0.9783	4.7035
0.0064	766	0.9967	0.9850	0.9781	4.6799
0.0128	766	0.9933	0.9697	0.9769	4.6360
0.0256	19,184	0.9864	0.9386	0.9731	4.5496
0.0512	116,003	0.9730	0.8741	0.9661	4.3672
0.1024	627,996	0.9427	0.7524	0.9373	3.9754
0.2048	921,291	0.8713	0.5116	0.8185	3.0808
0.4096	921,755	0.6518	0.1166	0.3824	0.9986
BFS	921,751				

Subset Sum

ϵ	Node Exps	ARS	ARN	INR	WIRE
0.0016	363,190	0.9985	0.9845	0.9399	3.7904
0.0032	427,384	0.9971	0.9689	0.9398	3.7649
0.0064	484,978	0.9940	0.9395	0.9352	3.7177
0.0128	528,999	0.9881	0.8832	0.9291	3.6237
0.0256	552,068	0.9757	0.7867	0.8968	3.4356
0.0512	562,801	0.9501	0.6374	0.8412	3.0607
0.1024	568,781	0.8956	0.4078	0.6967	2.3275
0.2048	569,621	0.7620	0.1387	0.3787	1.0798
0.4096	569,621	0.3591	0.0015	0.0138	0.0232
BFS	569,621				

Uncorrelated with Similar Weight

ϵ	Node Exps	ARS	ARN	INR	WIRE
0.0016	24,573	0.9987	0.9625	0.8817	3.1518
0.0032	40,951	0.9975	0.9524	0.8895	3.1368
0.0064	53,235	0.9950	0.9321	0.8821	3.1064
0.0128	65,517	0.9899	0.8913	0.8710	3.0449
0.0256	135,083	0.9793	0.8319	0.8598	2.9282
0.0512	236,660	0.9579	0.7233	0.8633	2.6829
0.1024	448,451	0.9116	0.5142	0.7417	2.1757
0.2048	681,705	0.7984	0.2125	0.4915	1.1238
0.4096	698,995	0.4592	0.0035	0.0241	0.0468
BFS	699,037				

Profit Ceiling

ϵ	Node Exps	ARS	ARN	INR	WIRE
0.0016	43,262	0.9967	0.9592	0.9034	2.4341
0.0032	69,291	0.9935	0.9188	0.8978	2.3952
0.0064	86,397	0.9869	0.8361	0.8795	2.3188
0.0128	94,178	0.9733	0.7187	0.8340	2.1689
0.0256	97,350	0.9457	0.5518	0.7504	1.8894
0.0512	99,497	0.8882	0.3356	0.5846	1.3916
0.1024	100,366	0.7615	0.1163	0.3171	0.6568
0.2048	100,351	0.4677	0.0072	0.0406	0.0669
0.4096	100,364	0	0	0	0
BFS	100,364				

5   Conclusions

This work provides evidence that the inconsistency and accuracy of heuristics are related. Theoretical evidence suggests that the heuristic accuracy could be upper-bounded by its level of inconsistency. Thus, requiring the heuristic to be consistent could limit the room to improve its accuracy. Empirical evidence with a family of practical admissible heuristics on Knapsack domains shows that the more accurate the heuristic, the more inconsistent it is. The experiments in this work also provide positive results about accurate heuristics and inconsistent admissible heuristics, that is, both the accuracy and the inconsistency of the heuristic can be used to improve the performance of A*.

References

[1] Dinh, H., Russell, A., Su, Y.: On the value of good advice: The complexity of A* with accurate heuristics. In: Proceedings of the Twenty-Second Conference on Artificial Intelligence (AAAI 2007), pp. 1140–1145 (2007)

[2] Dinh, H., Dinh, H., Michel, L., Russell, A.: The time complexity of A* with approximate heuristics on multiple-solution search spaces. Journal of Artificial Intelligence Research 45, 685–729 (2012)

[3] Felner, A., Zahavi, U., Schaeffer, J., Holte, R.C.: Dual lookups in pattern databases. In: Proceedings of the 19th International Joint Conference on Artificial Intelligence, IJCAI 2005, pp. 103–108. Morgan Kaufmann Publishers Inc., San Francisco (2005)

[4] Felner, A., Zahavi, U., Holte, R., Schaeffer, J., Sturtevant, N., Zhang, Z.: Inconsistent heuristics in theory and practice. Artificial Intelligence 175(9-10), 1570–1603 (2011)

[5] Gaschnig, J.: Performance measurement and analysis of certain search algorithms. PhD thesis, Carnegie-Mellon University, Pittsburgh, PA (1979)

[6] Hart, P., Nilson, N., Raphael, B.: A formal basis for the heuristic determination of minimum cost paths. IEEE Transactions on Systems Science and Cybernetics SCC-4(2), 100–107 (1968)

[7] Helmert, M., Röger, G.: How good is almost perfect? In: Proceedings of AAAI 2008 (2008)

[8] Ibarra, O.H., Kim, C.E.: Fast approximation algorithms for the knapsack and sum of subset problems. Journal of the ACM 22(4), 463–468 (1975) ISSN 0004-5411

[9] Korf, R., Reid, M.: Complexity analysis of admissible heuristic search. In: Proceedings of the National Conference on Artificial Intelligence (AAAI 1998), pp. 305–310 (1998)

[10] Korf, R., Reid, M., Edelkamp, S.: Time complexity of iterative-deepening-A*. Artificial Intelligence 129(1-2), 199–218 (2001)

[11] Korf, R.E.: Iterative-deepening-a: an optimal admissible tree search. In: Proceedings of the 9th International Joint Conference on Artificial Intelligence, IJCAI 1985, vol. 2, pp. 1034–1036. Morgan Kaufmann Publishers Inc., San Francisco (1985)

[12] Korf, R.E.: Recent progress in the design and analysis of admissible heuristic functions. In: Proceedings of the 17th National Conference on Artificial Intelligence (AAAI 2000), pp. 1165–1170. AAAI Press / The MIT Press (2000) ISBN 0-262-51112-6

[13] Huyn, J.P.N., Dechter, R.: Probabilistic analysis of the complexity of A*. Artificial Intelligence 15, 241–254 (1980)

[14] Pearl, J.: Heuristics: Intelligent Search Strategies for Computer Problem Solving. Addison-Wesley, MA (1984)

[15] Pisinger, D.: Where are the hard knapsack problems? Computers and Operations Research 32, 2271–2284 (2005) ISSN 0305-0548

[16] Pohl, I.: Practical and theoretical considerations in heuristic search algorithms. In: Elcock, W., Michie, D. (eds.) Machine Intelligence, vol. 8, pp. 55–72. Ellis Horwood, Chichester (1977)

[17] Sen, A.K., Bagchi, A., Zhang, W.: Average-case analysis of best-first search in two representative directed acyclic graphs. Artif. Intell. 155(1-2), 183–206 (2004)

[18] Vazirani, V.: Approximation Algorithms. Springer (2001)

[19] Zahavi, U., Felner, A., Schaeffer, J., Sturtevant, N.: Inconsistent heuristics. In: Proceedings of AAAI 2007, pp. 1211–1216 (2007)

[20] Zahavi, U., Felner, A., Burch, N., Holte, R.C.: Predicting the performance of IDA* using conditional distributions. J. Artif. Int. Res. 37(1), 41–84 (2010), http://dl.acm.org/citation.cfm?id=1861751.1861753 ISSN 1076-9757

[21] Zhang, Z., Sturtevant, N.R., Holte, R., Schaeffer, J., Felner, A.: A* search with inconsistent heuristics. In: Proceedings of the 21st International Joint Conference on Artificial Intelligence, IJCAI 2009, pp. 634–639. Morgan Kaufmann Publishers Inc., San Francisco (2009)

VMSP: Efficient Vertical Mining of Maximal Sequential Patterns

Philippe Fournier-Viger[1], Cheng-Wei Wu[2], Antonio Gomariz[3],
and Vincent S. Tseng[2]

[1] Dept. of Computer Science, University of Moncton, Canada
[2] Dept. of Computer Science and Information Engineering, National Cheng Kung
University, Taiwan
[3] Dept. of Information and Communication Engineering, University of Murcia, Spain
philippe.fournier-viger@umoncton.ca, silvemoonfox@gmail.com,
agomariz@um.es, tseng@mail.ncku.edu.tw

Abstract. *Sequential pattern mining* is a popular data mining task with
wide applications. However, it may present too many sequential patterns
to users, which makes it difficult for users to comprehend the results.
As a solution, it was proposed to mine *maximal sequential patterns*, a
compact representation of the set of sequential patterns, which is often several orders of magnitude smaller than the set of all sequential
patterns. However, the task of mining maximal patterns remains computationally expensive. To address this problem, we introduce a vertical
mining algorithm named *VMSP* (*Vertical mining of Maximal Sequential
Patterns*). It is to our knowledge the first vertical mining algorithm for
mining maximal sequential patterns. An experimental study on five real
datasets shows that VMSP is up to two orders of magnitude faster than
the current state-of-the-art algorithm.

Keywords: vertical mining, maximal sequential pattern mining, candidate pruning.

1 Introduction

Discovering interesting patterns in sequential data is a challenging task. Multiple
studies have been proposed for mining interesting patterns in sequence databases
[11,4]. *Sequential pattern mining* is probably the most popular research topic
among them. A subsequence is called *sequential pattern* or *frequent sequence*
if it frequently appears in a sequence database, and its frequency is no less
than a user-specified *minimum support threshold minsup* [1]. Sequential pattern
mining plays an important role in data mining and is essential to a wide range
of applications such as the analysis of web click-streams, program executions,
medical data, biological data and e-learning data [11].

Several algorithms have been proposed for sequential pattern mining such as
PrefixSpan [12], *SPAM* [2] and *SPADE* [14]. However, a critical drawback of
these algorithms is that they may present too many sequential patterns to users.

M. Sokolova and P. van Beek (Eds.): Canadian AI 2014, LNAI 8436, pp. 83–94, 2014.

A very large number of sequential patterns makes it difficult for users to analyze results to gain insightful knowledge. It may also cause the algorithms to become inefficient in terms of time and memory because the more sequential patterns the algorithms produce, the more resources they consume. The problem becomes worse when the database contains long sequential patterns. For example, consider a sequence database containing a sequential pattern having 20 distinct items. A sequential pattern mining algorithm will present the sequential pattern as well as its $2^{20} - 1$ subsequences to the user. This will most likely make the algorithm fail to terminate in reasonable time and run out of memory. For example, the well-known PrefixSpan [12] algorithm would have to perform 2^{20} database projection operations to produce the results.

To reduce the computational cost of the mining task and present fewer but more representative patterns to users, many studies focus on developing concise representations of sequential patterns. A popular representation that has been proposed is *closed sequential patterns* [13,6]. A closed sequential pattern is a sequential pattern that is not strictly included in another pattern having the same frequency. Several approaches have been proposed for mining closed sequential patterns in sequence databases such as *BIDE* [13] and *ClaSP* [6]. Although these algorithms mines a compact set of sequential patterns, the set of closed patterns is still too large for dense databases or database containing long sequences.

To address this problem, it was proposed to mine *maximal sequential patterns* [3,5,8,9,10,7]. A maximal sequential pattern is a closed pattern that is not strictly included in another closed pattern. The set of maximal sequential patterns is thus generally a very small subset of the set of (closed) sequential patterns. Besides, the set of maximal sequential patterns is representative since it can be used to recover all sequential patterns, and the exact frequency of these latter can also be recovered with a single database pass.

Maximal sequential pattern mining is important and has been adopted in numerous applications. For example, it is used to find the frequent longest common subsequences in texts, analysing DNA sequences, data compression and web log mining [5]. Although maximal sequential pattern mining is desirable and useful in many applications, it remains a computationally expensive data mining task and few algorithms have been proposed for this task. *MSPX* [9] is an approximate algorithm and therefore it provides an incomplete set of maximal patterns to users. *DIMASP* [5] is designed for the special case where sequences are strings (no more than an item can appear at the same time) and where no pair of contiguous items appears more than once in each sequence. *AprioriAdjust* [10] is an apriori-like algorithm, which may suffer from the drawbacks of the candidate generation-and-test paradigm. In other words, it may generate a large number of candidate patterns that do not appear in the input database and require to scan the original database several times. The *MFSPAN* [7] algorithm needs to maintain a large amount of intermediate candidates in main memory during the mining process. The most recent algorithm is *MaxSP* [3], which relies on a pattern-growth approach to avoid the problem of candidate generation from previous algorithms. However, it has to repeatedly perform costly database projection operations [3].

Given the limitations of previous work, we explore a novel approach, which is to mine maximal sequential pattern by using a depth-first exploration of the search space using a vertical representation. We propose a novel algorithm for maximal sequential pattern mining that we name *VMSP* (*Vertical mining of Maximal Sequential Patterns*). The algorithm incorporates three efficient strategies named EFN (Efficient Filtering of Non-maximal patterns), FME (Forward-Maximal Extension checking) and CPC (Candidate Pruning by Co-occurrence map) to effectively identify maximal patterns and prune the search space. VMSP is developed for the general case of a sequence database rather than strings and it can capture the complete set of maximal sequential patterns with a single database scan. We performed an experimental study with five real-life datasets to compare the performance of VMSP with MaxSP [3], the state-of-the-art algorithm for maximal sequential pattern mining. Results show that VMSP is up to two orders of magnitude faster than MaxSP, and perform well on dense datasets.

The rest of the paper is organized as follows. Section 2 formally defines the problem of maximal sequential pattern mining and its relationship to sequential pattern mining. Section 3 describes the VMSP algorithm. Section 4 presents the experimental study. Finally, Section 5 presents the conclusion and future works.

2 Problem Definition

Definition 1 (sequence database). Let $I = \{i_1, i_2, ..., i_l\}$ be a set of items (symbols). An *itemset* $I_x = \{i_1, i_2, ..., i_m\} \subseteq I$ is an unordered set of distinct items. The *lexicographical order* $\succ_{lex}$ is defined as any total order on I. Without loss of generality, we assume that all itemsets are ordered according to $\succ_{lex}$. A *sequence* is an ordered list of itemsets $s = \langle I_1, I_2, ..., I_n \rangle$ such that $I_k \subseteq I$ ($1 \leq k \leq n$). A *sequence database* SDB is a list of sequences $SDB = \langle s_1, s_2, ..., s_p \rangle$ having sequence identifiers (SIDs) $1, 2...p$. **Example.** A sequence database is shown in Fig. 1 (left). It contains four sequences having the SIDs 1, 2, 3 and 4. Each single letter represents an item. Items between curly brackets represent an itemset. The first sequence $\langle \{a, b\}, \{c\}, \{f, g\}, \{g\}, \{e\} \rangle$ contains five itemsets. It indicates that items a and b occurred at the same time, were followed by c, then f and g at the same time, followed by g and lastly e.

Definition 2 (sequence containment). A sequence $s_a = \langle A_1, A_2, ..., A_n \rangle$ is said to be *contained in* a sequence $s_b = \langle B_1, B_2, ..., B_m \rangle$ iff there exist integers $1 \leq i_1 < i_2 < ... < i_n \leq m$ such that $A_1 \subseteq B_{i1}, A_2 \subseteq B_{i2}, ..., A_n \subseteq B_{in}$ (denoted as $s_a \sqsubseteq s_b$). **Example.** Sequence 4 in Fig. 1 (left) is contained in Sequence 1.

Definition 3 (prefix). A sequence $s_a = \langle A_1, A_2, ..., A_n \rangle$ is a *prefix* of a sequence $s_b = \langle B_1, B_2, ..., B_m \rangle$, $\forall n < m$, iff $A_1 = B_1, A_2 = B_2, ..., A_{n-1} = B_{n-1}$ and the first $|A_n|$ items of B_n according to $\succ_{lex}$ are equal to A_n.

SID	Sequences
1	⟨{a, b},{c},{f, g},{g},{e}⟩
2	⟨{a, d},{c},{b},{a, b, e, f}⟩
3	⟨{a},{b},{f, g},{e}⟩
4	⟨{b},{f, g}⟩

Pattern	Sup.		Pattern	Sup.	
⟨{a}⟩	3	C	⟨{b},{g},{e}⟩	2	CM
⟨{a},{g}⟩	2		⟨{b},{f}⟩	4	C
⟨{a},{g},{e}⟩	2	CM	⟨{b},{f, g}⟩	2	CM
⟨{a},{f}⟩	3	C	⟨{b},{f},{e}⟩	2	CM
⟨{a},{f},{e}⟩	2	CM	⟨{b},{e}⟩	3	C
⟨{a},{c}⟩	2		⟨{c}⟩	2	
⟨{a},{c},{f}⟩	2	CM	⟨{c},{f}⟩	2	
⟨{a},{c},{e}⟩	2	CM	⟨{c},{e}⟩	2	
⟨{a},{b}⟩	2		⟨{e}⟩	3	
⟨{a},{b},{f}⟩	2	CM	⟨{f}⟩	4	
⟨{a},{b},{e}⟩	2	CM	⟨{f, g}⟩	2	
⟨{a},{e}⟩	3	C	⟨{f},{e}⟩	2	
⟨{a, b}⟩	2	CM	⟨{g}⟩	3	
⟨{b}⟩	4		⟨{g},{e}⟩	2	
⟨{b},{g}⟩	3	C			

C = Closed   M = Maximal

Fig. 1. A sequence database (left) and (all/closed/maximal) sequential patterns found (right)

Definition 4 (extensions). A sequence s_b is said to be an s-**extension** of a sequence $s_a = \langle I_1, I_2, ...I_h \rangle$ with an item x, iff $s_b = \langle I_1, I_2, ...I_h, \{x\} \rangle$, i.e. s_a is a prefix of s_b and the item x appears in an itemset later than all the itemsets of s_a. In the same way, the sequence s_c is said to be an i-**extension** of s_a with an item x, iff $s_c = \langle I_1, I_2, ...I_h \cup \{x\} \rangle$, i.e. s_a is a prefix of s_c and the item x occurs in the last itemset of s_a, and the item x is the last one in I_h, according to $\succ_{lex}$.

Definition 5 (support). The *support* of a sequence s_a in a sequence database SDB is defined as the number of sequences $s \in SDB$ such that $s_a \sqsubseteq s$ and is denoted by $sup_{SDB}(s_a)$.

Definition 6 (sequential pattern mining). Let $minsup$ be a threshold set by the user and SDB be a sequence database. A sequence s is a *sequential pattern* and is deemed *frequent* iff $sup_{SDB}(s) \geq minsup$. The *problem of mining sequential patterns* is to discover all sequential patterns [1]. **Example.** Fig. 1 (right) shows the 29 sequential patterns found in the database of Fig. 1 (left) for $minsup = 2$, and their support. For instance, the patterns $\langle \{a\} \rangle$ and $\langle \{a\}, \{g\} \rangle$ are frequent and have respectively a support of 3 and 2 sequences.

Definition 7 (closed/maximal sequential pattern mining). A sequential pattern s_a is said to be *closed* if there is no other sequential pattern s_b, such that s_b is a superpattern of s_a, $s_a \sqsubseteq s_b$, and their supports are equal. A sequential pattern s_a is said to be *maximal* if there is no other sequential pattern s_b, such that s_b is a superpattern of s_a, $s_a \sqsubseteq s_b$. The *problem of mining closed (maximal) sequential patterns* is to discover the set of closed (maximal) sequential patterns. **Example.** Consider the database of Fig. 1 and minsup = 2. There are 29 sequential patterns (shown in the right side of Fig. 1), such that 15 are closed (identified by the letter C) and only 10 are maximal (identified by the letter M).

Property 1. (Recovering sequential patterns). The set of maximal sequential patterns allow recovering all sequential patterns.

Proof. By definition, a maximal sequential pattern has no proper super-sequence that is a frequent sequential pattern. Thus, if a pattern is frequent, it is either a proper subsequence of a maximal pattern or a maximal pattern. Figure 2 presents a simple algorithm for recovering all sequential patterns from the set of maximal sequential patterns. It generates all the subsequences of all the maximal patterns. Furthermore, it performs a check to detect if a sequential pattern has already been output (line 3) because a sequential pattern may be a subsequence of more than one maximal pattern. After sequential patterns have been recovered, an additional database scan can be performed to calculate their exact support, if required.

RECOVERY (*a set of maximal patterns M*)
1.   **FOR** each sequential pattern *j* ∈ *M*
2.   **FOR** each subsequence *k* of *j*
3.   **IF** *k* has not been output
4.   **THEN** output *k*.

Fig. 2. Algorithm to recover all sequential patterns from maximal patterns

Definition 8 (horizontal database format). A *sequence database in horizontal format* is a database where each entry is a sequence. **Example.** Figure 1 (left) shows an horizontal sequence database.

Definition 9 (vertical database format). A *sequence database in vertical format* is a database where each entry represents an item and indicates the list of sequences where the item appears and the position(s) where it appears [2]. **Example.** Fig. 3 shows the vertical representation of the database of Fig. 1 (left).

a		b		c		d	
SID	Itemsets	SID	Itemsets	SID	Itemsets	SID	Itemsets
1	1	1	1	1	2	1	
2	1,4	2	3,4	2	2	2	1
3	1	3	2	3		3	
4		4	1	4		4	

e		f		g	
SID	Itemsets	SID	Itemsets	SID	Itemsets
1	5	1	3	1	3,4
2	4	2	4	2	
3	4	3	3	3	
4		4	2	4	2

Fig. 3. The vertical representation of the database shown in Fig. 1(left)

Vertical mining algorithms associate a structure named *IdList* [14,2] to each pattern. IdLists allow calculating the support of a pattern quickly by making join operations with IdLists of smaller patterns. To discover sequential patterns, vertical mining algorithms perform a single database scan to create IdLists of patterns containing single items. Then, larger patterns are obtained by performing the join operation of IdLists of smaller patterns (cf. [14] for details). Several works proposed alternative representations for IdLists to save time in join operations, being the bitset representation the most efficient one [2].

3   The VMSP Algorithm

We present VMSP, our novel algorithm for maximal sequential pattern mining. It adopts the IdList structure [2,6,14]. We first describe the general search procedure used by VMSP to explore the search space of sequential patterns. Then, we describe how it is adapted to discover maximal patterns efficiently.

3.1   The Search Procedure

The pseudocode of the search procedure is shown in Fig. 4. The procedure takes as input a sequence database SDB and the $minsup$ threshold. The procedure first scans the input database SDB once to construct the vertical representation of the database $V(SDB)$ and the set of frequent items F_1. For each frequent item $s \in F_1$, the procedure calls the SEARCH procedure with $\langle s \rangle$, F_1, $\{e \in F_1 | e \succ_{lex} s\}$, and $minsup$.

The SEARCH procedure outputs the pattern $\langle \{s\} \rangle$ and recursively explores candidate patterns starting with the prefix $\langle \{s\} \rangle$. The SEARCH procedure takes as parameters a sequential pattern pat and two sets of items to be appended to pat to generate candidates. The first set S_n represents items to be appended to pat by s-extension. The second set S_i represents items to be appended to pat by i-extension. For each candidate pat' generated by an extension, the procedure calculate the support to determine if it is frequent. This is done by the IdList join operation (see [2,14] for details) and counting the number of sequences where the pattern appears. If the pattern pat' is frequent, it is then used in a recursive call to SEARCH to generate patterns starting with the prefix pat'.

It can be easily seen that the above procedure is correct and complete to explore the search space of sequential patterns since it starts with frequent patterns containing single items and then extend them one item at a time while only pruning infrequent extensions of patterns using the anti-monotonicity property (any infrequent sequential pattern cannot be extended to form a frequent pattern)[1].

3.2   Discovering Maximal Patterns

We now describe how the search procedure is adapted to discover only maximal patterns. This is done by integrating three strategies to efficiently filter non-maximal patterns and prune the search space. The result is the VMSP algorithm, which outputs the set of maximal patterns.

PATTERN-ENUMERATION(*SDB, minsup*)
1.   Scan *SDB* to create $V(SDB)$ and identify S_{init}, the list of frequent items.
2.   **FOR** each item s $\in S_{init}$,
3.   **SEARCH**($\langle s \rangle$, S_{init}, the set of items from S_{init} that are lexically larger than s, *minsup*).

SEARCH(*pat*, S_n, I_n, *minsup*)
1.   Output pattern *pat*.
2. $S_{temp} := I_{temp} := \emptyset$
3.   **FOR** each item $j \in S_n$,
4.   **IF** the s-extension of *pat* is frequent **THEN** $S_{temp} := S_{temp} \cup \{i\}$.
5.   **FOR** each item $j \in S_{temp}$,
6.   **SEARCH**(the s-extension of *pat* with j, S_{temp}, elements in S_{temp} greater than j, *minsup*).
7.   **FOR** each item $j \in I_n$,
8.   **IF** the i-extension of *pat* is frequent **THEN** $I_{temp} := I_{temp} \cup \{i\}$.
9.   **FOR** each item $j \in I_{temp}$,
10.   **SEARCH**(i-extension of *pat* with j, S_{temp}, all elements in I_{temp} greater than j, *minsup*).

Fig. 4. The search procedure

Strategy 1. Efficient Filtering of Non-maximal patterns (EFN).

The first strategy identifies maximal patterns among patterns generated by the search procedure. This is performed using a novel structure named Z that stores the set of maximal patterns found until now. The structure Z is initially empty. Then, during the search for patterns, every time that a pattern s_a, is generated by the search procedure, two operations are performed to update Z.

- *Super-pattern checking.* During this operation, s_a is compared with each pattern $s_b \in Z$ to determine if there exists a pattern s_b such that $s_a \sqsubseteq s_b$. If yes, then s_a is not maximal (by Definition 7) and thus, s_a is not inserted into Z. Otherwise, s_a is maximal with respect to all patterns found until now and it is thus inserted into Z.
- *Sub-pattern checking.* If s_a is determined to be maximal according to super-pattern checking, we need to perform this second operation. The pattern s_a is compared with each pattern $s_b \in Z$. If there exists a pattern $s_b \sqsubseteq s_a$, then s_b is not maximal (by Definition 7) and s_b is removed from Z.

By using the above strategy, it is obvious that when the search procedure terminates, Z contains the set of maximal sequential patterns. However, to make this strategy efficient, we need to reduce the number of pattern comparisons and containment checks ($\sqsubseteq$). We propose three optimizations.

1. *Size check optimization.* Let n be the number of items in the largest pattern found until now. The structure Z is implemented as a list of heaps $Z = \{Z_1, Z_2, ...Z_n\}$, where Z_x contains all maximal patterns found until now having x items ($1 \le x \le n$). To perform sub-pattern checking (super-pattern checking) for a pattern s containing w items, an optimization is to only compare s with patterns in $Z_1, Z_2...Z_{w-1}$ (in $Z_{w+1}, Z_{w+2}...Z_n$) because a pattern can only contain (be contained) in smaller (larger) patterns.

2. *Sum of items optimization.* In our implementation, each item is represented by an integer. For each pattern s, the *sum of the items* appearing in the pattern is computed, denoted as $sum(s)$. In each heap, patterns are ordered by decreasing sum of items. This allows the following optimization. Consider super-pattern checking for a pattern s_a and a heap Z_x. If $sum(s_a) < sum(s_b)$ for a pattern s_b in Z_x, then we don't need to check $s_a \sqsubseteq s_b$ for s_b and all patterns following s_b in Z_x. A similar optimization is done for sub-pattern checking. Consider sub-pattern checking for a pattern s_a and a heap Z_x. If $sum(s_b) < sum(s_a)$ for a pattern s_b in Z_x, then we don't need to check $s_b \sqsubseteq s_a$ for s_b and all patterns following s_b in Z_x, given that Z_x is traversed in reverse order.

3. *Support check optimization.* This optimization uses the support to avoid containment checks ($\sqsubseteq$). If the support of a pattern s_a is less than the support of another pattern s_b (greater), then we skip checking $s_a \sqsubseteq s_b$ ($s_b \sqsubseteq s_a$).

Strategy 2. Forward-Maximal Extension checking (FME). The second strategy aims at avoiding super-pattern checks. The search procedure discovers patterns by growing a pattern by appending one item at a time by s-extension or i-extension. Consider a pattern x that is generated. An optimization is to not perform super-pattern checking if the recursive call to the SEARCH procedure generate a frequent pattern (because this pattern would have x has prefix, thus indicating that x is not maximal).

Strategy 3. Candidate Pruning with Co-occurrence map (CPC). The last strategy aims at pruning the search space of patterns by exploiting item co-occurrence information. We introduce a structure named *Co-occurrence MAP* (CMAP) defined as follows: an item k is said to *succeed by i-extension* to an item j in a sequence $\langle I_1, I_2, ..., I_n \rangle$ iff $j, k \in I_x$ for an integer x such that $1 \leq x \leq n$ and $k \succ_{lex} j$. In the same way, an item k is said to *succeed by s-extension* to an item j in a sequence $\langle I_1, I_2, ..., I_n \rangle$ iff $j \in I_v$ and $k \in I_w$ for some integers v and w such that $1 \leq v < w \leq n$. A CMAP is a structure mapping each item $k \in I$ to a set of items succeeding it.

We define two CMAPs named $CMAP_i$ and $CMAP_s$. $CMAP_i$ maps each item k to the set $cm_i(k)$ of all items $j \in I$ succeeding k by i-extension in no less than *minsup* sequences of SDB. $CMAP_s$ maps each item k to the set $cm_s(k)$ of all items $j \in I$ succeedings k by s-extension in no less than *minsup* sequences of SDB. For example, the $CMAP_i$ and $CMAP_s$ structures built for the sequence database of Fig. 1(left) are shown in Table 1. Both tables have been created considering a *minsup* of two sequences. For instance, for the item f, we can see that it is associated with an item, $cm_i(f) = \{g\}$, in $CMAP_i$, whereas it is associated with two items, $cm_s(f) = \{e, g\}$, in $CMAP_s$. This indicates that both items e and g succeed to f by s-extension and only item g does the same for i-extension, being all of them in no less than *minsup* sequences.

VMSP uses CMAPs to prune the search space as follows:

1. *s-extension(s) pruning.* Let a sequential pattern *pat* being considered for *s*-extension with an item $x \in S_n$ by the SEARCH procedure (line 3). If the last item a in *pat* does not have an item $x \in cm_s(a)$, then clearly the pattern resulting from the extension of *pat* with x will be infrequent and thus the join operation of x with *pat* to count the support of the resulting pattern does not need to be performed. Furthermore, the item x is not considered for generating any pattern by *s*-extension having *pat* as prefix, by not adding x to the variable S_{temp} that is passed to the recursive call to the SEARCH procedure. Moreover, note that we only have to check the extension of *pat* with x for the last item in *pat*, since other items have already been checked for extension in previous steps.

2. *i-extension(s) pruning.* Let a sequential pattern *pat* being considered for *i*-extension with an item $x \in I_n$ by the SEARCH procedure. If the last item a in *pat* does not have an item $x \in cm_i$, then clearly the pattern resulting from the extension of *pat* with x will be infrequent and thus the join operation of x with *pat* to count the support of the resulting pattern does not need to be performed. Furthermore, the item x is not considered for generating any pattern by *i*-extension(s) of *pat* by not adding x to the variable I_{temp} that is passed to the recursive call to the SEARCH procedure. As before, we only have to check the extension of *pat* with x for the last item in *pat*, since others have already been checked for extension in previous steps.

CMAPs are easily maintained and are built with a single database scan. With regards to their implementation, we define each one as a hash table of hash sets, where an hashset corresponding to an item k only contains the items that succeed to k in at least *minsup* sequences.

Table 1. $CMAP_i$ and $CMAP_s$ for the database of Fig. 1 and $minsup = 2$

CMAP_i		CMAP_s	
item	is succeeded by (i-extension)	item	is succeeded by (s-extension)
a	$\{b\}$	a	$\{b, c, e, f\}$
b	$\emptyset$	b	$\{e, f, g\}$
c	$\emptyset$	c	$\{e, f\}$
e	$\emptyset$	e	$\emptyset$
f	$\{g\}$	f	$\{e, g\}$
g	$\emptyset$	g	$\emptyset$

Lastly, since the VMSP algorithm is a vertical mining algorithm, it relies on IDLists. We implement IDLists as bitsets as it is done in several state-of-art algorithms [2,6]. Bitsets speed up the join operations. Algorithms using this representation were demonstrated to be much faster than vertical mining algorithms which do not use them.

4   Experimental Evaluation

We performed several experiments to assess the performance of the proposed algorithm. Experiments were performed on a computer with a third generation Core i5 64 bit processor running Windows 7 and 5 GB of free RAM. We compared the performance of VMSP with MaxSP, the current state-of-the-art algorithm for maximal sequential pattern mining. All algorithms were implemented in Java. All memory measurements were done using the Java API. Experiments were carried on five real-life datasets having varied characteristics and representing three different types of data (web click stream, text from a book and protein sequences). Those datasets are *Leviathan, Snake, FIFA, BMS* and *Kosarak10k*. Table 2 summarizes their characteristics. The source code of all algorithms and datasets used in our experiments can be downloaded from `http://goo.gl/hDtdt`.

Table 2. Dataset characteristics

dataset	sequence count	item count	avg. seq. length (items)	type of data
Leviathan	5834	9025	33.81 (std= 18.6)	book
Snake	163	20	60 (std = 0.59)	protein sequences
FIFA	20450	2990	34.74 (std = 24.08)	web click stream
BMS	59601	497	2.51 (std = 4.85)	web click stream
Kosarak10k	10000	10094	8.14 (std = 22)	web click stream

Experiment 1. Influence of the *minsup* parameter. The first experiment consisted of running all the algorithms on each dataset while decreasing the *minsup* threshold until an algorithm became too long to execute, ran out of memory or a clear winner was observed. For each dataset, we recorded the execution time and memory usage.

In terms of execution time, results (cf. Fig. 5) show that VMSP outperforms MaxSP by a wide margin on all datasets. Moreover, VMSP performs very well on dense datasets (about 100 times faster than MaxSP on Snake and FIFA).

In terms of memory consumption the maximum memory usage of VMSP (MaxSP) on BMS, Snake, Kosarak, Leviathan and FIFA was respectively 840 MB (403 MB), 45 MB (340 MB), 1600 MB (393 MB), 911 MB (1150 MB) and 611 MB (970 MB). Overall, VMSP has the lowest memory consumption for three out of five datasets.

Experiment 2. Influence of the strategies. We next evaluated the benefit of using strategies in VMSP. We compared VMSP with a version of VMSP without strategy CPC (VMSP_W3), a version without strategies FME and CPC (VMSP_W2W3), and a version of VMSP without FME, CPC and strategies in EFN (VMSP_W1W2W3). Results for the BMS and FIFA datasets are shown in Fig. 6. Results for other datasets are similar and are not shown due to space limitation. As a whole, strategies improved execution time by up to to 8 times, CPC being the most effective strategy.

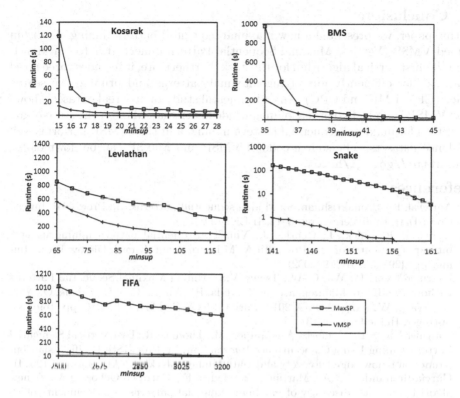

Fig. 5. Execution times

We also measured the memory used by the CPC strategy to build the CMAPs data structure. We found that the required amount memory is very small. For the BMS, Kosarak, Leviathan, Snake and FIFA datasets, the memory footprint of CMAPs was respectively 0.5 MB, 33.1 MB, 15 MB, 64 KB and 0.4 MB.

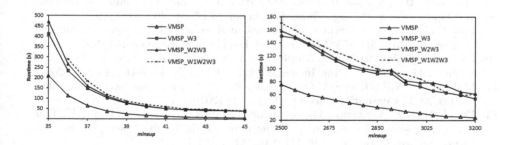

Fig. 6. Influence of optimizations for BMS (left) and FIFA (right)

5   Conclusion

In this paper, we presented a new maximal sequential pattern mining algorithm named VMSP (Vertical Maximal Sequential Pattern miner). It is to our knowledge the first vertical algorithm for this task. Furthermore, it includes three novel strategies for efficiently identifying maximal patterns and pruning the search space (EFN, FME and CPC). An experimental study on five real datasets shows that VMSP is up to two orders of magnitude faster than MaxSP, the state-of-art algorithm for maximal sequential pattern mining, and that VMSP performs well on dense datasets. The source code of VMSP and MaxSP can be downloaded from http://goo.gl/hDtdt.

References

1. Agrawal, R., Ramakrishnan, S.: Mining sequential patterns. In: Proc. 11th Intern. Conf. Data Engineering, pp. 3–14. IEEE (1995)
2. Ayres, J., Flannick, J., Gehrke, J., Yiu, T.: Sequential pattern mining using a bitmap representation. In: Proc. 8th ACM Intern. Conf. Knowl. Discov. Data Mining, pp. 429–435. ACM (2002)
3. Fournier-Viger, P., Wu, C.-W., Tseng, V.S.: Mining Maximal Sequential Patterns without Candidate Maintenance. In: Motoda, H., Wu, Z., Cao, L., Zaiane, O., Yao, M., Wang, W. (eds.) ADMA 2013, Part I. LNCS (LNAI), vol. 8346, pp. 169–180. Springer, Heidelberg (2013)
4. Fournier-Viger, P., Gomariz, A., Campos, M., Thomas, R.: Fast Vertical Sequential Pattern Mining Using Co-occurrence Information. In: Proc. 18th Pacific-Asia Conference on Knowledge Discovery and Data Mining. LNCS (LNAI). Springer (2014)
5. García-Hernández, R.A., Martínez-Trinidad, J.F., Carrasco-Ochoa, J.A.: A new algorithm for fast discovery of maximal sequential patterns in a document collection. In: Gelbukh, A. (ed.) CICLing 2006. LNCS, vol. 3878, pp. 514–523. Springer, Heidelberg (2006)
6. Gomariz, A., Campos, M., Marin, R., Goethals, B.: ClaSP: An Efficient Algorithm for Mining Frequent Closed Sequences. In: Pei, J., Tseng, V.S., Cao, L., Motoda, H., Xu, G. (eds.) PAKDD 2013, Part I. LNCS, vol. 7818, pp. 50–61. Springer, Heidelberg (2013)
7. Guan, E.-Z., Chang, X.-Y., Wang, Z., Zhou, C.-G.: Mining Maximal Sequential Patterns. In: Proc. 2nd Intern. Conf. Neural Networks and Brain, pp. 525–528 (2005)
8. Lin, N.P., Hao, W.-H., Chen, H.-J., Chueh, H.-E., Chang, C.-I.: Fast Mining Maximal Sequential Patterns. In: Proc. of the 7th Intern. Conf. on Simulation, Modeling and Optimization, Beijing, China, September 15-17, pp. 405–408 (2007)
9. Luo, C., Chung, S.: Efficient mining of maximal sequential patterns using multiple samples. In: Proc. 5th SIAM Intern. Conf. Data Mining, Newport Beach, CA (2005)
10. Lu, S., Li, C.: Apriori Adjust: An Efficient Algorithm for Discovering the Maximum Sequential Patterns. In: Proc. Intern. Workshop Knowl. Grid and Grid Intell. (2004)
11. Mabroukeh, N.R., Ezeife, C.I.: A taxonomy of sequential pattern mining algorithms. ACM Computing Surveys 43(1), 1–41 (2010)
12. Pei, J., Han, J., Mortazavi-Asl, B., Wang, J., Pinto, H., Chen, Q., Dayal, U., Hsu, M.: Mining sequential patterns by pattern-growth: the PrefixSpan approach. IEEE Trans. Known. Data Engin. 16(11), 1424–1440 (2004)
13. Wang, J., Han, J., Li, C.: Frequent closed sequence mining without candidate maintenance. IEEE Trans. on Knowledge Data Engineering 19(8), 1042–1056 (2007)
14. Zaki, M.J.: SPADE: An efficient algorithm for mining frequent sequences. Machine Learning 42(1), 31–60 (2001)

A Comparison of Multi-Label Feature Selection Methods Using the Random Forest Paradigm

Ouadie Gharroudi, Haytham Elghazel, and Alex Aussem

Université de Lyon, Université Lyon 1, LIRIS UMR CNRS 5205, F-69622, France
firstname.lastname@liris.cnrs.fr

Abstract. In this paper, we discuss three wrapper multi-label feature selection methods based on the Random Forest paradigm. These variants differ in the way they consider label dependence within the feature selection process. To assess their performance, we conduct an extensive experimental comparison of these strategies against recently proposed approaches using seven benchmark multi-label data sets from different domains. Random Forest handles accurately the feature selection in the multi-label context. Surprisingly, taking into account the dependence between labels in the context of ensemble multi-label feature selection was not found very effective.

Keywords: Feature Selection, Multi-label Learning, Ensemble Methods, Random Forest.

1 Introduction

The problem of single-label classification is concerned with learning from a set of examples $\mathcal{X}$, where each example is associated with a single label λ from a finite set of disjoint labels $\mathcal{L}$ of size L, with $L > 1$. If $L = 2$, then the learning task is called binary classification, while if $L > 2$, then it is called multi-class classification. On the other hand, the task of learning a mapping from an instance $x \in X$ to a set of labels $Y \in \mathcal{L}$ is referred to as a multi-label classification. Multi-label classification is a challenging problem that emerges in several modern applications such as text categorization, gene function classification, and semantic annotation of images [3,24]. The issue of learning from multi-label data has recently attracted significant attention from many researchers and a considerable number of approaches have been proposed [5,17,25]. Basically, they can be summarized into two categories: (a) algorithm adaptation methods and (b) problem transformation methods. Algorithm adaptation methods extend specific learning algorithms to handle multi-label data directly. Problem transformation methods, on the other hand, transform the multi-label learning problem into either several binary classification problems, such as the Binary Relevance (BR) approach, or one multi-class classification problem, such as the Label Powerset (LP) approach. The single-label classification problems are then solved with a commonly used single-label classification approach and the output is transformed back into a multi-label representation.

M. Sokolova and P. van Beek (Eds.): Canadian AI 2014, LNAI 8436, pp. 95–106, 2014.

The identification of relevant subsets of random variables among thousands of potentially irrelevant and redundant variables is a very important topic of pattern recognition research that has attracted much attention over the last few years. In traditional single-label learning, feature selection algorithms use information from labeled data to find the relevant subsets of variables, i.e., those that conjunctively prove useful to construct an efficient classifier from data. It enables the classification model to achieve good or even better solutions with a restricted subset of features [10]. As in the single-label case, multi-label feature selection had been widely studied and have encountered some success in many applications [9,16,20].

Although considerable attention has been given on the problem of using Random Forest (RF) to estimate the feature importance for traditional supervised [4], unsupervised [11,12,8] and semi-supervised [1] learning, little attention has been given to exploiting the power of this ensemble method with a view to identify and remove the irrelevant features in a multi-label setting. The way internal estimates are used to measure variable importance in RF paradigm [4] have been influential in our thinking. In this study, we propose and experimentally evaluate three wrapper multi-label feature selection methods, which use the RF paradigm. The main idea is to run three variants of RF for Multi-label learning (BRRF, RFLP and RFPCT) and then exploit the RF permutation importance measure [4] to evaluate the goodness of a feature. BRRF, for *Binary Relevance Random Forest* and RFLP, for *Random Forest Label power-Set*, consists of the two problem transformation approaches BR and LP, to previously transform the multi-label data into single-label data, which is then used to perform a Random Forest. However, RFPCT [15] (Random Forest of Predictive Clustering Trees) is another extension of RF that uses as base classifier PCT [2], a decision tree predicting multiple target attributes at once. We would like to mention that feature selection using RFPCT was initially proposed in [13], nonetheless, it was evaluated on a single biological data set and only compared to a trivial random feature ranking algorithm in [14]. To the best of our knowledge, this study is the first attempt to compare several RF-based feature selection methods in the context of multi-label classification.

Empirical results on seven multi-labeled datasets will be presented to answer the following questions: (1) Is there any benefit of exploiting label dependence structure in the context of multi-label feature selection as suggested by several authors [9,25]? (2) How can we extend the RF approach to address the multi-label feature selection problem? (3) Are these RF-based methods competitive with other state-of-the-art feature selection methods?

The rest of the paper is organized as follow: Section 2 reviews recent studies on multi-label feature selection and ensemble methods. Section 3 introduces the three RF-based multi-label feature selection methods and describes how variable importance used in RF can be extended in multi-label context. Experiments using conventional benchmark data sets are presented in Section 4. We raise several issues for future work in Section 5 and conclude with a summary of our contributions.

2 Related Work

In this section, we briefly review the multi-label feature selection and multi-label ensemble learning approaches that appeared recently in the literature.

2.1 Multi-Label Feature Selection

In multi-label classification, most feature selection tasks have been addressed by extending the techniques available for single-label classification using the bridge provided by multi-label transformations. These methods propose a previous transformation of multi-label data to single-label data, i.e., to binary data or multi-class data using respectively the BR or the LP approach. Thus, when the BR strategy is used, it is straightforward to employ a filter approach on each binary classification task, and then combining somehow the results (by an averaging for example). In this context, different feature importance measures have been used, such as Information Gain [20] and ReliefF [20]. Since each label is treated independently, these methods fail to consider the correlation among different labels. On the other hand, methods which perform feature selection using the same evaluation measures according to the LP approach take into account the label correlation [7,20]. Furthermore, the PMU approach in [16] is considered as the first filter approach that takes into account label interactions in evaluating the dependency of given features without resorting to problem transformation. The proposed method is presented as a multivariate mutual information-based feature selection method for multi-label learning that naturally derives from mutual information between a set of features and a set of labels.

In contrast to these previous filter approaches, Gu el al. propose an embedded-style feature selection method for multi-label learning called CMLFS [9]. CMLFS (for Correlated Multi-Label Feature Selection) is based on LaRank SVM, which is among state-of-the-art multi-label learning methods. In the proposed method, the goal is to find a subset of features, based on which the label correlation regularized loss of label ranking is minimized. Although this method considers correlation among labels, it optimizes a set of parameters during feature selection process to tune the kernel function of multi-label classifier making it impractical from the viewpoint of computational cost [16].

2.2 Multi-Label Ensemble Learning

The ensemble methods for multi-label learning are developed on top of the common problem transformation or algorithm adaptation methods. The most well known problem transformation ensembles are the RAkEL system by Tsoumakas et al. [21] and ensembles of classifier chains (ECC) [18]. RAkEL (for RAndom k-labELsets) is an ensemble of LP classifiers. It proposes breaking the initial set of labels into a number m of small-sized random subsets, called k-labelsets (k is the labelset size) and employing LP to train a corresponding classifier. A simple voting process determines the final classification set. In this manner, RAkEL take into account correlation between labels, and at the same time, avoid the weakness

of LP methods, by reducing the the number of labels handled by the LP classifiers. On the other hand, ECC are ensemble methods that are based on classifier
chain CC. The algorithm use p classifier chains $C_1, C_2, \ldots, C_p$; where in each C_i
a random subset of instances is chosen as training data and the order of learners
is performed using a random sequence of labels. For the multi-label classification of an unlabeled instance, the decisions of all CC classifiers are gathered and
combined. A threshold is used to choose the final multi-label set. Both RAkEL
and ECC are ensemble methods based on problem transformation algorithms. In
algorithm adaptation category, RFPCT [15] is a standard Random Forest, which
uses PCT [2] as base learner. PCT is an algorithm adaptation decision tree capable of predicting multiple target attributes at once.The induction process in
PCT uses the sum of the Gini indices throughout all labels to identify the best
separation at each node. In RFPCT, each tree makes multi-label predictions, and
then predictions are combined using a majority or a probability voting scheme.
The diversity among trees is promoted using two strategies; bootstrap sampling
of training data and random selection of feature subsets.

3   Ensemble Feature Selection for Multi-Label Learning

RF has several desirable characteristics for feature selection: It is robust, exhibits
high-quality predictive performance, does not overfit and handles simultaneously
categorical and continuous features [4]. Furthermore, RF have proved to be efficient in traditional supervised [4], semi-supervised [1], and unsupervised [8]
feature selection process. This section introduces and experimentally evaluates
three wrapper multi-label feature selection methods, which use the RF paradigm.
In this way, we discuss three variants of RF for Multi-label learning *Random
forest of predictive clustering trees* (RFPCT), *Binary Relevance Random Forest* (BRRF), and *Random Forest Label Power-set* (RFLP); and then exploit the
RF permutation importance measure [4] to evaluate the goodness of a feature.
Before introducing the proposed methods, we recall how RF with permutation
based out-of-bag (oob) measures feature importance.

The variable importance measure in RF is based on the decrease of predictive
performance when values of a descriptive variable in a node of a tree are permuted randomly. Basically, a bootstrap is used as training set to create trees in
the forest. In each bootstrapped data set, almost 33% are left oob, *i.e.*, they are
not used for the construction of the ith corresponding model h^i ($i \in \{1, \ldots, T\}$).
We refer to them as Oob_i. Thus, these patterns can be used to estimate non
biased feature relevancies. In every tree grown in the forest, the values of the fth
feature in the Oob_i data, is randomly permuted to form Oob_i^f, and the tree h^i is
used to predict the labels of the new oob patterns. The predictive performance
of each tree h^i is evaluated on the untouched oob data and the permuted versions of the oob data. The importance of the fth variable is then calculated as
the relative increase of the error that is obtained when its values are randomly
permuted (*c.f.* Equation 1). The average of this number over all trees in the

forest is the raw importance score for variable f. We note that the greater the value of the importance measure, the more relevant is the feature,

$$I(f) = \frac{1}{T} \sum_{i=1}^{T} \frac{e(h^i(Oob_i^f)) - e(h^i(Oob_i))}{e(h^i(Oob_i))} \tag{1}$$

where T is the size of the forest and e is the error measure function.

Given a label space $\mathcal{L} = (\lambda_1, \lambda_2, ..., \lambda_L)$ and a data set $\mathcal{D}$ that consists of N instances each taking the form (x_i, y_i) where $x_i = (x_{i1}, ..., x_{iM})$ is a vector of M descriptive attributes and $y_i \in \mathcal{L}$ is the subset of labels associated to x_i (represented by a binary feature vector $(y_{i1}, ..., y_{iL}) \in \{0,1\}^L$), we present, in the sequel, the three used variants of RF for multi-label learning and describe how variable importance used in RF can be extended in this context.

Binary Relevance Random Forest (BRRF) - This method transforms the multi-label dataset $\mathcal{D}$ into many single-label datasets, one for each individual label in $\lambda_i \in \mathcal{L}$. After this transformation, a RF is created for each label λ_i. The relevance of each feature according to each individual label is measured using the above Equation 1 for which e is the traditional single-label classification error. Finally, the average of the score of all features across all labels is considered. BRRF, focuses on each label individually and does not take into account label dependence.

Random Forest Label Power-set (RFLP) - In this method the multi-label feature selection problem is handled using Power-set strategy. This approach reduces the multi-label dataset $\mathcal{D}$ to a multi-class dataset by treating each distinct label set as an unique multi-class label. To avoid creating too many calsses with few instances, that may issue an overfitting and an imbalance multi-class problems the Pruned Problem Trans formation as in [7] was used; patterns with too rarely occurring labels are simply removed from the training set by considering label sets with a predefined minimum occurrence τ. A RF could be now performed and the above described feature selection procedure will be naturally applied using in Equation 1 the traditional single-label classification error e. In this way, this approach directly takes into account label correlation.

Random Forest Predictive Clustering Tree (RFPCT) - In contrast to both previous approaches (BRRF and RFLP) for which the RF grows many classification trees using a CART as a base classifier, RFPCT [15] is an extension of RF that use a randomized variant of the non Pruned Predictive Clustering Tree (PCT) [2], as a base classifier. In this approach, the multi-label data $\mathcal{D}$ is handled directly and is then able to provide an intuitive way for taking into account relationships between labels. Nevertheless, it is noteworthy that BRRF and RFPCT perform comparably for classification (see [15] for more details).

The feature selection problem with RFPCT follows the same procedure described above. Feature relevance is measured on each PCT tree, and then averaged over all the trees in the forest. However, since PCT is an adaptation method devoted to learning simultaneously all the labels, the RF-based feature evaluation procedure requires an appropriate multi-label error measure e in Equation 1 instead of the ordinary classification error used for BRRF and RFLP. As suggested in [13,14], the multi-label error measure e used in each tree is obtained by averaging the individual classification errors across the L labels.

4   Performances Analysis

In this section, we investigate the effectiveness of the aforementioned RF-based feature importance measures for multi-label feature selection and compare their performances against two recent multi-label feature selection methods on seven benchmark data sets.

4.1   Data Sets and Evaluation Protocol

Seven benchmark multi-label data sets, mostly obtained from the *Mulan's repository* [22], were used to assess the performance of feature selection algorithms. We selected these data sets as they have already been used in various empirical studies and cover different application domains: Biology, semantic scene analysis, music emotions and text categorization. Table 1 shows, for each data set, the number of examples (N), the number of features (M), where b indicates that the feature values are binary and n indicates that the feature values are numeric; the number of labels (L), the Label Cardinality (LC), which is the average number of single-labels associated with each example; the Label Density (LD), which is the normalized cardinality; and the number of Distinct Combinations (DC) of labels.

Table 1. Description of the multi-label data sets used in the experiments

Data	Domain	N	M	L	LC	LD	DC
Emotions	Music	593	72 n	6	1.869	0.311	27
Enron	Text	1702	1001 b	53	3.378	0.064	753
Genbase	Biology	662	1186 b	27	1.252	0.046	32
Medical	Text	978	1449 b	45	1.245	0.028	94
Scene	Image	2407	294 n	6	1.074	0.179	15
Slashdot-f	Text	3782	1079 b	22	1.180	0.041	156
Yeast	Biology	2417	103 n	14	4.237	0.303	198

We compared the three RF-based multi-label feature selection methods to two recently proposed ones: PPT-MI [7] and PMU [16]. PPT-MI is a multi-label feature selection method using the Pruned Problem Transformation (PPT) to improve the LP approach followed by a sequential forward selection with the

Mutual information (MI) as search criterion. PMU is a filter approach that takes into account label interactions in evaluating the dependency of given features without resorting to problem transformation. It is presented as a multivariate mutual information-based feature selection method for multi-label learning that naturally derives from mutual information between selected features and a set of labels. The classification performance of the five feature selection methods was measured using RAkEL multi-label classification algorithm [21]. We evaluated the performance of the methods using a 3-fold cross validation. In order to get reliable statistics over the performance metrics, the experiments were repeated 5 times. So the results obtained are averaged over 15 iterations which allows us to apply statistical tests in order to discern significant differences between the compared methods. Note that the LIBSVM (with linear kernel) is employed as the binary learner for classifier induction to instantiate RAkEL. The number of models m in RAkEL is set to $min(2 \times L, 100)$ for all datasets [17], the size of the label-sets k to half the number of labels $(L/2)$ [17] and the threshold value to 0.5. For PMU and PPT-MI, the numeric data sets are discretized using the Equal-width interval scheme, as suggested by the authors in [16]. Furthermore, the three variants of RF of multi-label learning (BRRF, RFLP and RFPCT) are tuned similarly. The number of variables to split on at each node and the committee size are set to $\sqrt{M}$, and 100, respectively.

In the multi-label classification problem, performance can be assessed by several evaluation measures. Here, we employed the subset accuracy measure (also called multi-label classification accuracy) defined as follow:

$$Subset\_Accuracy(h) = \frac{1}{|Te|} \sum_{i=1}^{|Te|} I(h(x_i) = \mathcal{Y}_i) \tag{2}$$

where $\mathcal{Y}_i$ is the set of true labels and $h(x_i)$ is the set of predicted labels. $|Te|$ is the cardinality of test data set and $I(true) = 1$ and $I(false) = 0$

This metrics takes values in the interval $[0; 1]$. The greater the value, the better the algorithm performance. Note that the subset accuracy implicitly takes into account the label correlations. It is therefore a very strict evaluation measure as it requires an exact match of the predicted and the true set of labels.

4.2   Comparison Results

This section presents the results obtained from our empirical study and concludes on the applicability and performance of RF for multi-label feature selection. Figure 1 plots the classification performance in terms of subset accuracy averaged over the 5x3 runs of the above five compared approaches against the 50 most important features. Due to space limitation, we only show experimental results for the four largest data sets. As may be observed, BRRF outperforms the other methods by generally achieving the highest subset accuracy values. On the other hand, RFLP perform the worst.

For the sake of completeness, we also averaged the subset accuracy averaged over the 15 runs for different numbers of selected features for an extensive

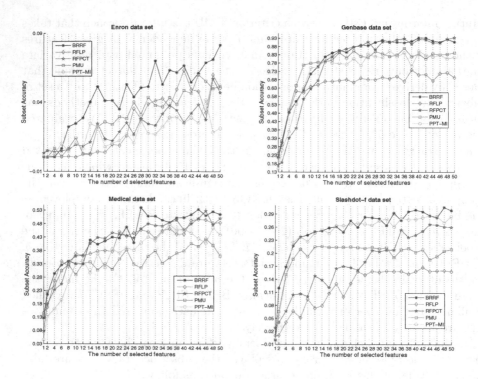

Fig. 1. Subset accuracy averaged over the 5x3 runs vs. different numbers of selected features on the four largest data sets

statistical analysis. The averaged metrics of the five feature selection methods over the top 50 features are depicted in Table 2. In order to better assess the results obtained for each feature selection algorithm on each metric, we adopt in this study the methodology proposed by [6] for the comparison of several algorithms over multiple datasets. In this methodology, the non-parametric Friedman test is firstly used to evaluate the rejection of the hypothesis that all the approaches perform equally well for a given risk level. It ranks the algorithms for each data set separately, the best performing algorithm getting the rank of 1, the second best rank 2 etc. In case of ties it assigns average ranks. Then, the Friedman test compares the average ranks of the algorithms and calculates the Friedman statistic. If a statistically significant difference in the performance is detected, we proceed with a *post hoc* test. The Nemenyi test is used to compare all the classifiers to each other. In this procedure, the performance of two classifiers is significantly different if their average ranks differ more than some critical distance (CD). The critical distance depends on the number of algorithms, the number of data sets and the critical value (for a given significance level p) that is based on the Studentized range statistic (see [6] for further details). In this study, based on the values in table 2, the Friedman test reveals statistically significant differences ($p < 0.1$) for each metric. Furthermore, we present the

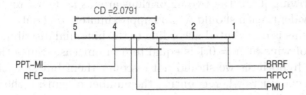

Fig. 2. Average rank diagram comparing the feature selection algorithms in terms of subset accuracy

result from the Nemenyi posthoc test with average rank diagram as suggested by Demsar [6]. This is given on Figure 2. The ranks are depicted on the axis, in such a manner that the best ranking algorithms are at the rightmost side of the diagram. The algorithms that do not differ significantly (at $p = 0.1$) are connected with a line. The critical difference CD is shown above the graph (here CD=2.0791).

Table 2. Subset accuracy averaged over the 5x3 runs and the 50 most important features for all algorithms and all data sets. Bottom row of the table presents the average rank of Subset accuracy mean used in the computation of the Friedman test

Data	BRRF	RFLP	RFPCT	PMU	PPT-MI
Emotions	0.266±0.05	0.253±0.04	0.260±0.05	0.245±0.04	0.248±0.05
Enron	0.046±0.01	0.024±0.00	0.022±0.01	0.029±0.01	0.019±0.01
Genbase	0.808±0.02	0.650±0.03	0.781±0.02	0.775±0.02	0.744±0.01
Medical	0.431±0.03	0.402±0.02	0.417±0.05	0.329±0.04	0.366±0.02
Scene	0.547±0.01	0.496±0.02	0.535±0.01	0.614±0.01	0.457±0.01
Slashdot	0.257±0.00	0.125±0.05	0.180±0.03	0.196±0.01	0.249±0.01
Yeast	0.157±0.01	0.155±0.01	0.155±0.01	0.139±0.01	0.152±0.01
Av Rank	1.1429	3.5714	2.8571	3.4286	4.0000

From Figure 2, we observe that BRRF performs significantly better than PMU, RFLP and PPT-MI, which seem to have equivalent performances. Although the average rank exhibit clear differences, the test doesn't allow us to conclude whether RFPCT is equivalent to BRRF or to the worst three methods. The Freidman test we use is known to be overly conservative. So to further exploit these rank comparisons, we compared, on each data set and for each pair of methods, the subset accuracy values obtained over the 15 iterations by using the paired t-test (with $\alpha = 0.1$). The results of these pairwise comparisons are depicted in Table 3 in terms of "Win-Tie-Loss" statuses of all pairs of methods; the three values in each cell (i, j) respectively indicate how times many the approach i is significantly better/not significantly different/significantly worse than the approach j. Following [6], if the two algorithms are, as assumed under the null-hypothesis, equivalent, each should win on approximately $n/2$ out of n data sets. The number of wins is distributed according to the binomial distribution and the critical number of wins at $\alpha = 0.1$ is equal to 6 in our case. Since tied matches support the null-hypothesis we should not discount them but split them evenly between the two classifiers when counting the number of wins; if there is an odd number of them, we again ignore one.

In Table 3, each pairwise comparison entry (i, j) for which the approach i is significantly better than j is boldfaced. The analysis of this table reveals that the approach that is never beaten by any other approach is BRRF.

Overall, these experiments confirm the ability of RF, that showed promising results for multi-label classification in [17], to rank the relevant features accurately in a multi-label context. More specifically, they suggest a relative superiority of the feature selection method built using the BRRF approach, compared with the ones that use RFPCT and RFLP. Indeed, it is more effective to use a RF that treats each label independently (*i.e.*, BRRF) rather than exploiting the underlying dependencies between labels (*i.e.*, RFPCT and RFLP) for evaluating feature importance in a multi-label setting. However, it was expected that methods that take the interaction among labels into consideration (*i.e.*, RFPCT and RFLP) would show better results than the ones using the BR approach (*i.e.*, BRRF). Nevertheless, this observation corroborate the previous finding in [20], namely that ignoring correlation among labels within the feature selection process doesn't affect the quality of the multi-label classification.

The superiority of BRRF compared to the remaining RF-based approaches (RFLP and RFPCT) in the feature selection process could be further motivated by the following reasons:

- The RFLP approach is based on the LP algorithm which suffers from class size issues, *i.e.*, the large number of label sets appearing in the training set (class values for the single-label classifier of LP), makes the learning task quite hard as many of these label sets are usually associated with very few training examples [21] giving rise to a poor feature importance estimation in a wrapper way. Although the Pruned Problem Transformation were used to avoid this problem of creating too many rarely classes in RFLP, the latter remains inefficient and does not give competitive results.
- With RFPCT, the classification error does not vary significantly when the values of a specific feature are randomly permuted. Indeed, we noticed that the label errors often compensate each other. This is why the classification error vary moderately after shuffling a variable. This issue worsen as the number of labels is increased. To confirm this observation from an experimental point of view, we analyzed the average gap between classification error before and after the variable shuffling in Equation 1. We observed error variations of the magnitude of 10^{-7} on the data sets with a large number of labels (*e.g.* Enron, Medical).

Table 3. Pairwise t-test comparisons of FS methods in terms of Subset accuracy. Bold cells (i, j) highlights that the approach i is significantly better than j according to the sign test at $\alpha = 0.1$.

	BRRF	RFLP	RFPCT	PMU	PPT-MI
BRRF	–	**7/0/0**	**5/2/0**	**6/0/1**	**7/0/0**
RFLP	0/0/7	–	0/4/3	2/1/4	4/1/2
RFPCT	0/2/5	3/4/0	–	2/2/3	5/1/1
PMU	1/0/6	4/1/2	3/2/2	–	3/1/3
PPT-MI	0/0/7	2/1/4	1/1/5	3/1/3	–

5 Conclusion

This work proposed and experimentally evaluated three wrapper multi-label feature selection methods, which use the RF paradigm: BRRF, RFLP and RF-PCT. These extensions differ in the way they consider label dependence within the feature selection process. The performance of the methods were compared against recently proposed approaches using seven benchmark multi-label data sets emerging from different domains. The result of this evaluation is three-fold: 1) Random Forest handles accurately the feature selection in a multi-label context and; 2) Surprisingly, BRRF appears more suitable for multi-label feature selection, as taking into account relationships between labels was not shown remarkably effective for multi-label feature selection using the RF paradigm.

Future work will be conducted to assess the stability of the feature selection methods [19] when noise is added to the data. We will also investigate the effectiveness of using label-specific feature selection [23] in the multi-label learning process. This is currently being undertaken and will be reported in due course.

References

1. Barkia, H., Elghazel, H., Aussem, A.: Semi-supervised feature importance evaluation with ensemble learning. In: ICDM 2010, pp. 31–40 (2011)
2. Blockeel, H., De Raedt, L., Ramon, J.: Top-down induction of clustering trees. In: ICML, pp. 55–63 (1998)
3. Boutell, M.R., Luo, J., Shen, X., Brown, C.M.: Learning multi-label scene classification. Pattern Recognition 37(9), 1757–1771 (2004)
4. Breiman, L.: Random forests. Machine Learning 45(1), 5–32 (2001)
5. Dembczynski, K., Waegeman, W., Cheng, W., Hüllermeier, E.: On label dependence and loss minimization in multi-label classification. Machine Learning 88(1-2), 5–45 (2012)
6. Demsar, J.: Statistical comparisons of classifiers over multiple data sets. Journal of Machine Learning Research 7, 1–30 (2006)
7. Doquire, G., Verleysen, M.: Feature selection for multi-label classification problems. In: Cabestany, J., Rojas, I., Joya, G. (eds.) IWANN 2011, Part I. LNCS, vol. 6691, pp. 9–16. Springer, Heidelberg (2011)

8. Elghazel, H., Aussem, A.: Unsupervised feature selection with ensemble learning. Machine Learning, 1–24 (2013)
9. Gu, Q., Li, Z., Han, J.: Correlated multi-label feature selection. In: CIKM, pp. 1087–1096 (2011)
10. Guyon, I., Elisseeff, A.: An introduction to variable and feature selection. Journal of Machine Learning Research 3, 1157–1182 (2003)
11. Hong, Y., Kwong, S., Chang, Y., Ren, Q.: Consensus unsupervised feature ranking from multiple views. Pattern Recognition Letters 29(5), 595–602 (2008)
12. Hong, Y., Kwong, S., Chang, Y., Ren, Q.: Unsupervised feature selection using clustering ensembles and population based incremental learning algorithm. Pattern Recognition 41(9), 2742–2756 (2008)
13. Kocev, D., Slavkov, I., Dzeroski, S.: More is better: Ranking with multiple targets for biomarker discovery. In: 2nd International Workshop on Machine Learning in Systems Biology, p. 133 (2008)
14. Kocev, D., Slavkov, I., Dzeroski, S.: Feature ranking for multi-label classification using predictive clustering trees. In: International Workshop on Solving Complex Machine Learning Problems with Ensemble Methods, in Conjunction with ECML/PKDD, pp. 56–68 (2013)
15. Kocev, D., Vens, C., Struyf, J., Dzeroski, S.: Tree ensembles for predicting structured outputs. Pattern Recognition 46(3), 817–833 (2013)
16. Lee, J.-S., Kim, D.-W.: Feature selection for multi-label classification using multivariate mutual information. Pattern Recognition Letters 34(3), 349–357 (2013)
17. Madjarov, G., Kocev, D., Gjorgjevikj, D., Dzeroski, S.: An extensive experimental comparison of methods for multi-label learning. Pattern Recognition 45(9), 3084–3104 (2012)
18. Read, J., Pfahringer, B., Holmes, G., Frank, E.: Classifier chains for multi-label classification. Machine Learning 85(3), 333–359 (2011)
19. Saeys, Y., Abeel, T., Van de Peer, Y.: Robust feature selection using ensemble feature selection techniques. In: Daelemans, W., Goethals, B., Morik, K. (eds.) ECML PKDD 2008, Part II. LNCS (LNAI), vol. 5212, pp. 313–325. Springer, Heidelberg (2008)
20. Spolaôr, N., Cherman, E.A., Monard, M.C., Lee, H.D.: A comparison of multi-label feature selection methods using the problem transformation approach. Electr. Notes Theor. Comput. Sci. 292, 135–151 (2013)
21. Tsoumakas, G., Katakis, I., Vlahavas, I.P.: Random k-labelsets for multilabel classification. IEEE Trans. Knowl. Data Eng. 23(7), 1079–1089 (2011)
22. Tsoumakas, G., Xioufis, E.S., Vilcek, J., Vlahavas, I.P.: Mulan: A java library for multi-label learning. Journal of Machine Learning Research 12, 2411–2414 (2011)
23. Zhang, M.-L.: Lift: Multi-label learning with label-specific features. In: IJCAI, pp. 1609–1614 (2011)
24. Zhang, M.-L., Zhou, Z.-H.: Multilabel neural networks with applications to functional genomics and text categorization. IEEE Transactions on Knowledge and Data Engineering 18(10), 1338–1351 (2006)
25. Zhang, M.-L., Zhou, Z.-H.: A review on multi-label learning algorithms. IEEE Transactions on Knowledge and Data Engineering 99(PrePrints):1 (2013)

Analyzing User Trajectories from Mobile Device Data with Hierarchical Dirichlet Processes

Negar Ghourchian and Doina Precup

McGill University
{negar,dprecup}@cs.mcgill.ca

Abstract. Mobile devices have become pervasive among users in both work environments as well as everyday life, and they sense a wealth of information that can be exploited for a variety of tasks, such as activity recognition, security or health monitoring. In this paper, we explore the feasibility of trajectory clustering, i.e., detecting similarities between moving objects, for an application related to workplace productivity improvement. We use Hierarchical Dirichlet Processes due to their ability to automatically extract appropriate trajectory segments. The application domain is the analysis of RSSI data, where this machine learning method proves successfully.

1 Introduction

Modern mobile devices are a ubiquitous part of modern society, and they are capable of recording a wide range of information about the users, including position (through GPS or Wifi localization), velocity and acceleration (through accelerometers), in addition to other sensors. There is an increased interest in trajectory data analysis, due in part to applications such as advertising (e.g. sending ads to a user that lingers in front of a particular store), surveillance and security, as well as health monitoring (e.g. recognizing if an elderly person living at home needs assistance). Characterizing trajectory data is of great importance for a variety of problems in urban planning, transportation engineering, public health, social behaviour analysis, etc. In this paper, we are exploring the use of trajectory analysis in a workplace context, in which one may want to understand the types of behaviors that employees use to fulfill their duties, and use this characterization to further optimize workplace organization. Many of the existing studies on trajectory analysis focus on the problem of computing similarity metrics between trajectories. A typical approach is to simplify the shape of the whole trajectory to a representative model, and define a distance or similarity measure between such models, or try to deform one trajectory onto another, e.g, using dynamic time warping. [3,4] compare several trajectory similarity measures (longest common subsequence, dynamic time warping and Hausdorff distance) in the problem of outdoor scene analysis and clustering of vehicle trajectories in a surveillance problem. These methods work well under the assumption that one can find a unique class of models for the shape of whole trajectories.

M. Sokolova and P. van Beek (Eds.): Canadian AI 2014, LNAI 8436, pp. 107–118, 2014.

However, in real data there are often complex, long trajectories, in which partial segments contain interesting information, instead of whole route. In such real-world application scenarios, a spatial segmentation is needed to generate discriminative features for trajectory classification or clustering. Frank et al define an alternative distance metric, called geometric template matching, which applies to time series containing one observation, and is geared towards segmenting trajectories that one could encounter into main "modes". However, extending this method to multidimensional time series is not obvious.

In this paper, we consider the setting in which trajectories have categorized, in an unsupervised setting, in which the length and structure of trajectories in the same category very greatly. We analyze a real data set recorded by an organization which required employees to use wearable electronic badges for a period of time. Such badges automatically capture a variety of signals such as physical activity and indoor location information for an individual, in order to study and analyze employees' well-being and organizational effectiveness [5]. This wearable computing platform produces several types of sensor data e.g., received signal strength indicator (RSSI), speech features and 3-axis accelerometer data. In this work, we are mainly interested in RSSI readings to anchor nodes with fixed positions, which convey instantaneous locations of employees in the workspace. Trajectories are defined by time-stamped RSSI readings. Employees with different job titles were given related tasks to complete while wearing the devices, and were asked to submit a report on their completed tasks. One important characteristic of this organization is that the tasks are information-sensitive and thus required the employees to discuss their problems with other employees in their group. Hence, fulfilling a task usually requires employees to follow fairly long trajectories. In this naturalistic setting, completing a task was not time-restricted so the length of an RSSI trajectory corresponding to one task would vary from minutes to hours and even days. We aim to perform task recognition based on these complex location trajectories.

We propose a hierarchical clustering approach that allows us to extract region-based features from location trajectories and exploit the correlation between the spatial movement of employees and their intended task. Our proposed framework is motivated by two main observations. First, in the presence of long trajectories, partitioning such trajectories into regions provides higher-level features that overcome the limitations of the whole trajectory shape analysis. Second, our goal is to achieve a semantic trajectory clustering that captures the different types of tasks. In many workplaces, including the one where our dataset was recorded, people with different jobs share common places for performing their assigned tasks. Hence, trajectories that belong to different clusters still share common passing regions. This suggests the need for a hierarchical clustering model that allows sharing components across clusters. The Hierarchical Dirichlet Process (HDP) [6] is a non-parametric Bayesian model for clustering problems involving multiple groups of data which share components. Hence, in principle it is ideally suited for this problem. The paper aims to be a proof of concept that such machine learning methods are indeed suitable for this type of important real problem.

The paper is organized as follows. In Section 2, we introduce the data set used, we describe the organizational dynamics and wearable computing platform, and present our proposed feature generation framework. Section 3 discusses our proposed HDP-based clustering approach. Section 4 presents the experimental results and Section 5 concludes the work.

2   Trajectory Feature Generation

This section presents the dataset we use, and also our proposed feature extraction method used for the problem of task recognition from location trajectories.

2.1   A Working Example

In this work we considered a dataset collected using Sociometric Badges [7] which records the behaviour and interactions of employees at Chicago-area server configuration firm over one month [5]. It includes multiple real and synthetic measures that reflect the performance and dynamics of the organization with varying temporal resolution. More precisely, the dataset documents the work of 23 participating employees in an IT facility that were given computer system tasks on a first-come, first-serve basis. Each employee was asked to take a client's IT configuration requirements and produce IT products according to these specifications, while wearing a badge. In total, 1,900 hours of data were collected with a median of 80 hours per employee. We utilized three main data types.

- *Workspace layout* is depicted in Figure 1. Participating employees with different departmental roles are indicated at their booths with different colors. The base stations (anchor node), on yellow squares, were placed at fixed positions throughout the place in order to locate badges and timestamps that data were collected by them. The booths indicated with letter "N" belong to employees who did not participate in the study. This floormap and the role indications on it will later be used for clustering results verification.
- *Behaviour data* includes the locations of employees estimated from Zigbee RSSI readings by the badges worn by each employee, representing to which fixed location (anchor node) they went. The synthetic coordinates for each employee were extracted from raw RSSI reading under two constraints. First, each badge at coordinate (x, y) should see RSSI records from at least three base stations for a particular instance in time. Second, the times of these readings should be at most 1 second apart. In [5] the authors claim that solving the optimization problem of finding the best station and corresponding distance under these two constraints led to the coordinate estimation with standard deviation being the radius of one booth. The coordinate trajectories that we used for our evaluation were computed per employee ID per minute.

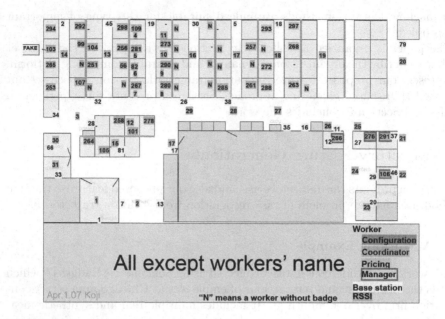

Fig. 1. Different branches in the organization have their own specific space. Each participating employee was assigned a unique ID and base stations with unique IDs were placed at fixed locations for RRSI records [5].

– *Performance data* includes the assigning time, closing time, difficulty level (basic, complex, or advanced), assigned-to, closed-by, number of follow-ups and role of the employee (pricing or configuration) of each completed task. In total there were 455 task reports in this dataset. We aim to cluster the location trajectories based on the tasks that we were doing at each specific time step. Therefore, we focused on assigning and closing time, closed-by employee ID and the role of employee. Also, for each observation we borrow the corresponding coordinate trajectory from the Instantaneous locations dataset.

Figure 2 shows two typical example of our observations. Although the employees from different jobs spend most of their time visiting their own group's specified region, they regularly pass through other regions and common areas as well (see Figure 1 for details of workplace layout). This shared areas, make it hard to learn the pattern of each "configuration" and "pricing" tasks. On the other hand, we need to extract location-based features instead of temporal or shape-based features in order to discover the correlation between tasks and visited places. Our framework aims to generate discriminative regional-based features and then carry out hierarchical non-parametric Bayesian clustering to achieve a distinguished group of trajectories.

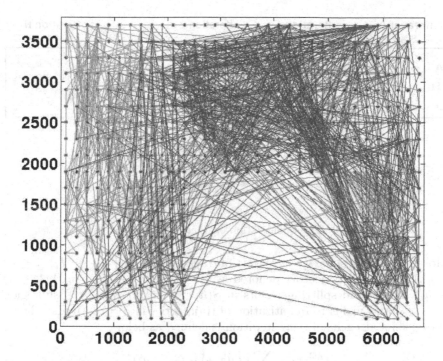

Fig. 2. Example of complex trajectories. The green and red trajectories are the tracks of two different employees performing "configuration" and "pricing" tasks, respectively.

2.2 Regional Quantization

For the feature generation phase, we take an approach motivated by partition-and-group framework proposed in [8]. Region segmentation aim to partition a 2-dimensional space into basic regions, either by identical or arbitrary small regions. We proposed to recursively split the original region into *four* identical regions and obtain a grid structure. Then each trajectory is quantized to a finite number of smaller subregions and will be represented by the labels of the traversed subregions. For ease of computation, each rectangular small subregion is represented by its centroid. Ideally, we would want to have homogeneous subregions in the sense that they would mostly contain trajectory parts from the same class distribution. However, a good quantization also possesses conciseness, which means that the number of subregions should stay as small as possible to avoid overfitting and feature complexity. Thus, it is essential to find a measure that offers a desirable compromise between these two properties. We decided to monitor the "quantization error" (QE) of subregions after each iteration of quad splitting to experimentally find the optimum number of splits. As shown in Figure 3, at each iteration of splitting, the average QE of each subregion is calculated by summation of the Euclidean distances of raw trajectory values enclosed in that subregion from the centroid.

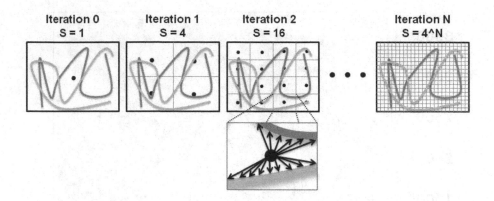

Fig. 3. The quantization error calculation process

More formally, assume we have a set of trajectories $\mathcal{T} = \{T_1, \cdots, T_i\}$ $(1 \leq i \leq num_{tra})$, with each trajectory denoted as $T_i = \{(x_1, y_1), \cdots, (x_\ell, y_\ell)\}$ $(1 \leq \ell \leq length_i)$. Each quad splitting results in $S(iter) = 4^{iter}$ $(itrer \geq 0)$ subregions $(r_1, \cdots, r_S)$ and yields to quantization of trajectories into S centroids. The QE of subregion r at each splitting iteration is defined as below,

$$QE^2(r) = \sum_{\hat{T}_j \in r} (\hat{T}_j(x, y) - C_r(x, y))^2, \tag{1}$$

where $\hat{T}_j$ represents all trajectory partitions enclosed in region r that are mapped to the centroid of r (C_r). The average QE of each iteration is then calculated as follows,

$$QE(iter) = \frac{\sum_{r=1}^{S(iter)} QE(r)}{S(iter)}. \tag{2}$$

Figure 4 demonstrates the QE for six iterations of recursive quad splitting of our working example workplace map. From this figure we can observe that the average QE significantly drops after 3rd iteration and almost stays unchanged from there. Based on this information we split the workspace to 64 identical regions and therefore, each trajectory is compound of this 64 distinct location features and corresponding number of passes from each subregion. In this approach, temporal aspect of the trajectories is not a concern since we only care about the most visited regions by each employee and not the sequence of regions.

3 Hierarchical Location Clustering

We developed and evaluated a statistical model to predict a user's task type given his/her location trajectory within the workplace. Although in our working example, the true labels of each employee's job and also the total number of job titles are known in advance, we are willing to solve the general case where this

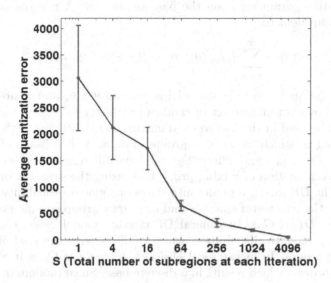

Fig. 4. Average quantization errors for different number of subregions

two parameter are unknown. We only use these ground truths for validating our clustering results.

Prior to clustering, we generated groups of data from raw data (trajectory of employees completing their tasks), e.i. each group is a mixture of components (subregions) with different mixing proportions (number of passes from subregions) specific to the group. Hierarchical Dirichlet process is a nonparametric Bayesian model for clustering problems involving multiple groups of data points where these points are exchangeable and are to be modeled with a mixture model [6].

3.1 Hierarchical Dirichlet Process

WWe are motivated by the problem of topic modeling in document corpora where a topic (i.e., a cluster) is a distribution across words and documents are viewed as distributions across topics while the goal is to discover topics that are common across multiple documents [9]. As mentioned before, the HDP works best in the scenarios where we have grouped or relational data points and rather than generating distinct cluster parameters, we prefer a more forgiving model that allows hierarchical discrimination. The HDP is a nonparametric prior, which means that the number of clusters is open-ended and at each step of generating data points a Dirichlet process (DP) mixture model can either assign the data point to a previously-generated cluster or start a new cluster.

The Dirichlet process $DP(\alpha, H)$ is a distribution over discrete probability measures on a potentially infinite parameter space Θ and, is uniquely defined by a concentration parameter α on the base measure H. A random draw $G \sim DP(\alpha, H)$ is equivalent to,

$$G(\theta) = \sum_{k=1}^{\infty} \pi_k \delta_{\theta_k}(\theta) \quad \theta_k \sim H, \quad k = 1, 2, \cdots \tag{3}$$

where the notation $\delta_{\theta_k}(\theta)$ indicates a Dirac delta at $\theta = \theta_k$ and π_θ are mixture weights defined over an infinite set of random parameter θ_k.

The DP can be used in the hierarchical mixture model setting. Let's consider a general setting in which we have J groups of data, each consists of n_j data points $(x_{j1}, x_{j2}, \cdots, x_{jn_j})$ and different groups have different characteristic given by a different combination of mixing proportion, using the same set of mixture components. The DP mixture model introduces one global probability measure $DP(\alpha_0, G_0)$ on the parameter space Θ and then draw group specific probability measures $G_j \sim DP(\alpha, G_0)$. In general DP mixture model cases, G_0 is often assumed to be a smooth distribution which conflicts with the goal of sharing clusters among groups. In hierarchical Dirichlet process the G_0 is itself a draw from Dirichlet process which results in a discrete base. Since random probability measures drawn from a Dirichlet process are discrete, there is a strictly positive probability of the group specific distributions having overlapping support (i.e. sharing parameters between groups) [6].

3.2 Region Sharing Clusters

Each observation of our working example, consist of a set of subregions (represented by its centroid), and the corresponding number of passes from each subregion for a particular location trajectory of task performance. We aim to compute the probability distribution over each subregion to be able to highlight important clusters or locations for each specific task. However the number of clusters is not known a priori and tends to grow as more data points are seen. Further, each subregion is been traveled by employees with various task description, which means that the clusters we compute have to share components. The number of components is itself an unbounded random variable, which is allowed to increase with the number of data points. In comparison with original topic modeling for documents, we consider each observation (set of subregions acquired by quantization of a trajectory) a *document* where each *document* is a collection of *words* (subregions), which are independent draws from underlying distribution. Also, the number of passes from each subregions corresponds to the word counts.

The HDP assigns to each trajectory a distribution over subregions which emphasize the locations of most popular subregions traversed by employees of each branch. However, the HDP model does not provide a unique cluster for each location. Instead, it computes cluster consisting of several subregions for each task and assigns to each trajectory a distribution over clusters(tasks).

In the language of Baysian modelling

$$P(Task|Traj) = P(Traj|Task) \times P(Task) \tag{4}$$

where $P(Traj)$ is assumed to be a constant and

$$P(Traj|Task) = \sum_{Sub} P(Traj, Sub|Task) = \tag{5}$$

$$\sum_{Sub} P(Traj|Sub, Task) \times P(Sub|Task).$$

Due to the nature of the HDP model, where number of clusters grows logarithmically with number of the data points, subregions that are passed more often tend to be represented by more clusters. As an advantage of this property, the HDP allows a better level of discrimination in location where the employee travel more. However, the drawback is that there is no one to one correspondence between clusters and subregions. To address this concurrent clustering, a greedy pruning phase seems to be critical for a semantic clustering task and a finer visualization of the average behavior of the clusters over the workspace map. Given the distribution $P(Subregion|Task)$, the cluster probability given a particular set of subregions, we computed pairwise L^2 norm regularization to learn similarity and correlation between clusters. The pairwise L^2 norm between each pair data points of $P(Subregion|Task)$ shows the structure of the underlying distribution of data points and is calculated as follows,

$$L^2_{P(i,j)} = \sum_{m=1}^{M} \|P_{(i,m)} - P_{(j,m)}\| \tag{6}$$

where $P_i, P_j \in P(Subregion|Task)$ and $P_i = \{p_{(i,1)}, \cdots, p_{(i,M)}\}$ represents cluster i with corresponding probability distributions over M subregions. Ideally, we would like to keep clusters with highest L^2 distances to others, which reveal the most unique clusters and helps to prune the rest, spurious ones. Ultimately, the significantly different clusters, detected by the HDP, are the *tasks* that we were interested to assign to each trajectory observation.

4   Evaluation

In this section we evaluate our approach on the introduced working example. The original raw trajectory were RSSI recordings mapped into a network of 502 grid points evenly distributed throughout the workspace using the algorithm described in [5]. The trajectories were quantized to 64 subregions based on the experimental results we achieved in section 2.2 and these region-based features (and their corresponding number of passes) were fed into HDP algorithm to produce task-dependent clusters. We expected the HDP models learn the mobility patterns that are intuitively correlated to the employees assignments and be able to predict correct labels based on spatial characteristics.

For the experiments, we used the HDP implementation provided by [10]. The HDPs requires setting a number concentration parameters that govern the a priori number of clusters, namely, α_0 and α, which we picked by initial exploratory

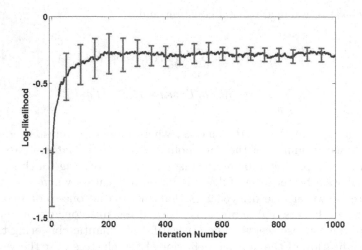

Fig. 5. Log-likelihood over training data, as the HDP algorithm progresses

experiments using a small subset of the data. Figure 5 presents the average log-likelihood over the training set for $\alpha_0 = 0.2$ and $\alpha = 0.2$ (determined by a line search procedure), as a function of the number of iterations of the HDP algorithm. As seen, the algorithm converge quickly and successfully to a solution with good log-likelihood.

As discussed in section 3.2, the HDP does not provide a one-to-one map between trajectories and clusters, hence we used L^2 norm to quantify the correlation between clusters and prune the repetitious and spurious. In our experiments, two main independent clusters remain after the pruning step, which had the most distinct probability distribution over the subregion space. Figure 6 presents a visualization of the "task" dependent clusters which reveals a semantic division of trajectories(into "configuration" and "pricing" branches) in comparison with the real layout of the organization. As it was expected, HDP allowed sharing subregions between clusters which explains why we have subregions that are assigned to multiple clusters with different colors. Eventually, we always can pick the dominant cluster (bigger probability) for each subregion.

In addition to visual verification, in order to verify the clustering performance with a ground truth, we used the independent trajectory labels provided in "Performance data". Also, we calculated the maximum likelihood of subregion assignment by adding up the maximum probability distributions of subregions $P(Trajectory|Subregion)$ over the winning cluster. The HDP obtained a 72% accuracy and the maximum log-likelihood of subregion assignments were -0.15. The results show that the algorithm mostly fails when the observation sample is extremely long. There were some tasks assignment where more than one working day were needed to complete the task. In this case, the employee would leave and then return the workplace multiple times during a single task performance

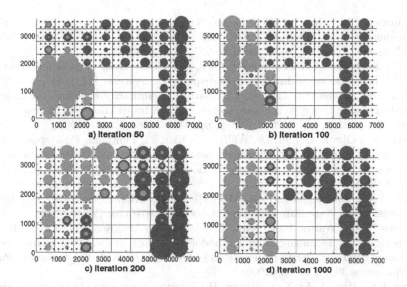

Fig. 6. Cluster visualization from different iterations of the HDP algorithm. The probability distribution over each sub-region is represented by a filled circle on its centroid where various colors denote different clusters and sizes correspond to the probability.

and so many non-relevant places could be repeatedly recorded in the mobility path. The results confirm that the non-parametric hierarchical framework has the ability of offering a high-level viewpoint for learning underlying structure of events, even in the presence of noisy and complex trajectories.

5   Conclusion

This project was conducted as an application for the "Badge Dataset" from "MIT Reality Mining Datasets", which attempt to investigate the influence of mobile-smart devices on individual, social and networks behaviour. We presented an application of Hierarchical Dirichlet Process to analysis of trajectories from wearable mobile devices. While ground truth is not available, the HDP computes cluster probabilities that can be used to accurately discriminate between various locations. Our results show that such methods are capable of interesting qualitative modelling. We plan to use this methodology in the future for other datasets as well.

References

1. Nascimento, J.C., Figueiredo, M., Marques, J.S.: Trajectory classification using switched dynamical hidden markov models. IEEE Transactions on Image Processing 19(5), 1338–1348 (2010)

2. Bennewitz, M., Burgard, W., Thrun, S.: Using em to learn motion behaviors of persons with mobile robots. In: IEEE/RSJ International Conference on Intelligent Robots and Systems, vol. 1, pp. 502–507. IEEE (2002)
3. Atev, S., Miller, G., Papanikolopoulos, N.P.: Clustering of vehicle trajectories. IEEE Transactions on Intelligent Transportation Systems 11(3), 647–657 (2010)
4. Zhang, Z., Huang, K., Tan, T.: Comparison of similarity measures for trajectory clustering in outdoor surveillance scenes. In: 18th International Conference on Pattern Recognition, ICPR 2006, vol. 3, pp. 1135–1138. IEEE (2006)
5. Dong, W., Olguin-Olguin, D., Waber, B., Kim, T., Pentland, P.: Mapping organizational dynamics with body sensor networks. In: 2012 Ninth International Conference on Wearable and Implantable Body Sensor Networks (BSN), pp. 130–135. IEEE (2012)
6. Teh, Y.W., Jordan, M.I., Beal, M.J., Blei, D.M.: Hierarchical dirichlet processes. Journal of the American Statistical Association 101(476) (2006)
7. Olguin, D.O., Waber, B.N., Kim, T., Mohan, A., Ara, K., Pentland, A.: Sensible organizations: Technology and methodology for automatically measuring organizational behavior. IEEE Transactions on Systems, Man, and Cybernetics, Part B: Cybernetics 39(1), 43–55 (2009)
8. Lee, J.-G., Han, J., Li, X., Gonzalez, H.: Traclass: Trajectory classification using hierarchical region-based and trajectory-based clustering. Proceedings of the VLDB Endowment 1(1), 1081–1094 (2008)
9. Blei, D.M., Ng, A.Y., Jordan, M.I.: Latent dirichlet allocation. The Journal of Machine Learning Research 3, 993–1022 (2003)
10. Teh, Y.W., Jordan, M.I.: Hierarchical Bayesian nonparametric models with applications. In: Hjort, N., Holmes, C., Muller, P., Walker, S. (eds.) Bayesian Nonparametrics: Principles and Practice. Cambridge University Press (2010)

Use of Ontology and Cluster Ensembles for Geospatial Clustering Analysis

Wei Gu[1], Zhilin Zhang[2], Baijie Wang[3], and Xin Wang[2]

[1] petroWeb, 1200 333-7th Avenue, Calgary, AB, Canada
victorgucanada@hotmail.com
[2] Department of Geomatics Engineering, University of Calgary
2500 University Drive NW, Calgary, AB, Canada, T2N 1N4
{zhi-lin.zhang,xcwang}@ucalgary.ca
[3] Husky Energy Inc. 707-8th Avenue, Calgary, AB, Canada
wangbaijie@gmail.com

Abstract. Geospatial clustering is an important topic in spatial analysis and knowledge discovery research. However, most existing clustering methods clusters geospatial data at data level without considering domain knowledge and users' goals during the clustering process. In this paper, we propose an ontology-based geospatial cluster ensemble approach to produce good clustering results with the consideration of domain knowledge and users' goals. The approach includes two components: an ontology-based expert system and a cluster ensemble method. The ontology-based expert system is to represent geospatial and clustering domain knowledge and to identify the appropriate clustering components (e.g., geospatial datasets, attributes of the datasets, and clustering methods) based on a specific application requirement. The cluster ensemble is to combine a diverse set of clustering results produced by recommended clustering components into an optimal clustering result. A real case study has been conducted to demonstrate the efficiency and practicality of the approach.

Keywords: Spatial analysis, Ontology, Cluster ensemble, Facility location analysis.

1 Introduction

Geospatial clustering is an important topic in spatial analysis and knowledge discovery research. It aims to partition similar objects into the same group (called a cluster) based on their similarity or connectivity in geographical space while placing dissimilar objects in different groups [1,2]. It can be used to find natural clusters (e.g., extracting the type of land use from the satellite imagery), identify hot spots (e.g., epidemics, crime, traffic accidents), and partition an area based on utility (e.g., market area assignment by minimizing the distance to customers).

Domain knowledge and users' goals play important roles during geospatial clustering [3,4,5]. The background knowledge concerning the domain described

M. Sokolova and P. van Beek (Eds.): Canadian AI 2014, LNAI 8436, pp. 119–130, 2014.

by the geospatial data is called domain knowledge. In geospatial clustering analysis, a user seeks to discover knowledge from geospatial data based on a particular goal by applying clustering methods. However, most existing clustering processes and clustering methods focus solely on the data itself without considering domain knowledge. Thus, clustering occurs at the data level instead of the knowledge level, which prevents the user from precisely understanding the clustering results and achieving his or her goals.

A few options for handling the problem seem apparent. One option is to develop new geospatial clustering methods exactly tailored for users' applications. The customized methods should consider which attributes of the geospatial data are needed and what kinds of the domain knowledge have to be exploited. Some customized clustering methods called constrained-based clustering methods, have been proposed [8,9,11]. Since they only consider limited knowledge concerning the domain and the user's goals, they are typically difficult to be reused. In particular, they usually have very restricted means of incorporating domain-related information from non-geospatial attributes.

The second option can be defined by considering the overall nature of the clustering process and building a knowledge-based system to support the integration of knowledge in the geospatial clustering process. The clustering process consists of all steps required to accomplish a clustering task given by a user. It starts from data preprocessing (including data cleaning, data integration, data selection, and data transformation), then applies clustering methods on the datasets, and finally presents the clustering results to the user [12]. Applying a clustering method is only one step of the overall process. Thus, the second option is to incorporate domain knowledge and users' goals into the clustering process, which allows an informed choice to be made from choosing the available datasets, suitable attributes of the datasets and clustering methods. However, this option can only get the best clustering results by using the existing most suitable clustering method and thus will be not helpful when none of the existing clustering method can provide good clustering results for a specific application.

The third option is to apply cluster ensembles [13,29] to geospatial clustering analysis. By applying available clustering methods to different attributes of datasets, cluster ensembles can obtain a large set of clustering results and finally combine them into a single consolidated clustering result. The result contains all information in the ensemble. However, it is time consuming to get a diverse set of clustering results. Previous research also shows that it is not always the best to include all available clustering results in the ensemble [14,28]. Thus, there is an emerging interest on reducing the number of clustering results in the ensemble.

In this paper, we propose a novel approach to geospatial clustering analysis, which combines an ontology-based expert system with a cluster ensemble method. Specifically, we first build an ontology to represent geospatial and clustering domain knowledge and then use an expert system to help identify appropriate geospatial datasets, attributes of the datasets and clustering methods for a specific application. Next, all the datasets, the attributes of the datasets and the clustering methods recommended by the ontology-based expert system are used to produce a diverse set of clustering results. Finally, with the help of the domain knowledge in the expert system, a subset of clustering results are selected and combined into a single clustering result. The approach can perform better than existing clustering methods because of the following reasons:

First, instead of developing a new clustering method for every specific application, the approach considers knowledge reuse. The domain knowledge learned from previous applications is formalized into the ontology-based expert system and would be reused for the similar applications in the future.

Second, with the help of the domain knowledge, the approach can identify appropriate components, including appropriate datasets, attributes of datasets, and clustering methods, according to users' goals.

Third, the approach combines the clustering results produced by the appropriate clustering components and results in one best clustering result using the cluster ensemble method. It provides more comprehensive clustering results when none of the clustering results produced by the appropriate clustering components is good enough.

The rest of the paper is organized as follows. The ontology-based geospatial clustering ensemble approach is proposed in the Section 2. A case study of the approach to do facility location analysis in Alberta, Canada is presented in Section 3. Section 4 summarizes the paper and discusses the future work.

2 An Ontology-Based Geospatial Cluster Ensemble Approach

In this section, we present an ontology-based cluster ensemble approach for geospatial clustering analysis. The approach includes two components: the ontology-based geospatial clustering system and a clustering ensemble method. In the following, we start with the general workflow of the approach and then we will introduce each component in detail.

2.1 Work Flow of the Approach

As shown in Fig. 1, users' clustering goals are first sent to the ontology-based geospatial clustering system, GEO_CLUST. The GEO_CLUST identifies appropriate geospatial datasets, attributes of the datasets and clustering methods according to the goals and the domain knowledge. Then, a diverse set of clustering results are produced by applying different clustering methods to different datasets and attributes of datasets multiple times.

For a better understanding of the approach, we treat each clustering result of a geospatial clustering application as a solution to that application and plotted them into a solution space. Fig. 2 shows the changes in a solution space when applying the ontology-based geospatial cluster ensemble approach under different statuses. At status 1, only the solutions within the circle are left for the following analysis. Second, a set of clustering results are sent the cluster ensemble method. In the first step of the method, a subset of the clustering results are selected with the criteria of high quality and diversity. The domain knowledge may be extracted from the GEO_CLUST to measure the quality. Thus at status 2, the solution space is further reduced, as shown the smaller rectangle in Fig. 2. Finally, the second step of the cluster ensemble method combines the selected clustering results into one optimal combined clustering result (as shown the black point in Fig. 2). According to the Equation (6), the combination may need the domain knowledge which could be extracted from the GEO_CLUST.

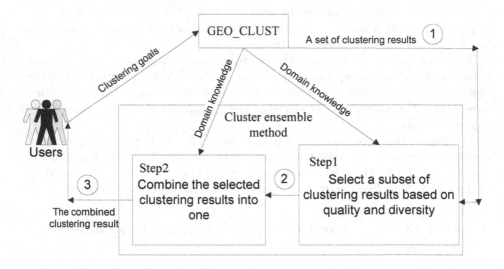

Fig. 1. Work flow of the approach

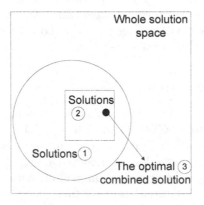

Fig. 2. Changes in solution space

2.2   An Ontology-Based Geospatial Clustering System

In this section, we present an ontology-based expert system, named GEO_CLUST, for performing geospatial clustering. The goal of the system is to make better use of geospatial and clustering knowledge to select proper methods and datasets to achieve clustering results that better meet users' requirements. The system consists of the GeoCO ontology for geospatial clustering and the Ontology Reasoner reasoning mechanism. The GeoCO ontology is used to represent geospatial and clustering domain knowledge. The Ontology Reasoner uses classification and decomposition techniques to specify users' tasks.

An **ontology** is a formal explicit specification of a shared conceptualization [15]. It provides domain knowledge relevant to the conceptualization and axioms for reasoning with it. For geospatial clustering, an appropriate ontology must include a rich set of geospatial and clustering concepts. Therefore, it can provide a knowledge source that supplements domain experts. Since the ontology in the geospatial domain is complex and varies according to the application [18], we build GeoCO at a high generic level such that it can be extended and materialized for specific applications. The GeoCO geospatial clustering ontology has been represented in using Protg-OWL [24] and the detail information about it can be found in [4].

The structure of the GEO_CLUST system for ontology-based clustering is shown in Fig. 3. It includes five components: the Geospatial Clustering Ontology, the Ontology Reasoner, the Clustering Methods, the Data, and the Graphical User Interface (GUI). The geospatial clustering ontology component is used when identifying the clustering problem and the relevant data. Within this component, the task model specifies the data and methods that may potentially be suitable for meeting the user's goals, and GeoCO includes all classes, instances, and axioms in a geospatial clustering domain. Through classification and decomposition conducted in the Ontology Reasoner, proper clustering data and methods can be indentified from the ontology.

The system works as follows: with the system, the user first gives his or her goals for clustering through the GUI. To be able to find proper data and clustering methods in ontology, the goal needs formalizing as a task instance. A task instance describes the specific problem to be solved. An example of a user's goal is to identify the best locations for five hospitals in Alberta. A task instance "determine best locations of five hospitals in Alberta" is created. The task instance is refined in the task model by using Ontology Reasoner and the refined elementary sub-tasks are used to search the domain ontology. For the example above, one of elementary tasks of "determine the best locations of five hospitals in Alberta" is "to find population data in Alberta," which can be implemented through database queries. The results of these queries identify the proper clustering methods and the appropriate data sets. Based on these results, clustering is conducted. The clustering results can be used for statistical analysis or be interpreted using the task ontology and the domain ontology.

In the system, the Ontology Reasoner is used to reason about knowledge represented in the ontology. In this component, classification are applied to detect the most specialized task node in a tree structure that the task instance belongs to, where the tree is used to organize all tasks hierarchically. Specifically, each task instance is classified according to its data and constraints, and thus the sub-task best fitting the characteristics of the task instance can be selected for the following processes. Decomposition describes the process of decomposing a task into simpler but more elementary subtasks, which presents a problem-solving strategy for tackling the task with a list of sub-tasks and operators (sequence, choice, iteration).

The input of the reasoner is the user's goals, and the output is a set of appropriate geospatial clustering methods and datasets. The reasoner performs the following steps. First, it builds a task instance [25] to associate the reasoning with the geospatial clustering ontology and the user's requirements. For the above example, the task instance "determine best locations of five hospitals in Alberta" is created based on a Partitioning-Clustering task in the task model, because the final results of clustering are to form five population clusters assigned to individual hospitals. Second, each available geospatial clustering method described in the ontology is considered either as an elementary task (which is accomplished by a simple primitive function) or as a complex task (which is accomplished via a task decomposition method represented in some problem solving strategy). The task "determine best locations of five hospitals in Alberta" is a complex task because we cannot solve the task by simply calling an existing primitive function. For the task, we first need to find the proper data, such as Alberta population data, and then identify the proper partitioning clustering method based on the characteristics of the data and the user's specification of "best locations". So the task needs to be decomposed. Finally each complex task is recursively decomposed into elementary sub-tasks [26]. The detailed description about the Ontological Reasoner component in GEO_CLUST is in [4], including task model, inference engine, and classification algorithm.

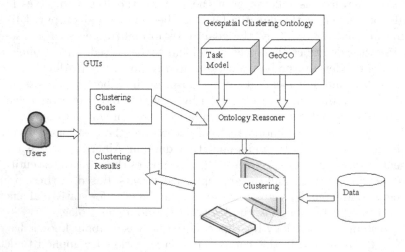

Fig. 3. The GEO_CLUST system for ontology-based clustering

2.3   A Cluster Ensemble Method

Given a set of clustering results, the cluster ensemble method used in the approach aims to select a subset of the clustering results and then combines them into a new clustering result which is better than the best result in the given result set. We extend cluster ensemble selection method in [14] by cooperating with the domain knowledge. The method includes two steps.

Step one: select a subset of clustering results with high quality and diversity, where quality measures the accuracy of clustering results in the subset and diversity measures the difference among the clustering results. According to the research did by Fern and Lin [14], selecting only a subset of clustering results based quality and diversity could improve the accuracy of the final ensemble result as well as reducing the execution time. Particularly, in this paper, given a set of clustering results C (i.e., $C = \{C_1, C_2, ..., C_r\}$) and a subset $C' \in C$, the way to measure the quality of C' is separated into two conditions:

(1) Without the domain knowledge. The quality of ($C_i \in C'$) is defined as the similarity between it and the other clustering results in C (as shown in Equation (1)) and the quality of C' is defined as the sum of the quality of all the clustering results in it (as shown in Equation (2)).

$$Quality(C_i) = \sum_{k=1}^{r} NMI(C_i, C_k) \tag{1}$$

$$Quality(C') = \sum_{C_i \in C'} Quality(C_i) \tag{2}$$

Where $NMI(C_i, C_k)$ is the normalized mutual information[1] between clustering C_i and C_k. We adopt the Equation 3 in [13] to estimate the NMI value between two clustering results. According to [13], if two clusterings are completely independent partitions, their NMI value is 0, vice versa. Thus, the larger is the value, the higher is the quality.

(2) With the domain knowledge. The quality of C_i ($C_i \in C'$) is measured by the external objective function according to the domain knowledge, and the quality of C' is defined as the sum of the quality of all the clustering results in it (as shown in Equation (3)).

$$Quality(C') = \sum_{C_i \in C'} EF(C_i) \tag{3}$$

Where $EF(C_i)$ is the external objective function value of C_i. For instance, when applying geospatial clustering analysis for the facility location planning [7], all the demand nodes in the target region are clustered into different groups and the demand nodes in each group are served by one facility. According to the domain knowledge in facility location planning, the external objective function

---

[1] Mutual information is a symmetric measure to quality the statistical information shared between two distributions.

value of a clustering result is the total travelling distance from the demand nodes to their assigned facilities.

The diversity of C' is defined as the sum of all pairwise similarities in the set (as shown the Equation (4)). The lower the value, the higher is the diversity. We measure the diversity as follows because it has been proved to efficiently affect the cluster ensemble performance [14].

$$Diversity(C') = \sum_{i \neq j, C_i, C_j \in C'} NMI(C_i, C_j) \tag{4}$$

Since high quality and diversity are two objectives to be achieved during the clustering result selection, we treat it as a bi-objective optimization problem [6] and solve it by using a bi-objective function. Specifically, given a set of clustering results, a subset of clustering results C' is selected from the whole set that minimizes the value of $BOF(C')$ in the following.

$$BOF(C') = \alpha Quality(C') + (1 - \alpha)Diversity(C') \tag{5}$$

In the Equation (5), α is defined as a co-efficient for balancing the quality objective and diversity objective, which is within [0,1]. The parameter α is a constant value to control the weight of each objective.

Selecting a subset of solutions to minimize the value of $BOF(C')$ is a NP-hard problem. In the research, we adopt a greedy procedure to perform the selection. It begins with a C' that only contains the single solution of the lowest linear normalized cost value and then incrementally adds one solution at a time into C' to minimize the value of $BOF(C')$. The procedure stops when the size of C' reaches the predefined number.

Step two: combine the selected clustering results into one optimal combined clustering result. The optimal clustering result C_{opt} should reach the objective function (Equation (6)), which asks for the optimal result has maximal mutual information with other selected clustering results and achieves higher external objective function value if domain knowledge is applied. We adopt the greedy optimization approach in [13] to solve the Equation (6) by defining our objective function $\Gamma(C_{opt})$.

$$\Gamma(C_{opt}) = \arg \max_{C_{opt}} \sum_{C_i \in C'} \left(\beta EF(C_{opt}) + (1 - \beta)NMI(C_{opt}, C_i) \right) \tag{6}$$

Where C_{opt} is a single labeling of clustering results which maximizes $\Gamma(C_{opt})$.

3   Case Study

In this section, we apply the approach to a real case study, finding the best locations for breast cancer screening clinics in Alberta (AB), Canada.

A population-based program to increase the number of Alberta women screened regularly for breast cancer was implemented in 1990 and today the Alberta Breast Cancer Screening Program (ABCSP) recommends Alberta women between the ages of 50 and 69 have a screening mammogram at least once every two years [10]. A key challenge is to determine the best locations for screening clinics to minimize the average demand-weighted distance from demand nodes to their assigned clinics. In this case, we perform the approach to separate the whole province into 26^2 screening clusters and for each cluster find a suitable location for building a clinic to serve the people within it.

In the approach, the first step is to find appropriate datasets, attributes of datasets, and clustering methods by sending the clustering goal to the ontology-based geospatial clustering system. The user interface and parameter setting of the ontology-based system can be found in [4]. Here we only list the results as below:

- Dataset: Census dataset in AB
- Attributes of datasets: women from 50 to 69 in DA level
- Similarity Measurement: Euclidean distance
- Clustering method: Capability K-MEANS

The dataset is the 2006 Canadian census data in AB. Estimates of the screening population (Alberta women aged 50 to 69 years) were derived from census data at the Dissemination Area (DA) level [16]. There are 327830 women within the target age in Alberta. A total of 5180 DAs were used in the research. Their values range from 0 to 920. The system adopts Euclidean distance to measure the similarity among the locations. The recommended clustering method is Capability K-MEANS [17]. In order to meet the overall demands in the province, the capacity of each clinic is set to 15,000. We generate 30 clustering results by applying Capability K-MEANS with different initializations.

The second step is to select a subset of clustering results based on quality and diversity. The Equation (3) in 3.1 is chosen to measure the quality. The external objective function value is the total travelling distance from the DAs to their assigned clinics. The diversity is measured by the Equation (4) in 3.1. α in the Equation (5) is set to 0.5. The number of the selected solutions is set to 5.

In the third step, the selected 5 results are combined into one optimal result. Since the case has a clear clustering goal, minimizing the average demand-weighted distance from demand nodes to their assigned clinics, the way to choose the optimal result is only based on the goal and thus the value of β in the Equation (6) is set to 1. Fig. 4 shows the clinic locations in the optimal result. We compare the optimal result with the one produced by the Interchange algorithm [27], the most popular algorithm used in facility location planning. The average demand-weighted distance are 24.12 km for the optimal result and 26.64 for the one produced by the Interchange algorithm. In this case, the approach achieves better results than by simply picking a standard facility location solution approach.

Finally, the knowledge obtained from this case study can be incorporated into some other applications in the future. For example, the same results can be used for planning the locations of pharmacies in the communities.

---

[2] The number of current cancer care facilities in Alberta is 26.

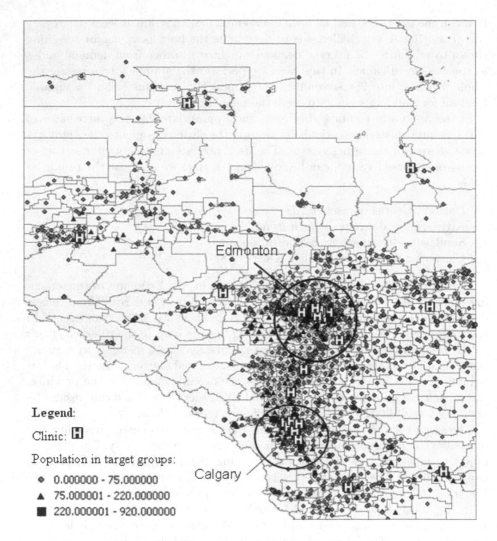

Fig. 4. The The optimal clinics distribution in AB

4   Conclusion and Future Work

In this paper, we present an ontology-based cluster ensemble approach to produce good clustering results for geospatial applications. The approach includes two components: an ontology-based expert system for formalizing the domain knowledge in the geospatial clustering, and a cluster ensemble method for combing good clustering results into an optimal result. To our best knowledge, it is the first research work to combine geospatial ontology and cluster ensembles for geospatial clustering analysis. The real case study did in Alberta, Canada shows that it is practical to combine the ontology and the cluster ensembles together for geospatial analysis. In the future, we will investigate how to further improve clustering results by integrating clustering ensemble with domain knowledge in theory and apply the presented approach on more application scenarios.

References

1. Ng, R., Han, J.: Efficient and Effective Clustering Method for Spatial Data Mining. In: Proc. of 20th International Conference on Very Large Data Bases, pp. 144–155. Morgan Kaufmann, San Francisco (1994)
2. Shekhar, S., Chawla, S.: Spatial Databases: A Tour. Prentice Hall (2003)
3. Graco, W., Semenova, T., Dubossarsky, E.: Toward knowledge-driven Data Mining. In: Proc. of International Workshop on Domain Driven Data Mining at 13th ACM SIGKDD, pp. 49–54. ACM, New York (2007)
4. Wang, X., Gu, W., Ziebelin, D., Hamilton, H.: An Ontology-based Framework for Geospatial Clustering. International Journal of Geographical Information Science 24, 1601–1630 (2010)
5. Wang, X., Hamilton, H.J.: Towards an Ontology-based Spatial Clustering Framework. In: Kégl, B., Lee, H.-H. (eds.) Canadian AI 2005. LNCS (LNAI), vol. 3501, pp. 205–216. Springer, Heidelberg (2005)
6. Mitropoulos, P., Mitropoulos, I., Giannikos, I., Sissouras, A.: A Biobjective Model for the Locational Planning of Hospitals and Health Centers. Health Care Management Sci. 9, 171–179 (2006)
7. Liao, K., Guo, D.: A Clustering-Based Approach to the Capacitated Facility Location Problem. Trans GIS 12, 323–339 (2008)
8. Prabakara Raj, S.R., Ravindran, B.: Incremental Constrained Clustering: A Decision Theoretic Approach. In: Li, J., Cao, L., Wang, C., Tan, K.C., Liu, B., Pei, J., Tseng, V.S. (eds.) PAKDD 2013 Workshops. LNCS, vol. 7867, pp. 475–486. Springer, Heidelberg (2013)
9. Wang, X., Rostoker, C., Hamilton, H.J.: Density-based Spatial Clustering in the Presence of Obstacles and Facilitators. In: Boulicaut, J.-F., Esposito, F., Giannotti, F., Pedreschi, D. (eds.) PKDD 2004. LNCS (LNAI), vol. 3202, pp. 446–458. Springer, Heidelberg (2004)
10. Alberta Breast Cancer Screening Program website, http://www.cancerboard.ab.ca/abcsp/program.html
11. Thiago, F.C., Eduardo, R.H., Joydeep, G.: A Study of K-Means-based Algorithms for Constrained Clustering. J. Intelligent Data Analysis 17, 485–505 (2013)
12. Han, J.W., Kamber, M., Pei, J.: Data Mining: Concepts and Techniques, 3rd edn. Morgan Kaufmann, San Francisco (2011)
13. Strehl, A., Ghosh, J.: Cluster Ensembles A Knowledge Reuse Framework for Combining Multiple Partitions. Machine Learning Research 3, 583–617 (2002)
14. Fern, X.Z., Lin, W.: Cluster Ensemble Selection. Journal of Statistical Analysis and Data Mining 1, 128–141 (2008)
15. Gruber, T.R.: A Translation Approach to Portable Ontologies. Knowledge Acquisition 5, 199–220 (1993)
16. Data quality index for census geographies, http://www12.statcan.ca.ezproxy.lib.ucalgary.ca/census-recensement/2006/ref/notes/DQ-QD_geo-eng.cfm
17. Ng, M.K.: A Note on Constrained k-means Algorithms. Pattern Recognition 33, 515–519 (2000)

18. Fonseca, F., Egenhofer, M., Agouris, P., Cmara, G.: Using Ontologies for Integrated Geographic Information Systems. Transactions in GIS 6, 231–257 (2002)
19. Maedche, A., Zacharias, V.: Clustering Ontology-based Metadata in the Semantic Web. In: Elomaa, T., Mannila, H., Toivonen, H. (eds.) PKDD 2002. LNCS (LNAI), vol. 2431, pp. 348–360. Springer, Heidelberg (2002)
20. Worboys, M.F.: Metrics and Topologies for Geographic Space. In: Advances in Geographic Information Systems Research II: International Symposium on Spatial Data Handling (1996)
21. Egenhofer, M.J., Clementini, E., di Felice, P.: Topological Relations between Regions with Holes. International Journal of Geographical Information Systems 8, 129–142 (1994)
22. Papadias, D., Egenhofer, M.: Hierarchical Spatial Reasoning about Direction Relations. GeoInformatica 1, 251–273 (1997)
23. Egenhofer, M.J., Franzosa, R.D.: Point-Set Topological Spatial Relations. International Journal of Geographical Information Systems 5, 161–174 (1991)
24. Protg web site, http://protege.stanford.edu/index.html
25. Crubzy, M., Musen, M.: Ontologies in Support of Problem Solving. In: Staab, S., Studer, R. (eds.) Handbook on Ontologies, pp. 321–341. Springer, Heidelberg (2004)
26. Parmentier, T., Ziébelin, D.: Distributed Problem Solving Environment Dedicated to DNA Sequence Annotation. In: Fensel, D., Studer, R. (eds.) EKAW 1999. LNCS (LNAI), vol. 1621, pp. 243–258. Springer, Heidelberg (1999)
27. Teitz, M.B., Bart, P.: Heuristic methods for estimating the generalized vertex median of a weighted graph. Oper. Res. 16, 955–961 (1968)
28. Naldi, M.C., Carvalho, A.C.P.L.F., Campello, R.J.G.B.: Cluster Ensemble Selection Based on Relative Validity Indexes. Data Mining and Knowledge Discovery 27, 259–285 (2013)
29. Sarumathi, S., Shanthi, N., Santhiya, G.: A Survey of Cluster Ensemble. International Journal of Computer Applications 65, 8–11 (2013)

Learning Latent Factor Models of Travel Data for Travel Prediction and Analysis

Michael Guerzhoy[1] and Aaron Hertzmann[1,2]

[1] University of Toronto
[2] Adobe Research
{guerzhoy,hertzman}@cs.toronto.edu

Abstract. We describe latent factor probability models of human travel, which we learn from data. The latent factors represent interpretable properties: travel distance cost, desirability of destinations, and affinity between locations. Individuals are clustered into distinct styles of travel. The latent factors combine in a multiplicative manner, and are learned using Maximum Likelihood.

We show that our models explain the data significantly better than histogram-based methods. We also visualize the model parameters to show information about travelers and travel patterns. We show that different individuals exhibit different propensity to travel large distances. We extract the desirability of destinations on the map, which is distinct from their popularity. We show that pairs of locations have different affinities with each other, and that these affinities are partly explained by travelers' preference for staying within national borders and within the borders of linguistic areas. The method is demonstrated on two sources of travel data: geotags from Flickr images, and GPS tracks from Shanghai taxis.

1 Introduction

Understanding human travel and mobility is a key component of understanding individuals and societies. In the past few years, researchers in many different disciplines have looked to online social media datasets as new sources of information about how we move from place to place. Good models of this data could yield new scientific insights into human behavior [1,8,11]. Furthermore, tools for analysis and prediction of movement can be useful for numerous applications. For example, travel models can help in predicting the spread of disease [6,15]; surveying tourism [9,10], traffic [2,14], and special events mobility for urban planning [3]; geolocating with computer vision [4,16]; interpreting activity from movements [19,22]; and recommending travel [7,17]. Analysis of travel data can also play a role within the larger context of analysis of social network data. More recent work has explored the value of social network and taxi trajectory data for predicting mobility [5] and knowledge discovery [23].

Previous work has generally employed two kinds of models to describe human travel probabilities. First, scientists have employed the Lévy flight, a stochastic process model [1,11]. This model captures the heavy-tailed nature of travel, but ignores the actual locations being visited. Second, one can build probability tables based on empirical travel histograms [4,6,9,10,15,16]. Such models are more accurate than simple stochastic process models, but require enormous datasets while revealing little about the underlying

M. Sokolova and P. van Beek (Eds.): Canadian AI 2014, LNAI 8436, pp. 131–142, 2014.

structure of the data. Detailed models have also been developed for some specialized applications (e.g., [14,19]).

This paper describes an approach to building structured models of human travel, using spatially-varying and individual-varying *latent factors*. The main idea is to model travel probabilities as functions of spatially-varying latent properties of locations and the travel distance. The latent factors represent interpretable properties: travel distance cost, desirability of destinations, and affinity between locations. Moreover, individuals are clustered into clusters with distinct travel models. The latent factors combine in a multiplicative manner, and are learned using Maximum Likelihood. The resulting models exhibit significant improvements in predictive power over previous methods, while also using far fewer parameters than histogram-based methods.

We present our approach as a family of models of increasing generality, similar to the use of multiplicative models in social network analysis [13]. While there are many factors in travel that we have not incorporated into the model, the multiplicative formulation easily lends itself to incorporating additional latent factors and sources of information.

Our model allows us to analyze the different "factors" separately. We visualize and analyze the propensity to travel long distance of different user clusters, independently of the locations to which they travel. We show that the desirabilities of the locations on the map are distinct from the number of transitions to those locations, partly since some desirable locations are more remote. We analyze the affinities between the different locations, independently of the desirability of the locations and the distances between locations. In particular, the affinities our model learns are stronger, on average, for pairs of locations inside national borders, and for some languages, for pairs of locations within the same linguistic areas.

2   Background

Suppose we discretize the world into J locations, and we observe an individual at location i. We wish to predict where that individual is likely to be after a certain time interval τ: $P_{ij\tau} \equiv P(L_{next} = j | L_{current} = i, \Delta T = \tau)$.

Like most previous work, we make the simplifying assumption that human travel can be treated as a Markov process. We also discretize all continuous quantities: τ represents one of a fixed set of time interval bins, and i and j represent location bins (quads) on the map.

To date, two general kinds of models have been applied to human travel. First, scientists have employed the Lévy flight, a stochastic process model [1]. In particular, the marginal probability distribution of traveling a distance d in some fixed time interval is given by a truncated power law:

$$p(d) \propto d^{-b} \tag{1}$$

for a parameter b. This model captures the heavy-tailed nature of travel: long-distance travel is rare but not surprising. A truncated power law [11] can be used to model limits in travel. These simple parametric models can be interpreted to yield insights into travel behavior; however, they assume that travel probabilities depend only on the travel distance. Destinations that are equidistant from one's current location are equally likely, regardless of whether they are in New York City or in the Arctic.

Second, one can build probability tables based on empirical travel histograms [4,6,9,10,15,16]. For example, Kalogerakis et al. [16] use counts of actual transitions in a database of 6 million travel records. Letting $N_{ij\tau}$ be the number of observed transitions from location i to location j in time interval τ, and $N_{d\tau}$ be the number of observed transitions over distance d in time interval τ, they set

$$P_{ij\tau} = \frac{N_{ij\tau} + \lambda q_{d_{ij}\tau}}{\sum_{\ell}(N_{i\ell\tau} + \lambda q_{d_{i\ell}\tau})} \qquad (2)$$

where d_{ij} is the distance between locations i and j. The histogram is regularized by the term $q_{d\tau} = \frac{N_{d\tau}}{\sum_{\delta} N_{\delta\tau}}$, which corresponds to the probability of traveling a given distance d in time interval τ in the training set, with a constant factor λ set using cross-validation. Such models are more accurate than simple stochastic process models, but require enormous datasets while revealing little about the underlying structure of the data.

Our work aims to combine the advantages of the two approaches described above, while extending them to yield new generality.

3   The Datasets

We consider two datasets. The first dataset comprises the publicly-available Flickr.com image streams of 75,250 distinct individuals, collected by Kalogerakis et al. [16]; there are about 6 million images in total in the dataset. Each image is accompanied by a timestamp, and a geotag specifies the location where the picture was taken. We consider only the geotags and the time intervals between consecutive geotags. As done by Kalogerakis et al., we discretize the Earth into 3186 bins, each of which is approximately 400km × 400km in size. An individual's trajectory is the sequence of transitions between the locations of the individual's photos.

The second dataset comprises taxi travel data, collected by Peng et al. [20]. We use the taxi trips recorded on Feb. 1, 2007. A trip (transition) begins or ends when the taxi changes its status from "occupied" to "vacant" or vice versa; we calculate the duration of each trip. The dataset, which we split into a training set and a test set, consists of 2,000 taxis which take about 287,000 trips in total. We discretize the area around Shanghai served by the taxis in the dataset (30°N to 32°N, 120°E to 122°E) into 50 × 50 quads using a rectangular grid.

The locations on the map differ in popularity (see Fig. 2). In the Flickr dataset, we observe there are more transitions to locations in the coastal United States, Western Europe, Australia, and Japan than to locations elsewhere on the map. In the taxi dataset, there are more transitions to locations in downtown Shanghai. The variation in the probability of transitioning from a given source to a given destination is not determined by the distance between the source and the destination alone. Individuals in our datasets exhibit travel patterns that are distinct from one another; see Fig. 1.

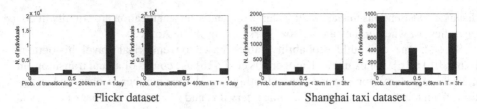

Flickr dataset   Shanghai taxi dataset

Fig. 1. Histogram of different individuals travel propensities. For a given distance and timeframe, and for a given probability, each bar in a histogram represents the number of individuals who travel the distance within the timeframe with the given probability.

4   Basic Model: Distance Cost and Desirability

The model can be motivated by observing that one is usually more likely to travel to nearby destinations than to further ones, but that certain destinations are more popular than others. For example, New York is almost always a local peak compared to its neighbors, whereas the ocean is never a destination. We note that there is a "logical-AND" relationship between these factors: undesirable destinations (like the ocean) are never visited no matter how close; highly-desirable destinations (like New York) are never visited if they are simply too far to be reached within the given time interval. This relationship suggests the use of a multiplicative model to combine these factors.

In particular, we consider the following multiplicative model:

$$P_{ij\tau} = \frac{r(d_{ij}, \tau)\, a_j}{\sum_\ell r(d_{i\ell}, \tau)\, a_\ell} \tag{3}$$

The model consists of two terms. First, the *distance factor* $r(d, \tau)$ captures the dependence of travel on the distance d_{ij} between locations i and j and the transition time τ. For a given τ, this term mimics the power-law terms used in previous work (Eq. 1). Second, the spatially-varying a_j term represents the *desirability* of the destination j. While not representing literal "desirability," it reflects that fact that some destinations attract more travel than others. For convenience, we define the parameters $\rho(d, \tau) = \log r(d, \tau)$ and $\alpha_j = \log a_j$. And we write the log-probability of traveling to location j from location i as $\log P_{ij\tau} \doteq \rho(d_{ij}, \tau) + \alpha_j$. (The symbol $\doteq$ indicates that we have omitted the normalization term.)

The quantities d and τ are discretized, and the functions $\rho(d, \tau)$ and α_j are each represented as 2D look-up tables. This method can be viewed as a form of "logistic PCA" [21] applied to travel data, taking the structure of the problem into account. The model can also be viewed as a special case of logistic regression, because the probability $P_{ij\tau}$ is linear in the parameters r and a. In particular, the model can be written as $P_{ij\tau} \propto \exp(\theta^T \phi_{ij\tau})$, where $\theta = [\text{vec}(\rho)^T, \text{vec}(\alpha)^T]^T$, and $\phi_{ij\tau}$ is a vector of indicator variables.

For our problem, the size of the dataset dwarfs the number of model parameters, which suggests that Maximum Likelihood should be sufficient for learning. We estimate the α and ρ parameters from the data, by minimizing the negative log-likelihood using the conjugate gradient method. The negative log-likelihood of the data is:

$$NLL = -\log \prod_{ij\tau}(P_{ij\tau})^{N_{ij\tau}} = -\sum_{ij\tau} N_{ij\tau} \log P_{ij\tau} \qquad (4)$$

Here, $N_{ij\tau}$ is the number of observed transitions from i to j in time interval τ. The equivalence to logistic regression shows that the negative log-likelihood is convex.

5 Location Affinities

We expect that similarities between locations would play a role in travel. For example, we expect that people are more likely to stay in their own countries than to cross borders, and, when leaving their country, they are more likely to visit countries where the same

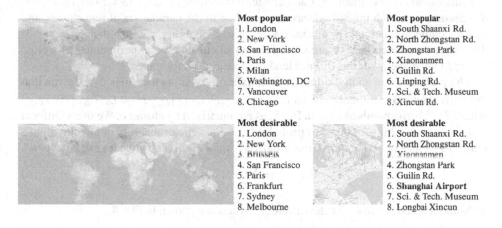

Most popular
1. London
2. New York
3. San Francisco
4. Paris
5. Milan
6. Washington, DC
7. Vancouver
8. Chicago

Most popular
1. South Shaanxi Rd.
2. North Zhongstan Rd.
3. Zhongstan Park
4. Xiaonanmen
5. Guilin Rd.
6. Linping Rd.
7. Sci. & Tech. Museum
8. Xincun Rd.

Most desirable
1. London
2. New York
3. Brussels
4. San Francisco
5. Paris
6. Frankfurt
7. Sydney
8. Melbourne

Most desirable
1. South Shaanxi Rd.
2. North Zhongstan Rd.
3. Xiaonanmen
4. Zhongstan Park
5. Guilin Rd.
6. **Shanghai Airport**
7. Sci. & Tech. Museum
8. Longbai Xincun

Fig. 2. The popularity (count of inbound transitions) and desirability (our learned parameter α) of destinations, with the top locations displayed. Desirability and popularity are related but distinct. On the Shanghai map, strong red means very large and strong green means very small. Quads in Shanghai identified by Metro stations, except for Shanghai Airport.

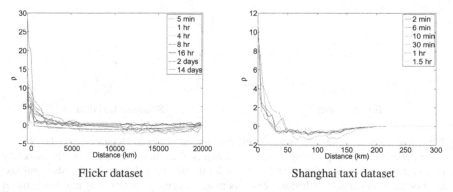

Flickr dataset

Shanghai taxi dataset

Fig. 3. The learned distance factors for the Flickr and Shanghai taxi datasets, for selected time intervals out of a total of 23 and 13 respectively

language is spoken as in their home country. We can express this by assigning a low-dimensional latent parameter vectors $\mathbf{u}_j$, $\mathbf{v}_j$ to each location j (for a discussion of why an asymmetric factor is useful, see [12]) . A low-dimensional vector is used, typically $\mathbf{u}_j, \mathbf{v}_j \in R^4$. The transition probability is then given by $\log P_{ij\tau} \doteq \rho(d_{ij}, \tau) + \alpha_j + \mathbf{u}_i^T \mathbf{v}_j$. The log-likelihood of a transition is then:

$$\log P_{ij\tau} \doteq \rho(d_{ij}, \tau) + \alpha_j + \mathbf{u}_i^T \mathbf{v}_j \tag{5}$$

The learned models are shown in Figs. 2 and 3.

6   Clustering Individuals

We incorporate the observation that travel varies between individuals [11] by clustering individuals into C clusters. Each cluster has its own distance factor functions $\rho^{(c)}$. We assume that each individual travels according to the parameters of one cluster, and that an individual belongs to cluster c with prior probability π_c, such that $\sum_c \pi_c = 1$. The other parameters $\alpha, \mathbf{u}, \mathbf{v}$ are shared across all clusters; we experimented with clustering them as well, but this did not lead to increased performance.

The probability of an individual's travel trajectory visiting locations $L_{1:K}$, given that the individual is a member of cluster c, is $P_{1:K}^{(c)} = P(L_1) \prod_{k=2}^{k} P^{(c)}(L_k | L_{k-1}, \tau_{k-1}, c)$ where $P^{(c)}$ corresponds to $P_{ij\tau}$ in Eq. 5, as parametrized by cluster c. We use a uniform distribution over bins for the start probability $P(L_1)$. The probability of an individual's entire trajectory of K locations, marginalized over the possible cluster assignments is

$$P_{1:K} = \sum_c \pi_c P_{1:K}^{(c)} \tag{6}$$

Some learned ρ values for the different clusters are shown in Fig. 4.

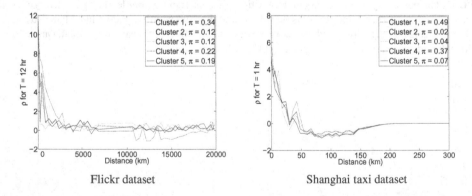

Flickr dataset   Shanghai taxi dataset

Fig. 4. Sample learned ρ values for different clusters for the Flickr and Shanghai taxi datasets for a set time period. The prior probability of each cluster is given in the legend. Note the markedly different probabilities for traveling less than 3km for the different clusters for the Shanghai taxi dataset. For the Flickr dataset, Cluster 1 exhibits more propensity for traveling mid-range travel distances than the other models, and Cluster 2 exhibits less propensity for traveling mid-range travel distances than the other models.

7 Fitting the Models

Our single-cluster model is given by Eq. 5. The Negative Log-Likelihood (NLL) of the data under the single-cluster model is then

$$NLL = -\log \prod_{ij\tau} (P_{ij\tau})^{N_{ij\tau}} =$$
$$-\sum_{\tau d} N_{\tau d}\rho(d, \tau) - \sum_j N_j \alpha_j + \sum_{ij\tau} N_{ij\tau} \log \sum_\ell \exp(\rho(d_{i\ell}\tau) + \alpha_\ell).$$

Taking derivatives with respect to the model parameters yields:

$$\frac{\partial NLL}{\partial \alpha_j} = -N_j + \sum_{i\tau} N_{i\tau} \frac{\exp(\rho(d_{ij},\tau)+\alpha_j)}{\sum_\ell \exp(\rho(d_{i\ell},\tau)+\alpha_\ell)} = -N_j + \sum_{i\tau} N_{i\tau} P_{ij\tau},$$

$$\frac{\partial NLL}{\partial \rho_{\tau d}} = -N_{\tau d} + \sum_{ij} N_{ij\tau} \frac{\sum_{\ell:d_{i\ell}=d} \exp(\rho(d_{i\ell},\tau)+\alpha_\ell)}{\sum_\ell \exp(\rho(d_{i\ell},\tau)+\alpha_\ell)} = -N_{\tau d} + \sum_i N_{i\tau} P_{i\tau d},$$

$$\frac{\partial NLL}{\partial u_i} = -\sum_{j\tau} N_{ij\tau} v_j + \sum_{j\tau} N_{ij\tau} \sum_\ell v_\ell P_{i\ell\tau}.$$

We optimize the NLL of the training data using the conjugate gradient method to fit the model.

For the Cluster model (Equation 6), we optimize the NLL using a generalized EM algorithm [18], using exact E-steps and gradient descent M-steps.

8 Results

We first analyze our models quantitatively and show that they demonstrate improved explanatory power as compared to the baseline model. We then analyze the parameters of the models.

8.1 The Predictive Power of the Models

We test the algorithm on the Flickr dataset by splitting individuals into test and training sets, with 52,769 individuals in the training set and 11,149 individual in the test set; we set the dimensionality of the u and v vectors to 4. We discretize time into 23 intervals. We use 737 taxis from the Shanghai taxi dataset for training and 1,018 taxis for the test set; the dimensionality of u and v is 4, and we discretize time into 13 intervals. We obtain the results shown in Table 1. Throughout, we compute the negative log-likelihood (NLL) of each test individual's transitions according to the learned models. For the Cluster model, we compute the likelihood of each individual trajectory (see Eq. 6); for comparison purposes, we compute the NLL of each transition separately by marginalizing over clusters and present the test NLL obtained by summing the NLLs of all the transitions, as in Eq. 4:

$$NLL = -\sum_{ij\tau} N_{ij\tau} \log \sum_c \pi_c P_{ij\tau}^{(c)} \tag{7}$$

Note that, for the non-Cluster models, the NLL of an individual's trajectory is always simply the sum of the NLLs of all the transitions in the trajectory (cf. Eq. 4).

Our models yield significantly better NLLs than the histogram models, while using orders-of-magnitude fewer parameters. We see performance improve as we use more sophisticated models, with the Cluster model giving the best predictions. Note that, for the Cluster model, the gain is obtained due to the fact that we marginalize the entire travel trajectory of each individual over cluster assignments. If we ignore the fact that all the transitions in a trajectory were generated by a single individual ("Cluster, indep."), there is no gain in the train likelihood over the Affinity model, and the test NLL is slightly worse. This is to be expected, since we have a non-parametric (binned) ρ, and model averaging does not increase the effective number of parameters much.

Table 1. NLL per user for the Shanghai and Flickr datasets. The number of parameters is calculated by excluding parameters that pertain to map quads that are never visited.

Model	Taxi dataset			Flickr dataset		
	Parameters	Train NLL	Test NLL	Parameters	Train NLL	Test NLL
Histogram (Eq. 2)	1,004,692	236.1	511.8	26,332,700	15.76	47.27
Reg. Histogram (Eq. 2)	1,004,692	251.0	279.0	26,332,700	17.36	21.53
Latent model (Eq. 3)	1,578	272.2	276.1	3,370	19.90	20.00
Affinity model (Eq. 5)	3,802	264.6	270.1	11,930	18.95	19.10
Cluster model (Eq. 6)	9,002	232.0	231.8	21,130	18.43	18.66
Cluster, indep. (Eq. 7)	9,002	263.2	271.1	21,130	18.95	19.28

The main reason for the improvement due to the parametric methods is that the histogram-based methods are very statistically inefficient, whereas the factored models can generalize. For the Flickr dataset, the full histogram contains 231 million entries, which would require enormous datasets to accurately estimate. Indeed, in the learned histogram, only 97,190 of the entries are non-zero. To see the benefit of the parametric model, consider all the possible self-transitions (i.e., transitions from bin i to i) in the test set that do not occur in the training set. The contribution from these transitions to the test NLL is $1.18e3$ under the Regularized Histogram model and 495.08 under the Affinity model. The differences in the transition probabilities can be quite significant. For example, the probability of the Tabriz $\rightarrow$ Tabriz transition in 30 to 60 days is 0.04 under the Regularized Histogram model and 0.14 under the Affinity model. Transitions that take more than 30 days are rare, but our model is able to infer a better transition probability by using data from shorter transitions.

8.2   Popularity vs. Desirability

We learned the *desirability* for each location on the maps. The desirabilities and the popularities (the number of inbound transitions to the location) of a location are related but distinct: see Fig. 2. A partial explanation of the difference is the remoteness of some locations: Sydney and Melbourne are in the top ten most desirable destinations, but are less popular because of their remoteness. The Shanghai Airport is also remote and more desirable than popular.

8.3 Affinity Factors

Analyzing affinity factors ($u_i^T v_j$) between pairs of locations allows us to discover affinities between locations that are not accounted for by proximity and desirability. For example, there is obviously more traffic than the average between Russian-speaking cities because some of them are desirable and they are all located in or near Russia. However, by analyzing our learned model we find that there is more traffic between Russian-speaking cities than we would expect based on the proximity between them and their popularity alone.

For the Flickr dataset, we show that our model captures the fact that locations within national borders have higher affinity with each other than the average. We show that the same holds for some linguistic areas. The mean log-affinity factors within and across nation boundaries are plotted in Fig. 5. As expected, the model learns that, on average, locations within the same national boundaries have higher affinities with each other.

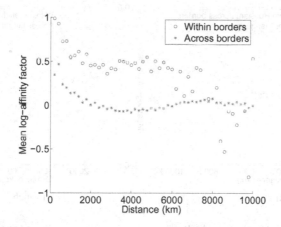

Fig. 5. The mean log-affinity factors within and across national boundaries. For distaces larger than 6000 km, very few location pairs are available.

We explore the affinities for pairs of locations which share an official language, but which are located in different countries. In Fig. 6, we observe a clear pattern where affinities are higher for pairs of Spanish-speaking locations in different countries. There is a clear pattern for Russian. For other languages, there is no clear pattern.

High affinities between locations that are not within the same national or subnational unit may indicate an interesting connection between the locations. For example, the second highest affinity for the Flickr dataset is between Mauritius and Réunion, two nearby islands with similar histories. Other, not as easily explainable, examples from the list of the highest 100 affinities are Hilo (Hawaii)→Vancouver, Glasgow→Miami, and Tunis→Sassari (Sardinia).

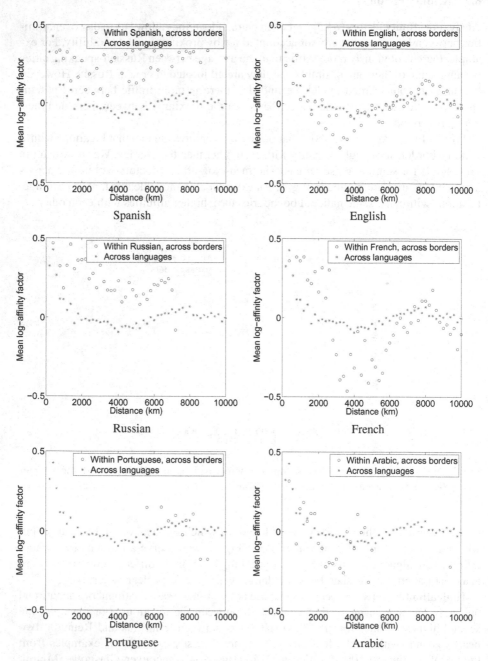

Fig. 6. Mean log-affinity factors for individual official languages, compared to the mean log-affinity factors for locations where the official languages are different. Note that the data is very noisy for larger distances because the mean is based on very few datapoints.

8.4   Travel Distance Cost

The ρs for the different clusters, for a single time interval, are shown in Fig. 4. We have obtained clusters that are different from one another: for the Flickr dataset, one cluster (Cluster 1 in Fig. 4) assigns much higher probability to mid-range distances than the other clusters, and one cluster (Cluster 2) assigns them lower probability than the other clusters. In the Shanghai taxi dataset, the clusters have different probabilities of staying roughly in place: the ρ values for the 5 different clusters for a distance smaller than 3km and a time interval of 1 hour are 5.8, 4.8, 7.7, 4.6, and 5.0; for the location centered at 31.22°N 121.46°E (selected to serve as an example), the corresponding probabilities under the five cluster models are 0.32, 0.25, 0.76, 0.24, and 0.22.

9   Conclusions

We have described several latent factor models of human travel. In contrast to previous approaches to modeling human travel data, our model represents interpretable properties of human travel: travel distance cost, desirability of destinations, and affinity between locations. The resulting models exhibit significant improvements in predictive power over previous methods, while also using far fewer parameters than histogram-based methods.

The parameters of our model are interpretable. We are able to describe the propensity to travel long distances of different clusters of individuals. We contrast the popularity and the desirability of locations, and show that they are distinct. By "factoring out" the desirability of locations and the distance to them, we can analyze the affinities between locations. Our model successfully learns the fact that locations that are located within the same country have higher affinity than the average. We explore the affinities of locations located across national borders, but which share an official language. We see that for pairs of locations where the official language is Spanish and Russian, the affinities are higher than average.

Statistical modeling of human society has become a major theme in recent years, yielding both new insights and predictive abilities. Travel is one of the pieces of this puzzle, and the statistical models here can be used in concert with models of other properties. For example, when modeling photo-collections in online datasets, the speed and location of travel will be correlated to the types of pictures taken, as well as to individual identity, both of which can be useful cues for recognition and tagging of image content (see, e.g., [16]). Furthermore, the use of latent models can be useful for making predictions in new situations. For example, the effects of world events on world travel could be predicted by adjusting the latent factors accordingly, which could be useful in epidemic forecasting [6,15].

References

1. Brockmann, D., Hufnagel, L., Geisel, T.: The Scaling Laws of Human Travel. Nature 439(7075), 462–465 (2006)
2. Calabrese, F., Di Lorenzo, G., Liu, L., Ratti, C.: Estimating Origin-Destination Flows Using Mobile Phone Location Data. Pervasive 10(4), 36–44 (2011)

3. Calabrese, F., Pereira, F.C., Di Lorenzo, G., Liu, L., Ratti, C.: The geography of taste: Analyzing cell-phone mobility and social events. In: Floréen, P., Krüger, A., Spasojevic, M. (eds.) Pervasive 2010. LNCS, vol. 6030, pp. 22–37. Springer, Heidelberg (2010)

4. Chen, C.-Y., Grauman, K.: Clues from the Beaten Path: Location Estimation with Bursty Sequences of Tourist Photos. In: Proc. CVPR (2011)

5. Cho, E., Myers, S.A., Leskovec, J.: Friendship and Mobility: User Movement in Location-Based Social Networks. In: Proc. KDD (2011)

6. Colizza, V., Barrat, A., Barthelmy, M., Valleron, A.-J., Vespignani, A.: Modeling the Worldwide Spread of Pandemic Influenza: Baseline Case and Containment Interventions. PLoS Med. 4(1) (2007)

7. Dearman, D., Sohn, T., Truong, K.N.: Opportunities Exist: Supporting the Development of Spatial Knowledge with Continuous Place Discovery for Activities. In: Proc. CHI (2011)

8. Eagle, N., Pentland, A.: Reality mining: sensing complex social systems. Pers. Ubiquit. Comput. 10, 255–268 (2006)

9. Girardin, F., Calabrese, F., Fiore, F., Ratti, C., Blat, J.: Digital Footprinting: Uncovering Tourists with User-Generated Content. Pervasive Computing 7(4), 36–43 (2008)

10. Girardin, F., Fiore, F.D., Ratti, C., Blat, J.: Leveraging explicitly disclosed location information to understand tourist dynamics: A case study. J. of Location Based Services 2(1), 41–56 (2008)

11. González, M.C., Hidalgo, C.A., Barabási, A.-L.: Understanding individual human mobility patterns. Nature 453(7196), 779–782 (2008)

12. Guerzhoy, M., Hertzmann, A.: Learning latent factor models of human travel. In: NIPS Workshop on Social Network and Social Media Analysis: Methods, Models and Applications (2012)

13. Hoff, P.D.: Multiplicative latent factor models for description and prediction of social networks. Comp. & Math. Org. Theory 15, 261–272 (2009)

14. Horvitz, E., Apacible, J., Sarin, R., Liao, L.: Prediction, Expectation, and Surprise: Methods, Designs, and Study of a Deployed Traffic Forecasting Service. In: Proc. IJCAI (2005)

15. Hufnagel, L., Brockmann, D., Geisel, T.: Forecast and control of epidemics in a globalized world. PNAS 101(24), 15124–15129 (2004)

16. Kalogerakis, E., Vesselova, O., Hays, J., Efros, A., Hertzmann, A.: Image Sequence Geolocation with Human Travel Priors. In: Proc. ICCV (2009)

17. Kurashima, T., Iwata, T., Irie, G., Fujimura, K.: Travel route recommendation using geotags in photo sharing sites. In: Proc. CIKM (2010)

18. Neal, R.M., Hinton, G.E.: A view of the EM algorithm that justifies incremental, sparse, and other variants. In: Jordan, M.I. (ed.) Learning in Graphical Models, pp. 355–368. Kluwer Academic Publishers (1998)

19. Patterson, D.J., Liao, L., Fox, D., Kautz, H.: Inferring High-Level Behavior from Low-Level Sensors. In: Dey, A.K., Schmidt, A., McCarthy, J.F. (eds.) UbiComp 2003. LNCS, vol. 2864, pp. 73–89. Springer, Heidelberg (2003)

20. Peng, C., Jin, X., Wong, K.-C., Shi, M., Liò, P.: Collective Human Mobility Pattern from Taxi Trips in Urban Area. PLoS ONE 7(4) (2012)

21. Schein, A.I., Saul, L.K., Ungar, L.H.: A Generalized Linear Mmodel for Principal Component Analysis of Binary Data. In: Proc. AISTATS (2003)

22. Sohn, T., et al.: Mobility Detection Using Everyday GSM Traces. In: Dourish, P., Friday, A. (eds.) UbiComp 2006. LNCS, vol. 4206, pp. 212–224. Springer, Heidelberg (2006)

23. Yuan, J., Zheng, Y., Xie, X.: Discovering regions of different functions in a city using human mobility and pois. In: Proc. KDD (2012)

Task Oriented Privacy Preserving Data Publishing Using Feature Selection

Yasser Jafer[1], Stan Matwin[1,2,3], and Marina Sokolova[1,2,4]

[1] School of Electrical Engineering and Computer Science, University of Ottawa, Canada
[2] Institute for Big Data Analytics, Dalhousie University, Canada
[3] Institute for Computer Science, Polish Academy of Sciences
[4] Faculty of Medicine, University of Ottawa, Canada
{yjafe089,sokolova}@uottawa.ca,
stan@cs.dal.ca

Abstract. In this work we show that feature selection can be used to preserve privacy of individuals without compromising the accuracy of data classification. Furthermore, when feature selection is combined with anonymization techniques, we are able to publish privacy preserving datasets. We use several UCI data sets to empirically support our claim. The obtained results show that these privacy-preserving datasets provide classification accuracy comparable and in some cases superior to the accuracy of classification of the original datasets. We generalized the results with a paired t-test applied on different levels of anonymization.

Keywords: Privacy, Data Publishing, Data Mining, Classification, Feature Selection.

1 Introduction

A tremendous growth of datasets containing personal information has necessitated a need in methods designed to protect the privacy of individuals. In general, efforts to protect privacy have lead to two related research areas, namely, Privacy Preserving Data Mining (PPDM) and Privacy Preserving Data Publishing (PPDP). PPDM is mainly concerned with building and using data mining models while controlling the privacy of individuals in a dataset. PPDP focuses on anonymizing and releasing datasets. However, if the data publishing techniques are customized according to particular types of analysis, i.e. "task", better results can be obtained [1].

Thus, when the ultimate usage of the data is known, PPDM and PPDP can be combined into a new privacy preserving technique. We call it Task Oriented Privacy Preserving (TOP) Data Publishing. In such setting, the data mining task guides the data publishing process in order to preserve the privacy of individuals in the dataset, and yet maintain high quality of information for analysis. We assume a scenario that consists of a Data Holder (DH) who holds the original data on the on hand and, a Data Recipient (DR) who wants the data in order to apply certain data mining task on the other hand. This task could be classification, regression, clustering, or association rule

M. Sokolova and P. van Beek (Eds.): Canadian AI 2014, LNAI 8436, pp. 143–154, 2014.
© Springer International Publishing Switzerland 2014

mining. In our work, we assume that the DR wants to classify the dataset. In a typical privacy preserving data publishing process, the DH would use a given privacy model such as K-anonymity [2] to anonymize the data and to publish the anonymized dataset regardless of the data mining task. In our scenario, the DH publishes a customized dataset which takes the intended analysis task of the DR into consideration. Since the dataset is going to be published, in practice, it will also be available to the attacker. However, since the dataset is tailored for a given analysis task, while guaranteeing privacy according to the privacy model, if the attacker uses the same dataset to do clustering analysis instead, the results would be misleading and of less or no value.

The process in our scenario is initiated by the DR requesting the dataset. The DH subsequently provides the DR with dataset's metadata (including number of instances, number of attributes, name and type of attributes). Such information will be used by the DR to decide on selecting a particular classification algorithm and identifying the target class. These preferences will be communicated to the DH. The DH, according-ly, uses feature selection and publishes a dataset that is tailored for the intended pur-pose. The high-level illustration of the interaction between the DH and the DR is shown in Figure1.

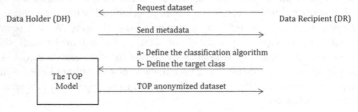

Fig. 1. The TOP data publishing scenario

In our work we show that, well-known machine learning techniques such as feature selection can be used to solve privacy preserving data publishing problem. We present the TOP data publishing model in which feature selection is integrated into the ano-nymization process. We show that such integration results in an anonymized dataset that, while preserving the privacy, does not have negative impact on the performance of the resulting models. We generalized the accuracy results through a paired t-test applied to four levels of anonymization. For every data set and every algorithm, the test compared results obtained on the original and pre-processed data set.

2   Related Work

Different methods have been proposed in the past which aim at protecting individu-als' privacy while preserving the data utility for certain data mining tasks. These works mainly focused on optimizing anonymization for classification applications. Lefevre et. al, [3] proposed a suite of greedy algorithms in order to address the K-anonymization problem for a number of analysis tasks such as classification and regression analysis for single/multiple categorical and numerical target attribute(s) respectively. The work in [3] argues that it is best to judge quality with regard to the workload for which the data will be eventually used. Other works such as top-down

[4] and bottom-up [5] greedy heuristic searches and genetic algorithm [6] used the target classification model in order to evaluate the recoding.

Byun et al. [7] proposed a comprehensive approach for privacy preserving access control based on the notion of *purpose*. Xiong and Rangachari [8] presented an application-oriented approach for data anonymization which considers the relative attribute importance for the target applications. The authors proposed a prioritized anonymization scheme in which the attributes are prioritized based on how critical and important they are for a given application. One main shortcoming associated with [8] is the fact that such attribute priority assignment by the user is impractical in many scenarios and "it is not always possible for users to specify that attribute priorities beforehand" [8]. In a relevant work, Sun et al [1] used the notion of purpose in terms of applications queried by different data requesters. In order to address the shortcoming associated with [8], the work in [1] devises a method to automatically derive the attribute priorities by adopting the concept of entropy to measure the independency among attributes. However, one main disadvantage associated with this technique is that it requires the most useful and the least useful attribute to be determined by the user.

We do not make assumptions about the priority of the attributes. Our work uses feature selection which automatically identifies the subset of features that are most relevant to the classification task at hand. We then use the resulting selected attributes to produce anonymized dataset.

3   Feature Selection

Feature selection aims at removing irrelevant and/or redundant attributes in order to improve the quality of data. It is also considered an effective dimensionality reduction method. There are two broad categories of feature selection techniques, namely, *fil ters* and *wrappers*. Filter approach attempts to assess the merits of features from the data without considering the induction algorithm. Therefore, it does not take into account the effects of the selected feature subset on the performance of the induction algorithm.

The wrapper model, on the other hand, uses a target learning algorithm in order to estimate the worth of attribute subsets. Wrappers assess subsets of variables according to how useful they are with respect to predicting the target class. Previous works have shown that the wrapper feature selection achieves better classification accuracy compared with filter feature selection techniques [9]. In the TOP data publishing, the classification algorithm is pre-determined and therefore wrapper feature selection technique was considered a perfect fit for our purpose. Wrapper feature selection essentially conducts a search for a good subset. However, the problem of choosing a search method is known to be NP-hard [10]. BestFirst is a well-known search method which searches the space of attribute using greedy strategies. Two main advantages of such search include its computational speed and its robustness against overfitting [11]. BestFirst may start with empty set of attributes and then attributes are progressively incorporated into larger subsets, a.k.a. *forward selection*, or may start with the full set of attributes and progressively eliminates the least important ones, a.k.a. *backward selection*. From the classification performance's point of view, these selection techniques usually result in very similar results. The question we want to ask is whether the search option impacts the elimination of potentially harmful attributes or not.

4   Data Anonymization

Raw data usually does not satisfy a specified privacy requirement and needs to be modified before release. The modification is done via a number of anonymization operations such as *generalization* and *suppression* of attributes [12].

With respect to privacy, data attributes can be categorized into (i) explicit identifier, (ii) quasi identifiers, (iii) sensitive and (iv) non-sensitive attributes. Explicit identifiers refer to a set of attributes such as name or SIN with information that explicitly identifies individuals. Quasi Identifiers (QI) refer to a set of attributes that when "combined", could be linked to external datasets and potentially breach the privacy. Sensitive attributes correspond to person-specific private information such as salary, disease, and so on. Finally, non-sensitive attributes consists of attributes that do not fall into any of the above categories. While the explicit identifiers are removed from the table, the QI set is transformed into a less specific form (QI') by applying anonymization operations. For example, a table is considered K-anonymous if the QI values of each tuple are indistinguishable from K-1 other tuples. In other words, if a record in the table has some QI value, there is at least K-1 other records which have the same QI value.

K-anonymity belongs to the syntactic anonymity approaches known to be susceptible to various attacks such as attribute linkage, table linkage, and probabilistic attacks. There are common limitations associated with these approaches such as information loss, ad hoc assumption on auxiliary information, and sub-optimality [13]. In order to responds to the need of firm foundation for privacy preserving data publishing, differential privacy [14] was proposed by Dwork. Differential privacy ensures that adding or removing a single dataset item does not substantially influence the outcome of any analysis. It is a rigorous notion of privacy; however, a study of its utility is still in its infancy [15]. One main limitation associated with differential privacy is the fact that it is a purturbative method and usually noise needs to be added in order to satisfy this privacy model. In privacy preserving data mining while protecting the privacy is the main goal, the other goal is to keep the data useful for data mining purposes. Therefore, until the usefulness of data for the data miner analysts is clarified, differential privacy is not considered rich enough to replace the other existing models and as such, enhancement of the existing syntactic models remains to be important. A fruitful research direction is to combine the benefits associated with syntactic anonymity approaches and differential privacy. A recent attempt [16] combines K-anonymity and differential privacy to improve the utility of the data using the very notion of indistinguishability offered by K-anonymity. In another work, Li et al. show that "safe" K-anonymization followed by random sampling could achieve differential privacy [17]. Integrating the ultimate analysis task and feature selection into the anonymization process reduces the amount of anonymization, improves the k-anonymization results, and is a step forward in this direction.

5   TOP Data Publishing

The TOP data publishing incorporates the ultimate usage of the data in order to generate a customized dataset specifically constructed for the intended analysis task. It uses a well-known machine learning techniques such as feature selection to achieve this goal. Figure 2 illustrates the intersection between all, selected, and QI attributes after applying feature selection to the original dataset. {A} refers to the complete set of attributes after removing the identifiers and the class attribute. From this list, we first identify a subset of attributes selected by the wrapper feature selection algorithm i.e. {B}. We identify {C} which represents the set of QI attributes. We then find {D'} which presents the intersection between {B} and {C}.

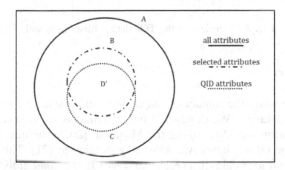

Fig. 2. The intersection between all, selected, and QI attributes

Depending on the number of attributes that fall in {D'}, the following outcomes are expected:

- **Case 1**: {D'} = {}. In this case, the features selected by feature selection do not include any the quasi-identifier features.
- **Case 2**: {D'} ≠ {}, B - {D'} ≠ {}, and C - {D'} ≠ {}. That is, {D'} includes some of the quasi-identifier features and excludes some others.
- **Case 3**: {D'} = {C}. In this case, it is possible that all of the quasi-identifier attributes are selected.

Recall from Section 4 that a dataset needs to be modified in order to satisfy a given privacy models and therefore, anonymization is applied to the dataset to achieve that. To the best of our knowledge, no previous study has combined feature selection and anonymization to release a task oriented privacy preserving dataset. We investigate such an impact by anonymizying the selected QI attributes, rather than all of the QI attributes and releasing them along other selected attributes according to a specific induction algorithm.

Figure 3 shows the TOP data publishing model. The first block corresponds to the wrapper feature selection algorithm and is borrowed from [18]. It, virtually, represents a black box from which, we first identify a subset of attributes that gives the best performance according to a predetermined induction algorithm X (being C4.5, N.B., SVM, etc). The TOP processing block constructs a new dataset with a feature vector equivalent to the selected features. Depending on the ultimate task, the *TOP Processing* block may include selecting a proportion of the dataset. For example, the analysis may include analyzing the records of male patients who are 20-40 years old. In such case only these records are selected.

The resulting dataset is then anonymized using the *Anonymization* block, released, and sent to the data recipient.

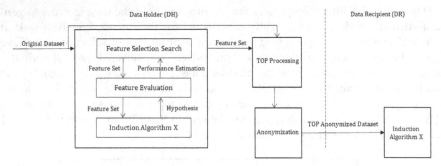

Fig. 3. Overview of the TOP data publishing model

6   Experiments

In this section, we study the impact of employing feature selection on the privacy of individuals in the dataset. We conducted our experiments using seven UCI [19] datasets. Anonymization was done using the Mondrian [20] technique. For some datasets, a subset of the QI attributes was selected according to [21]. This includes {age, sex} in the *heart stat logs* dataset, {preg, mass, age} in the *pima diabetes* dataset, and {purpose, credit_amount, personal_status, residence_since, age, job} in the *german credit* dataset. In the *liver patients* dataset, we identified {age, sex} as quasi-identifiers. For the remaining datasets, all of the attributes except the target class were selected as the QI set.

We first find the intersection between attribute sets according to Section 5. This includes {B}, {C}, {D'}, and {C} – {D'}. The results show that the list of selected features identified by the wrapper feature selection algorithm and, consequently, the list of included/excluded QI attributes depends on the choice of the base classifier.

6.1   Exclusion of the QI Attributes and Anonymization

In the case of *heart stat logs* and *liver patients* (which belong to Case 1 in Section 5) no anonymization at all is needed since none of the QI attributes are selected by feature selection. For the remaining cases, since feature selection excludes some of the QI attributes, rather than anonymizing all of the attributes in the QI set, only a subset of those attributes needs to be anonymized. This leads to less anonymization and less exposure of potentially harmful attributes.

In general, K-anonymity privacy model addresses the record linkage attack in which the QI attributes are linked to external datasets resulting in privacy breach. Figure 4 presents a sample of the *pima diabetes* dataset. Recall from Section 5 that, the QI attributes associated with this dataset are preg, mass, and age. If the DH publishes this dataset as is, it is possible for an attacker who has access to the external table to link the records in Figure 4(a) and Figure 4(b) and to figure out that a patient named 'Marilyn' who is 28 years old, whose BMI is 35.6, and have been pregnant

3 times in the past, has diabetes. Now, consider the case where Wrapper Feature Selection (WFS) is applied to the dataset using C4.5 as the base classifier. In this case out of the three QI attributes, only mass and age are retained and preg is excluded (Figure 4(c)). We may argue that, even without anonymization, such elimination of some of the QI attributes results in more privacy. Assume that the original table has other records x and y with the following QI values: x{preg = 5, mass = 29.8, age = 30, class = 'negative'} and y{preg = 6, mass = 29.8, age = 30, class = 'positive'}.

(a) Original pima diabetes records

preg	plas	pres	skin	insu	mass	pedi	age	class
2	112	66	22	0	25	0.307	24	negative
1	80	55	0	0	19.1	0.258	21	negative
4	123	80	15	176	32	0.443	34	negative
7	81	78	40	48	46.7	0.261	42	negative
3	83	58	31	18	34.3	0.336	25	negative
5	124	74	0	0	34	0.22	38	positive
0	105	84	0	0	27.9	0.741	62	positive

(b) External table

name	age	mass	preg
Alice	62	27.9	0
Elizabeth	21	19.1	1
Suzanne	34	32	4
Cathy	38	34	5
Jennifer	25	34.3	3
Helen	24	25	2
Nicole	42	46.7	7
Marilyn	28	35.6	3
Samantha	30	29.8	6

(c) WFS (C4.5)

plas	pres	mass	age	class
112	66	25	24	negative
80	55	19.1	21	negative
123	80	32	34	negative
81	78	46.7	42	negative
83	58	34.3	25	negative
124	74	34	38	positive
105	84	27.9	62	positive

(d) 3-anonymous (Mondrian) pima diabetes records

preg	plas	pres	skin	insu	mass	pedi	age	class
[0.0:3.0]	112	66	22	0	[0:68)	0.307	[0:82)	negative
[0.0:3.0]	80	55	0	0	[0:68)	0.258	[0:82)	negative
(3.0:17.0]	123	80	15	176	[0:68)	0.443	[0:82)	negative
(3.0:17.0]	81	78	40	48	[0:68)	0.261	[0:82)	negative
[0.0:3.0]	83	58	31	18	[0:68)	0.336	[0:82)	negative
(3.0:17.0]	124	74	0	0	[0:68)	0.22	[0:82)	positive
[0.0:3.0]	105	84	0	0	[0:68)	0.741	[0:82)	positive

(e) WFS (C4.5) + 3-anonymous (Mondrian)

plas	pres	mass	age	class
112	66	[0.0:32.0]	[0:82)	negative
80	55	[0.0:32.0]	[0:82)	negative
123	80	[0.0:32.0]	[0:82)	negative
81	78	(32.0:68.0)	[0:82)	negative
83	58	(32.0:68.0)	[0:82)	negative
124	74	(32.0:68.0)	[0:82)	positive
105	84	[0.0:32.0]	[0:82)	positive

Fig. 4. Example of pima diabetes records

An attacker can link this record with the external table and conclude with 100% accuracy that Samantha has diabetes. However, when preg is excluded, the following QI values are published instead: x{mass = 29.8, age = 30 where class = 'negative'} and y{mass = 29.8, age = 30, class = 'positive'}. In such case, the attacker can no longer link this record to that of Samantha in the external table confidently. We can argue that, since feature selection excludes some of the QI attributes, such exclusion disrupts the very meaning of the quasi-identifier set. We can make a hypothesis that, since the original QI set is being disrupted, the remaining attributes are no longer considered quasi-identifiers and may be released without any anonymization, although such statement needs to be investigated. To see the impact of feature selection on anonymization consider Figure 4(d): In order to satisfy the 3-anonymity requirement the dataset needs to undergo coarse generalization. As such, all values of the mass attribute are generalized to [0:68) and all values of the age attribute are generalized to [0:82). By anonymizing only, firstly, fewer attributes need to be anonymized, and secondly, the less coarse generalization (of the mass attribute) is required which leads to preserving more details for data mining purposes.

Results in Table 1 show that feature selection does not have any negative impact on the performance. In fact, in the case of N.B., feature selection even improves the classification accuracy and this improvement is statistically significant (based on paired t-test at 0.05). It was mentioned in Section 3 that, two commonly used search direction in the BestFirst search strategy are *forward selection* and *backward elimination*.

Table 1. Comparison of classification accuracy of models built using all attributes (original) vs. only selected attributes using WFS. ⊕ indicates statistically significant higher performance. Higher classification accuracy is shown in bold.

Dataset	C4.5			N.B.		
	Original	WFS	P value	Original	WFS	P value
heart stat logs	76.67	**85.19**⊕	0.0004	83.7	**86.3** ⊕	0.0248
pima diabetes	73.83	**75.78**	0.0767	76.3	**77.73**⊕	0.0481
German credit	70.5	**73.1**	0.1100	75.4	**76.1**	0.4716
liver patients	68.78	**71.012**	0.1852	55.74	**71.87**⊕	1.651e-05
CRX	85.29	**86.37**	0.325	78.25	**87.29**⊕	0.0006
CMC	52.14	**54.72**	0.1078	50.78	**55.39**⊕	0.01261
Winconsin Breast Cancer	93.41	**95.17**	0.0051	97.36	**97.80**	0.1944

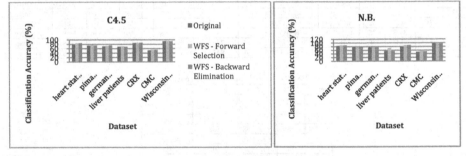

Fig. 5. Comparison of the performance of original vs. WFS with BestFirst (both forward selection and backward elimination) when the base classifier is C4.5 and N.B.

It is beneficial to investigate whether the search direction has any impact on the outcome from the performance and privacy perspectives. We conducted an experiment to compare the impact of search direction on the classification accuracy and the number of retained QI attributes. These results are shown in Figure 5 and Table 2 respectively. The results show that performance of the models built using forward selection or backward elimination are comparable. Following the definition of QI attributes, we assume that all of these attributes have the same weight with respect to their impact on privacy. If that is the case, since the performance of forward selection and backward elimination is equivalent, we need to select the one that excludes a larger number of QI attributes. As such, our results are in favor of forward selection search direction. We may argue that the strategy of giving equal weights to all of the attributes that constitutes the QI set may need to be reconsidered and an interesting research can be conducted to prioritize the attributes within a QI set according to their impact on privacy.

Table 2. Number of QI retained, the lower, the better. The lower value is shown in bold font.

		Heart stat logs	Pima diabetes	German credit	Liver patients	CRX	CMC	W breast cancer
C4.5	Forward Selection	0	2	2	0	1	4	2
	Backward Elimination	0	1	2	2	8	4	6
N.B.	Forward Selection	1	2	3	1	6	3	6
	Backward Elimination	1	2	4	2	7	5	8

6.2   Performance of the TOP Anonymized Dataset

We next compare the performance of the models built based on the original data vs. the anonymized data. We used paired t-test to compare the results obtained for different levels of anonymization. Let us consider the C4.5 results in Table 3. In discussing the results, Original and Mondrian refer to the classification accuracy of the models built using the original dataset and the dataset anonymized with Mondrian technique respectively."WFS + Mondrian" refers to the classification accuracy of the models built using the TOP anonymized dataset in Figure 2.

The performance of Mondrian in the case of *heart stat logs* ($K = 10$ and 20), *liver patients* ($K = 50$), and *Winconsin breast cancer* ($K = 10$) is higher than Original and this difference is statistically significant. In general, we expect anonymized dataset to result in lower performance due to the fact that generalization leads to hiding some of the details and potential information loss. This is, however, not always the case. In justifying such results, we follow the observation made in [4], that is, when the ultimate analysis task involves classification, anonymization may actually lead to better classification accuracy, or at least may not negatively impact the classification accuracy as we see in our results. Data usually consists of redundant structures for classification. Although generalization eliminates some useful structures, other structures emerge and could be helpful [4].

There are, however, other datasets such as *pima diabetes*, *CRX*, and *CMC* which consistently show lower accuracies of Mondrian compared with Original. This is specially the case with *CRX* dataset where the lower performance is statistically significant for all different anonymization levels. In the case of *Winconsin breast cancer*, anonymization with higher K values shows lower performance. *Liver patients* shows lower performance of anonymization at K values equal to 10 and 20, however, these results are not statistically significant.

Let us compare the accuracy of the models resulted from our approach "Mondrian + WFS". *Heart stat logs* dataset shows higher classification accuracy compared with Original and Mondrian for all different levels of anonymization and this increase in classification accuracy is statically significant. Compared with Mondrian results, "Mondrian + WFS" shows statistically significant improvement even at $K = 30$ and 50. The same is applied for the *Winconsin breast cancer* dataset. "Mondrian + WFS" outperforms Mondrian and Original. Such increase in performance is seen even for the cases where Mondrian resulted in lower classification accuracy compared with Original. Consider *CRX* dataset in which Mondrian significantly reduces the performance. With "Mondrian + WFS" approach (for all K values), compared with Original, better accuracies are obtained. For the *CMC* dataset, although both Mondrian and "Mondrian + WFS" show lower performance than Original, however, the p values are in favor or "Mondrian + WFS".

Table 3. Comparison of classification accuracy of the original, WFS, Mondrian, and Mondrian + WFS using C4.5 induction algorithm. $\oplus$ and $\ominus$ indicates statistically significant results in favor of anonymized and original datasets respectively. Higher accuracy is shown in bold. K refers to the anonymization level.

C4.5 Dataset	Original	WFS	Mondrian				Mondrian + WFS			
			K =10	K = 20	K = 30	K = 50	K =10	K = 20	K = 30	K = 50
			Classification Accuracy (%)							
Heart stat logs	76.67	85.18	**80.37**⊕	**80.37**⊕	**80.00**	79.63	**85.1**⊕	**85.1**⊕	**85.1**⊕	**85.1**⊕
Pima diabetes	73.83	75.78	72.66	73.44	71.74	72.53	69.4	73.44	**74.22**	73.83
German credit	70.7	73.1	**70.1**	**71.4**	**70.9**	**73.3**⊕	69.6	**72.4**	**72.5**	**73.5**
Liver patients	68.78	71.012	68.74	67.18	**70.12**	**72.19**	71.012	71.012	71.012	71.012
CRX	85.29	86.37	66.15⊖	61.10⊖	60.49⊖	58.8⊖	**86.37**	**86.37**	**86.37**	**86.37**
CMC	52.14	54.72	50.24	51.18	50.98	51.32	**54.18**	52.07	51.32	51.73
Wisconsin breast cancer	93.41	**95.17**	**95.75**⊕	94.29	93.85	90.6⊖	**96.0**⊕	94.29	**95.4**⊕	**95.3**⊕

Consider the results shown in Table 4 which corresponding to the N.B. classifier. With the exception of two cases, i.e. (*liver patients* when $K=20$) and (*CMC* when $K = 50$), Mondrian results in lower performance compared with Original. In the case of *CRX* this lower performance is statistically significant for all different levels of anonymization. The same is applied for *liver patients* ($K = 10, 30,$ and 50), and *Winconsin breast cancer* ($K = 30$ and 50). *German credit* shows statistically significant lower performance for Mondrian for $K = 20, 30,$ and 50.

Now, let us consider the results of "Mondrian + WFS" where we do not see such consistent lower performance. *Heart stat logs*, shows higher performance for all K values compared with Original. The same is applied to the *pima diabetes* and *German credit* datasets. Although this higher performance is not statistically significant it is consistently higher than the performance Mondrian. In the case of *German credit*, we observe that while Mondrian performance is statistically lower compared with Original, "Mondrian + WFS" resulted in comparable accuracies. Similarly, in the case of *liver patient,* where Mondrian shows statically significant lower performance compared with Original, "Mondrian + WFS" shows statistically significant higher performance across all K values. We also obtain better performance of "Mondrian + WFS" compared with the Original and Mondrian for *CRX* and *CMC. Wisconsin breast cancer* shows relatively similar results.

In addition to the privacy gains from using feature selections (Section 5.1), the above results show that combining wrapper feature selection and anonymization in the context of TOP data publishing, has a positive impact the performance. By integrating the intended analysis task into the data publishing process we are able to customize the datasets so that they are best fitted to our analysis purpose. The resulting customized dataset lead to better models and our results support this. Many real world datasets consist of large number of records and when the QI set consists of large number of attributes, anonymization becomes a challenging process. By anonymizing the selected attributes in the QI set only, we reduce the amount of generalization required, and this eventually leads to less computational cost and promising direction towards optimizing the anonymization process in the light of exponential increase of high dimensional datasets.

Table 4. Comparison of classification accuracy of the original, WFS, Mondrian, and Mondrian + WFS using N.B. induction algorithm. $\oplus$ and $\ominus$ indicates statistically significant results in favor of anonymized and original datasets respectively. Higher accuracy is shown in bold.

N.B. Dataset	Original	WFS	Mondrian				Mondrian + WFS			
			K=10	K=20	K=30	K=50	K=10	K=20	K=30	K=50
			Classification Accuracy (%)							
Heart stat logs	83.7	86.29	82.96	82.96	83.33	82.96	**84.81**	**84.81**	**84.81**	**84.81**
Pima diabetes	76.3	77.73	73.69	72.92	74.48	75.26	**76.56**	76.04	76.17	76.04
German credit	75.4	76.2	72.9	73.3⊖	74.20⊖	74.30⊖	**76.4**	**76.4**	**76.4**	**76.4**
Liver patients	55.74	71.87	55.61⊖	**55.78**	56.13⊖	55.96⊖	**71.35⊕**	70.49⊕	**71.35⊕**	**71.35⊕**
CRX	78.25	87.29	64.31⊖	60.4⊖	61.41⊖	58.96⊖	**84.99⊕**	82.54	77.03	77.03
CMC	50.78	55.39	50.37	49.08	50.03	**51.73**	52.74	**51.19**	50.03	**51.05**
Wisconsin breast cancer	97.36	97.80	95.75	93.7	84.63⊖	81.11⊖	95.90	92.09⊖	90.33⊖	90.33⊖

7   Conclusion and Future Work

In this study, we showed that, when the ultimate analysis task is predetermined, feature selection can be utilized to preserve privacy. We showed that, by anonymizing the selected features only, we release a task oriented dataset that results in higher classification accuracy compared to general purpose anonymization techniques. Our t-test results show that in many cases, such higher performance is comparable and even superior to the models built based on the original datasets.

In our experiments we showed that, feature selection resulted in excluding some or ull of the QI attributes. We will consider datasets where feature selection does not exclude any of the QI attributes. If that is the case, we can use the ranking property of feature selection in order to rank the QI attributes according to their importance and relevance to the target class. In general, the typical goal of supervised learning is to maximize the classification accuracy when classifying unseen data. A promising research direction is to consider feature selection that is not only guided by the classification accuracy but a trade-off between classification accuracy and privacy. We call it privacy preserving feature selection. This is related to the privacy by design concept [22] in which privacy is taken into consideration within the data mining process itself.

It is known that wrapper feature selection approaches result in higher performance compared with filter approaches. This performance is achieved at higher computational cost. By including another factor, i.e., the privacy cost, filter approaches may provide very useful results. A comparative study of the two techniques with respect to privacy is another interesting dimension for future research.

References

1. Sun, X., Wang, H., Li, J., Zhang, Y.: Injecting purpose and trust into data anonymization. Computers & Security 30, 332–345 (2011)
2. Sweeney, L.: Achieving k-anonymity privacy protection using generalization and suppression. International Journal of Uncertainty, Fuzziness and Knowledge-Based System 10(5), 571–588 (2002)
3. LeFevre, K., DeWitt, D.J., Ramakrishnan, R.: Workload-aware anonymization. In: Proceedings of the 12th ACM SIGKDD International Conference on Knowledge Discovery and Data Mining, Philadelphia, USA, pp. 277–286 (2006)

4. Fung, B., Wang, K., Yu, P.: Top-down specialization for information and privacy preservation. In: Proceedings of the 21st International Conference on Data Engineering, pp. 205–216 (2005)
5. Wang, K., Yu, P., Chakraborty, S.: Bottom-up generalization-A data mining solution to privacy protection. In: Proceedings of the 4th IEEE International Conference on Data Mining, Brighton, UK, pp. 249–256 (2004)
6. Iyengar, V.: Transforming data to satisfy privacy constraints. In: Proceedings of the 8th ACM SIGKDD Iinternational Conference on Knowledge Discovery and Data Mining, Edmonton, Alberta, Canada, pp. 279–288 (2002)
7. Byun, J.W., Bertino, E., Li, N.: Purpose based access control of complex data for privacy protection. In: The 10th ACM Symposium on Access Control Models and Technologies, Stockholm, Sweden, pp. 102–110 (2005)
8. Xiong, L., Rangachari, K.: Towards Application-Oriented Data Anonymization. In: First SIAM International Workshop on Practical Privacy-Preserving Data Mining, Atlanta, US, pp. 1–10 (2008)
9. Hall, M., Holmes, G.: Benchmarking attribute selection techniques for discrete class data mining. IEEE Transactions on Knowledge and Data Engineering 15(6), 1437–1447 (2003)
10. Amaldi, E., Kann, V.: On the approximation of minimizing non zero variables or unsatisfied relations in linear systems. Theoretical Computer Science 209, 237–260 (1998)
11. Guyon, I., Elisseff, A.: An introduction to variable and feature selection. Journal of Machine Learning Research 3, 1157–1182 (2003)
12. Fung, B., Wang, K., Chen, R., Yu, P.: Privacy-Preserving Data Publishing - A Survey of Recent Development. ACM Computing Surveys 42(4), Article 14 (2010)
13. Nguyen, H.H., Kim, J.: Differential Privacy in Practice. Journal of Computing Science and Engineering 7(3), 177–186 (2013)
14. Dwork, C.: Differential privacy. In: Proceedings of 33rd International Colloquium on Automata, Languages and Programming, Venice, Italy, pp. 1–12 (2006)
15. Clifton, C., Tassa, T.: On syntactic anonymity and differential privacy. Transactions of Data Privacy 6, 161–183 (2013)
16. Soria-Comas, J., Domingo-Ferrer, J., Sanchez, D., Martinez, S.: Improving the utility of differentially private data releases via k-anonymity. CoRR abs/1307.0966 (2013)
17. Li, N., Qardaji, W.: Su, Dong.: Provably private data anonyization: Or, k-anonymity meets differential privacy, CoRR abs/1101.2604 (2011)
18. Kohavi, R., John, G.H.: Wrappers for feature subset selection. Artificial Intelligence 97, 273–324 (1997)
19. UCI repository, http://archive.ics.uci.edu/ml/
20. Lefevre, K., DeWitt, D.J., Ramakrishnan, R.: Mondrian multidimensional k-anonymity. In: Proceedings of the 22nd International Conference on Data Engineering, Washington DC, USA, pp. 25–36 (2006)
21. Lin, K., Chen, M.: On the design and analysis of the privacy-preserving SVM classifier. IEEE Transaction on Knowedge and Data Engineering 23(11), 1704–1717 (2011)
22. Monreale, A.: Privacy by design in data mining. PhD dissertation, universit'a degli studi di pisa (2011)

A Consensus Approach for Annotation Projection in an Advanced Dialog Context

Simon Julien[1], Philippe Langlais[2], and Réal Tremblay[3]

[1] DIRO, Université de Montréal
juliensi@iro.umontreal.ca
[2] DIRO, Université de Montréal
felipe@iro.umontreal.ca
[3] Nuance Communications
real.tremblay@nuance.com

Abstract. Data annotation is a common way to improve the reliability of advanced dialog applications. Unfortunately, since those annotations are highly language-dependent, the universalization can become a very lenghty process. Even though some projection methods exist, most of them require a deeper level of annotation than the one used for advanced dialogs. In this paper, we present a consensus approach that exploits the specificities of a sparse annotation in order to do the data projection.

1 Introduction

Conversing with a computer can be tedious, which is why historically, many constraints were imposed on the dialog to ensure a viable (yet not very natural) conversation: short answers, step-by-step value collection, systematic confirmation, etc. Progress in computer technology lifted most of those constraints and, with current "advanced dialog" applications, it is now possible to use longer sentences, along with much more complex dialog patterns.

A common way to create such an application, or to enhance one, is to annotate a considerable amount of data. The annotations provide the desired parse results for typical sentences, from which new parses can be deducted, and then, from all the annotated data, a parsing grammar can be automatically generated. A major drawback of this otherwise easy method is to increase the development time, specially when it comes to universalization: all annotations are language-dependent, so the work done for a first language (say English), needs to be renewed for any further language (e.g. French, Spanish, Russian, Chinese. . .).

While many resource projection methods exist (Hwa et al., 2004; Santa-holma, 2008; Kim et al., 2010; Bouillon et al., 2006), none of them are fit to be used directly on the characteristic sparse annotation of advanced dialogs. For example, (Kim et al., 2003) uses a full-tree annotation hierarchy for its multilingual grammar development. Similarly, (Santaholma, 2005) needs access to the grammar syntax of each sentence to perform its speech-to-speech translation. That syntax has to be pinpointed in some way, usually through extra annotation. In all cases, the detailed annotation creates a deep level of abstraction that can be exploited by the proposed algorithms for the resource projection.

M. Sokolova and P. van Beek (Eds.): Canadian AI 2014, LNAI 8436, pp. 155–166, 2014.

Unfortunately, meta-information is too sparse when it comes to advanced dialogs, where most of the sentence is left unannotated (even though the whole sentence is used to determine the context). For instance, in "I would like to go to Newark, not New York", the annotation would typically be targeted on "Newark" and "New York", while "I would like to go to" and "not" would be left unannotated (their modifying action would however be accounted for in the way "Newark" and "New York" are annotated).

In this paper, we present a consensus method that projects a sparse annotation from a source language (English) to a target one (French), using Moses (Koehn et al., 2007), a public Statistical Machine Translation system (SMT – Koehn et al., 2003; Chiang, 2005). That method is directly aimed towards advanced dialog applications. We also provide a quantitative evaluation of the method against other intuitive solutions, using real data from two domains (Airline and Insurance). The datasets are described in Section 2, while the different methods are presented in Section 3. The comparative analysis is then conducted in Section 4, just before the conclusion of Section 5.

2   Data and Protocol

Prior to the deployment of an application, lots of representative utterances are typically gathered and manually annotated. The number of utterances varies according to each domain, but it is usually no less than a few thousands. Then, from those tagged utterances, a semantic parser is trained and can be used to parse new and unseen utterances. Our goal is to alleviate the annotation process for any further language, using previous annotation in the source language and machine translation. Since manual annotation is very time-consuming, this task is a real concern and a real challenge.

2.1   Datasets

In this work, we considered two datasets: the first one comes from a booking application in the airline domain, and the second one comes from a more general "how may i help you" application in insurance. English utterances have been internally collected at Nuance (duplicates were removed) and randomly split into training and testing sets, before they were manually translated in French. French duplicates, if any, were also removed. That way, we ensured that the translation of any source utterance in the train set is not in the target test set, where it could artificially boost the performance metrics. Typical examples from one of the application we studied are reported in Table 1.

For ease of understanding, we must define and illustrate the notions of tag, bounding tag and pattern:

Definition 1. *A* tag *is a conceptual entity that identifies the semantical role of contiguous words in a sentence. A* tag *can also identify the semantical role of contiguous tags, or of a mix of words and tags.*

Table 1. Examples of utterances from the Airline application, along with their tags and patterns

utterance	tags	pattern
Monday, I'm taking a plane bound for Christmas Island	**DAY_OF_WEEK**(Monday) **DATE**(DAY_OF_WEEK) **DEPARTURE_DATE**(DATE) **TERRITORY**(Christmas Island) **LOCATION**(TERRITORY) **DESTINATION**(LOCATION)	DEPARTURE_DATE, I'm taking a plane bound for DESTINATION
for Christmas, I need to take a flight between Melbourne and Sydney	**HOLIDAY**(Christmas) **DATE**(HOLIDAY) **DEPARTURE_DATE**(DATE) **CITY**(Melbourne) **LOCATION**(CITY) **ORIGIN**(LOCATION) **CITY**(Sydney) **LOCATION**(CITY) **DESTINATION**(LOCATION)	for DEPARTURE_DATE, I need to take a flight between ORIGIN and DESTINATION

Definition 2. *A pattern is a utterance in which all tagged words are replaced by their more general bounding tag. If a tag encompasses more than one word, a single instance of the tag replaces all the words.*

Take, for example, the utterance "leaving June 6th". A tag MONTH would target the literal "June", a tag DAY would target "6th" and a tag DATE would target those two previous tags (MONTH and DAY). Also, because of the context introduced by the keyword "leaving", a final tag DEPARTURE_DATE would encompass the lower tag DATE, in order to identify more throughly the se-mantical role of the underlying words. The DEPARTURE_DATE tag, which is not part of any other tag is considered the most general one. A common way to illustrate all those relations is with the use of a tree: "leaving DEPAR-TURE_DATE(DATE(MONTH(June) DAY(6th)))". As for Definition 2, the pat-tern of the sentences "leaving June 6th" and "leaving tomorrow" would likely be the same: "leaving DEPARTURE_DATE".

The notion of pattern is important, because the annotation (along with a list of recognized literals) is often sufficient to parse very close utterances. For instance, it's easy to parse "somewhere in Honduras" once you've seen "some-where in Belgium". Furthermore, that pattern redundancy will later be used in our consensus projection method.

The main characteristics of the two datasets we gathered are reported in Table 2 and 3 along with the number of different patterns and the percentage of the test patterns not seen at training time (OOVP%).

The number of utterances in both tasks is roughly similar (about 2500 utterances, two thirds of which are being used for training). We also observe that the tagset of the airline application is much larger than the one for insurance. This is because insurance data is more action driven (file a claim, deposit, withdraw, transfer ...), while airline data is more about targeting concepts (origin, destination, flight class ...). While distinct tags are needed for the latter type of data, an action usually only modifies existing tags, without creating new ones (for example, an amount becomes a deposit). This is also why the number of patterns and the OOVP% are more considerable in the insurance domain. Since more words are left unannotated, more different patterns arise, and with more different patterns to see, more are still unseen when it comes to testing.

Table 2. Data from the Airline domain

	English			**French**		
	total	train	test	total	train	test
Utterances	2449	1835	614	2233	1693	540
Tags	4163	3140	1023	3849	2938	911
Patterns	910	733	289	969	767	285
OOVP%			29.8%			38.2%

Table 3. Data from the Insurance domain

	English			**French**		
	total	train	test	total	train	test
Utterances	2459	1844	615	2313	1715	598
Tags	2663	2011	652	2426	1793	633
Patterns	1874	1454	538	1891	1459	551
OOVP%			69.3%			70.4%

2.2 Statistical Machine Translation

As for the bitexts used to train the SMT engine, we willingly chose them out-of-domain. Even if in-domain training provides better results (Koehn and Schroeder, 2007), the existence of in-domain bitexts is unsure in a real-life situation. Therefore, we used public corpora that anyone could find, from official records of the Canadian Parliament[1] (Hansard) and from movie subtitles[2] (OpenSubtitles2011 – Tiedemann, 2009). The main characteristics of the bitexts we used are reported in Table 4.

---

[1] http://www.isi.edu/natural-language/download/hansard/
[2] http://opus.lingfil.uu.se

Table 4. Corpora used to train the SMT engine

corpus	docs	phrases	tokens (en)	tokens (fr)
OpenSubtitles2011 (fr-ca)	24116	19.7×10^6	119.0×10^6	114.5×10^6
Hansard	2	1.2×10^6	19.8×10^6	21.2×10^6

2.3 Evaluation

The evaluation is pretty straightforward: for all the utterances of the test set, the retrieved tags are compared to the expected ones, using precision, recall and F_1 score (micro-averaged). The tags of a given utterance are directly inferred from the annotated sentences of the training set, with possibly some inner-filler words (that is, words between two recognized patterns that can be ignored). Basically, each tagged utterance becomes a recognized parsing rule, which is directly applied if possible (longer rules are preferred over shorter ones). Better parsing methods exist, but this simpler approach is useful to quickly zero in a better annotation projection algorithm.

3 Experiments

A short summary may be helpful here: what we are trying to do is to use n tagged utterances in a source language (English) to generate automatically, via SMT, m tagged utterances in a target language (French). Those m utterances, along with t "true" tagged target utterances, are then used to parse test sentences, from which precision, recall and F_1 are calculated. Given that the t "true" utterances from the French train set are never altered in any way, an algorithm that translates the n source utterances into the m ones will be considered better than another one if its F_1 score is higher. Below, we present three of those algorithms.

3.1 Baseline

For the baseline experiment, no SMT is used at all (all source utterances are ignored, i.e. $m = 0$). Instead, only samples from the French train set are randomly selected and directly used to train the semantic parser for the target language. This is a way to draw the learning curve of the applications over time, when $t = \{0, 50, 100, 150, \ldots\}$ samples are available in the current language, regardless of any other language.

3.2 Translation and Annotation

In this method, all source utterances are translated using the SMT. The literal expression encompassed by each tag is also translated. Then, the global annotation of a translation is inferred by searching each local literal translation, on which the original tag is restored. See Figure 1 for a simple example.

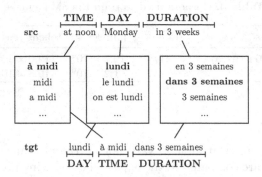

Fig. 1. Annotation projection through local search

Usually, more than one translation candidate is returned (but they are ranked according to their SMT score). Therefore, only the first candidate for which all tags were successfully restored is retained. For instance, in Figure 1, if "lundi dans 3 semaines" would have preceded "lundi à midi dans 3 semaines", the latter would still have be retained, due to the missing TIME tag ("à midi") in the former. Likewise, a tag will be considered missing if none of its 3 first translation candidates could be found. Even with those restrictions, there is almost always an adequate translation candidate for each source utterance, so $m \approx n$ ($m \leq n$) with this method.

This annotation projection method was preferred over a more intuitive approach, where the annotation alignment is combined with the translation alignment to retrieve a given target annotation. This is mainly because ghost concepts (see Figure 2) are sometimes wrongfully retrieved, a problem avoided with the projection through local search.

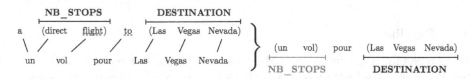

Fig. 2. Erroneous annotation projection when a local concept is missing from a translation

3.3   Consensus Approach

The consensus approach we present uses the same algorithm to project annotation between a source utterance and its translation, except it first focuses on identifying good translation candidates. In order to do so, source examples are no longer considered individually, but in clusters, using patterns as classifiers. For instance, the cluster "from ORIGIN to DESTINATION" would comprise utterances such as "from Toronto to Denver", "from Calgary Alberta to France", "from Italy to somewhere in Europe" and so on.

Because the SMT computes its transfer probabilities on each token individually, two utterances with the same source pattern will not necessarily produce the same n-best list of translation patterns. In fact, because a same concept (e.g. DATE) can be expressed in numerous ways ("today", "next weekend", "June 18th", "on Christmas", ...), significative differences arise in the translation candidates of yet very-close source utterances. On that matter, the example of Table 5 is very representative.

Table 5. Various n-best translation patterns from the source pattern "I'd like to go from ORIGIN to DESTINATION"

src utterance	n-best translation patterns	
I'd like to go from **Boston** to **New York**	je voudrais aller **à** ORIGIN **à** DESTINATION	✗
	je voudrais aller **de** ORIGIN **à** DESTINATION	✓
	voudrais aller **de** ORIGIN **à** DESTINATION	✓
	je voudrais aller **à** ORIGIN **pour** DESTINATION	✗
I'd like to go from **Montreal** to **New York**	je voudrais aller **de** ORIGIN **à** DESTINATION	✓
	je voudrais aller **à** ORIGIN **pour** DESTINATION	✗
	je voudrais **y** aller de ORIGIN DESTINATION	✗
	je voudrais aller **à** ORIGIN **de** DESTINATION	✗
I'd like to go from **Chicago Illinois** to **London UK**	je voudrais aller **de** ORIGIN **en** DESTINATION	✗
	je voudrais aller **à** ORIGIN **de** DESTINATION	✗
	je voudrais aller **de** ORIGIN **pour** DESTINATION	✓
	je voudrais aller **de** ORIGIN **à** DESTINATION	✓

Given those differences, our hypothesis is that recurrent translation patterns must be more reliable than unfrequent ones. A simple poll-voting algorithm with the 10 first translation candidates proved us right, at least on our datasets. And because the SMT is not completely wrong after all, a more effective way to pinpoint good translation patterns is to weight each one according to their translation rank. An inversely linear relationship turned out to be the best option.

Therefore, our consensus approach works as follows: each source utterance is sorted according to its pattern. Then, for each cluster, the best translation patterns are determined through a weighted-polling method. The 10-best translation candidates are considered for each utterance, then the score of each associated translation pattern is incremented according to the translation rank.

For example, the first example of Table 5 would increase the score of "je voudrais aller **à** ORIGIN **à** DESTINATION" by 10, the score of "je voudrais aller **de** ORIGIN **à** DESTINATION" by 9, "voudrais aller **de** ORIGIN **à** DESTINATION" by 8, etc. Once each example from a cluster voted, the 2 translation patterns with the highest scores are retained and considered "reliable". Finally, only pairs [src utt ; translation] in which the translation conforms to a reliable pattern are used to become the m tagged utterances.

Our consensus approach shares similarities with one of (Bangalore et al., 2002) with two important differences. First, we do not combine the translations of various off-the-shelf translation engines, but use a single system that we fully trained. We believe this is an easier setting to deploy, but further investigations are needed to compare the two approaches. Also, we directly use the annotation available in the source language, making our approach better tailored to our needs.

4   Results and Analysis

The three methods have been implemented and tested according to a growing number of real target utterances (examples from the French train set, which were manually annotated). For robustness purposes, the results shown in Figure 3 and 4 correspond to the average result over 5 repetitions (new manually-annotated samples were chosen at random each time).

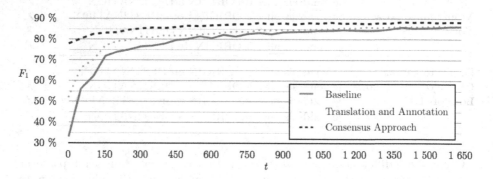

Fig. 3. F_1 variation on the Airline test set by the nb of first-hand utterances (t)

A first noticeable thing is that the F_1 baseline curves do not start at 0%, but near 30%. That is because roughly 30% of the tags in each test set can be identified with minimal knowledge. Those tags are usually basic concepts that can be listed[3] (e.g. list of countries). All other tags ($\pm 70\%$) are modified by the context (e.g. a COUNTRY that becomes a DESTINATION in "going to Finland").

The consensus approach curve clearly stands out, whether in the Airline or the Insurance domain. Furthermore, the performance of our method when only a few target annotated examples are being used is very close to the approach

---

[3] Main exceptions being DATE and TIME concepts, which cannot be listed per se, but are frequent enough to get a special treatment.

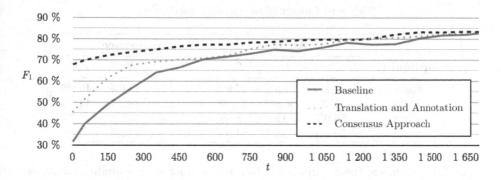

Fig. 4. F_1 variation on the Insurance test set by the nb of first-hand utterances (t)

obtained when all target examples are considered, which means the projected utterances are pretty good estimations of the real data. Therefore, few real target examples bring patterns that were not deducted from the original language, and few projected patterns are wrong (which is arguably the main problem of the "Translation and Annotation" approach). Because more than one target pattern can be kept for each source pattern, the consensus approach even keeps its edge over the baseline when all the Airline training data is considered (see Figure 3). A higher OOVP% explains why the approach scored lower at $t = 0$ on the Insurance test set than on the Airline test set.

4.1   Cardinality and Efficiency

Given that the consensus approach relies on various neighbour utterances, it is tempting to artificially increase the number of examples per cluster. One easy way to do so is to exploit the source annotation. That annotation is hierarchical, so from a common annotation node, new examples can be automatically generated. For instance, from the example "my departure date is on the 4th of July", where the annotation looks like "my departure date is on DEPARTURE_DATE(DATE(the DAY(4th) of MONTH(July)))", the subconcept DATE can easily be replaced by any other encountered DATE, creating new tagged examples for the cluster "my departure date is on DEPARTURE_DATE".

Over-generating data is promising, although real-life experiments in our domains showed very shy improvements. More importantly, over-generation gave us a way to measure how the clusters' cardinality impacted the results through our consensus approach. In order to do so, over-generation was used to ensure each source pattern had at least 4 associated examples. Then, the consensus approach was repeatedly used, always considering all the available source patterns, but with more and more underlying examples (first, a single example was used per pattern, then 2 examples per pattern, then 3 and then 4). Once again, experiments where run 5 times to ensure a minimal robustness.

Table 6. Impact of the clusters' cardinality

cardinality	Airline			Insurance		
	P	R	F_1	P	R	F_1
$n = 1$	44.1%	51.8%	47.6%	37.4%	43.6%	40.2%
$n = 2$	51.2%	56.5%	53.7%	42.1%	48.2%	44.9%
$n = 3$	56.0%	60.8%	58.3%	50.4%	55.0%	52.6%
$n = 4$	59.5%	63.2%	61.3%	54.3%	56.0%	55.2%

As Table 6 shows, the results get better as the clusters' cardinality increases. This seems to confirm that over-generation could be useful in cases where meaningful clusters are under-represented in the training set. It also confirms that the average translation of close utterances is more reliable than each utterance considered alone, at least in the cases of interest. Over-generation was not used in the results of Figure 3 and 4, where each clusters had a various number of associated examples.

4.2 Proximity with the SMT

At this point, another reasonable hypothesis would be that source examples "close" to the SMT probably produce overall better translation candidates. In other words, instead of mixing various translation candidates through our consensus approach, maybe it would be more efficient to identify a single source utterance per pattern, that source utterance being the one which is the most familiar to the SMT. After all, it is the SMT that performs the translation. So if, for example, "Munich" is much more frequent in the training bitexts than "Albuquerque", we should replace the latter with the former everywhere.

In order to test this intuition, we trained a bigram language model on the source side of the SMT training bitexts. We were then able to compute a proximity score on the training data of our two domains. The best candidate for each cluster was retained alone, and used to do the data projection (through the Translation and Annotation approach). The following results were obtained.

A small increase can be observed over the "Translation and Annotation" method, but the results are far from the ones obtained with the consensus approach. It would seem that examples close to the SMT are good translation candidates afterall, but not as good as an average translation over more examples. That is mainly because the SMT itself is not trained with data relevant to the current domain (in our case, airline or insurance), but with general data from various sources (in our case, parliament debates and movie subtitles). Therefore, an example that is close to the SMT will most likely be translated as if it came from a debate or a movie, rather than from an airline or an insurance application. Still, an efficient weighting of examples prior to their translation appears like a promising way to improve our current algorithm.

Table 7. Impact of the proximity with the SMT

method	Airline			Insurance		
	P	R	F_1	P	R	F_1
Consensus	74.9%	81.1%	77.9%	64.3%	70.2%	67.1%
Transl & Annot	49.5%	53.8%	51.6%	43.8%	44.7%	44.2%
SMT	52.6%	57.1%	54.7%	47.0%	49.0%	48.0%

5 Conclusion

From our experience, the consensus approach is an easy and effective way to project data for advanced dialog purposes. However, the reader must bear in mind that two main limitations could arisen if our method were to be used in another context. First, with the current algorithm, a source pattern can have no target pattern at all, should some concepts be completely absent from each translation candidates. In some (rare) cases, clusters can be ignored entirely. Also, the algorithm widely exploits the fact that wrong translations are often harmless in the current task, because no user will ever uses them in a real-life situation. Therefore, it is a minor error to deduct a rule from a completely wrong translation (it will simply become a dead rule in the grammar), but it is a major one to deduct a false rule from a meaningful translation.

Nonetheless, it appears that the projection of annotations in an advanced dialog context, even without a dedicated translation system, is perfectly feasible. In this regard, the consensus approach we presented is very effective. The projection is done automatically, and no further tagging is required. In the future, an automatic way to weight each examples according to their prior reliability could be helpful. Our method has also yet to be tested on a bigger scale. From a more general point of view, we believe that any algorithm meant to address this specific problem of data projection should exploit the redundancy of the source annotation, and our consensus approach is an easy and viable solution to do so.

References

1. Bangalore, S., Murdock, V., Riccardi, G.: Bootstrapping Bilingual Data Using Consensus Translation for a Multilingual Instant Messaging System. In: Proceedings of the 19th International Conference on Computational Linguistics, pp. 1–7 (2002)
2. Bouillon, P., Rayner, M., Novellas, B., Nakao, Y., Santaholma, M., Starlander, M., Chatzichrisafis, N.: Une grammaire multilingue partagée pour la traduction automatique de la parole. In: Proceedings of Traitement Automatique des Langues Naturelles, pp. 155–173 (2006)
3. Chiang, D.: A Hierarchical Phrase-Based Model for Statistical Machine Translation. In: ACL 2005 Proceedings of the 43rd Annual Meeting on Association for Computational Linguistics, pp. 263–270 (2005)
4. Rebecca, H., Resnik, P., Weinberg, A., Cabezas, C., Kolak, O.: Bootstrapping Parsers via Syntactic Projection across Parallel Texts. In: Natural Language Engineering, vol. 11(3), pp. 311–325 (2004)

5. Kim, R., Dalrymple, M., Kaplan, R., King, T.H., Masuichi, H., Ohkuma, T.: Multi-lingual Grammar Development via Grammar Porting. In: Proceedings of the ESS-LLI 2003 Workshop on Ideas and Strategies for Multilingual Grammar Development, pp. 49–56 (2003)

6. Kim, R., Jeong, M., Lee, J., Lee, G.G.: A Cross-lingual Annotation Projection Approach for Relation Detection. In: Proceedings of the 23rd International Conference on Computational Linguistics, pp. 564–571 (2010)

7. Koehn, P., Och, F.J., Marcu, D.: Statistical Phrase-Based Translation. In: Proceedings of the 2003 Conference of the North American Chapter of the Association for Computational Linguistics on Human Language Technology, vol. 1, pp. 48–54 (2003)

8. Koehn, P., Schroeder, J.: Experiments in Domain Adaptation for Statistical Machine Translation. In: Proceedings of the Second Workshop on Statistical Machine Translation, StatMT 2007, pp. 224–227 (2007)

9. Koehn, P., Hoang, H., Birch, A., Callison-Burch, C., Federico, M., Bertoldi, N., Cowan, B., Shen, W., Moran, C., Zens, R., Dyer, C., Bojar, O., Constantin, A., Herbst, E.: Moses: Open Source Toolkit for Statistical Machine Translation. In: Annual Meeting of the Association for Computational Linguistics, pp. 177–180 (2007)

10. Santaholma, M.: Linguistic Representation of Finnish in a Limited Domain Speech-to-Speech Translation System. In: Proceedings of the 10th Conference on European Association of Machine Translation, pp. 226–234 (2005)

11. Santaholma, M.: Multilingual Grammar Resources in Multilingual Application Development. In: Proceedings of Grammar Engineering Across Frameworks Workshop, pp. 25–32 (2008)

12. Tiedemann, J.: News from OPUS – A Collection of Multilingual Parallel Corpora with Tools and Interface. In: Recent Advances in Natural Language Processing, vol. V, pp. 237–248 (2009)

Partial Satisfaction Planning under Time Uncertainty with Control on When Objectives Can Be Aborted

Sylvain Labranche and Éric Beaudry

Université du Québec à Montréal
sylvain.labrance@courrier.uqam.ca, beaudry.eric@uqam.ca

Abstract. In real world planning problems, it might not be possible for an automated agent to satisfy all the objectives assigned to it. When this situation arises, classical planning returns no plan. In partial satisfaction planning, it is possible to satisfy only a subset of the objectives. To solve this kind of problems, an agent can select a subset of objectives and return the plan that maximizes the net benefit, i.e. the sum of satisfied objectives utilities minus the sum of the cost of actions. This approach has been experimented for deterministic planning. This paper extends partial satisfaction planning for problems with uncertainty on time. For problems under uncertainty, the best subset of objectives can not be calculated at planning time. The effective duration of actions at execution time may dynamically influence the achievable subset of objectives. Our approach introduces special abort actions to explicitly abort objectives. These actions can have deadlines in order to control when objectives can be aborted.

1 Introduction

Classical planners consider a goal as a non-separable set of objectives, i.e. a goal can be either satisfied or not satisfied in a particular state. In many real world applications, this binary state of goal satisfaction is not practical. Situations may arise where it is impossible to satisfy every single objective, either because of limited resources, limited time or even mutual exclusions. For these situations, classical planners simply return no solution because the goal cannot be satisfied.

This limitation of traditional planners motivated a branch in planning, which is often called 'over-subscription planning' or 'partial satisfaction planning'(PSP), where a goal is considered as a set of soft objectives, i.e. objectives that may be unsatisfied in a final state. Few works address the PSP problem [4,9,10,11,12].

A typical example of an application which requires partial satisfaction planning is the planning activity for Mars rovers that must gather scientific data. Since such an expedition represents important costs, time must be effectively spent. Therefore, it makes sense to input the rover more objectives than it is possible to satisfy and leave to automated planning the task of finding the best subset. Also, given that communication with a rover on Mars is a long and difficult task and since it is impossible to predict the effective duration of actions,

M. Sokolova and P. van Beek (Eds.): Canadian AI 2014, LNAI 8436, pp. 167–178, 2014.

it is crucial to be able to compute long-term plans that can assess many different outcomes. It has been estimated that sub-optimal plans and plan failures caused the Sojourner Mars rover to be idle between 40 % and 75 % of the time during the Pathfinder mission [5].

Approaches for solving PSP directly are utility based [4,10,11,12]. In order to generate a plan, these planners look to select the subset of objectives which maximizes the net benefit. The net benefit is defined by the sum of the utilities of satisfied objectives minus the costs of actions needed to satisfy them. A research note [9] criticized these approaches. They show that soft objectives can be compiled away by translating PSP problems to equivalent classical planning problems.

However, the proposed compilation in [9] has some limitations for PSP planning with uncertainty on the duration of actions. When planning under uncertainty, the best subset of objectives may not be calculated at planning time, because the effective duration of actions during execution may dynamically influence the optimal achievable subset of objectives. Thus, the decision to select or not an objective must sometimes be postponed at execution time. A way to treat these cases would be to allow dynamically the abortion of objectives.

In addition to allowing the abortion of objectives, it can also be required to force a deadline on the abortion time. In some situations, it is practical to know that past a certain time, an objective cannot be aborted any more and will be satisfied. This is the case when planning and execution overlap, like for Mars rovers missions where during execution, planning for the next day needs to already be started. It is also desirable in situations where commitment before a certain time is needed, like in the delivery business. With deadlines on abortion, it would be possible to know which objectives will be satisfied, because past abortion deadlines soft objectives would become hard.

We present a new approach to solve PSP problems under uncertainty on the duration of actions. Our approach is based on a compilation similar as in [9], giving the possibility to plan the abortion of objectives by introducing special actions to explicitly abort objectives. The planner then searches for a plan that satisfies all objectives, where aborting an objective is viewed as a way to satisfy it. Our approach also enables to set deadlines on when the abortion can be done. We experiment our method in the Action Contingency Time Uncertainty (ActuPlan) planner [2,3].

The rest of this paper is organized as follows. Section 2 formalizes the problem and reviews the state of the art. Then, in Section 3, we present ActuPlan, the abort actions method and a practical example. Empirical results are discussed in Section 4. Finally, a conclusion is presented.

2   State of the Art

2.1   Problem Definition

A planning problem is given by :

$$\mathcal{P} = \langle \mathcal{F}, \mathcal{I}, \mathcal{A}, \mathcal{G} \rangle$$

where $\mathcal{F}$ is a finite set of state variables, $\mathcal{I}$ an initial state, $\mathcal{A}$ a finite set of actions and $\mathcal{G} = \{g_1, g_2, .., g_n\}$ a goal consisting of a finite set of objectives. Every action in $\mathcal{A}$ has a cost, which is given here by the statistical distribution of the expected time it will take to apply it. Objectives in $\mathcal{G}$ can have deadlines, after which the objective cannot be satisfied any more. In classical planning, a plan π is a sequence of actions that leads from $\mathcal{I}$ to $\mathcal{G}$ and a final state is defined by a state where all predicates of $\mathcal{G}$ are true. In PSP, we are interested in finding a plan which maximizes the result of actions, therefore there might be only some, or even none, of the predicates of $\mathcal{G}$ that are true in a final state. This maximization is relative to how the problem is considered. Direct approaches either used the Net-Benefit method where quantitative values are assigned to objectives or introduced qualitative values.

2.2   Current Approaches

In Net-Benefit, utilities are assigned to objectives as a way to represent how important they are compared to each other. Then, the subset of objectives yielding the highest net benefit (utilities of objectives satisfied minus the costs to achieve them) is generated. Two classes of utility-based approaches have been developed, based on prior heuristic calculations asserting objective dependencies.

First, in [12], the problem is abstracted as an *orienteering* one, in which cities are attributes that create dependency between objectives or that are very important for their satisfaction. This abstraction is based on a natural PSP hierarchical decomposition that was later exposed in [1]. The idea was pushed further in [10] by considering stochastic action results and separating the PSP into a set of multiple small (compared to the whole problem) Markov Decision Processes (MDPs) and in which a main MDP process controls the lower hierarchically ranked ones. They also use the term *Abort Action*, but it has not the same meaning as in the method we introduce in Section 3. Their *Abort Action* means that a sub-process returns the control to the main process.

On the other hand, in [4,11], heuristic search methods for PSP have been proposed. The main idea is to heuristically estimate which subset of objectives yields the highest net benefit. This task is hard to achieve since, most of the times, objectives are not independent. Such dependency is treated either by incrementally building the optimal subset or by iteratively removing objectives from $\mathcal{G}$. This is done by using relaxed plan heuristics and assessing how dependant objectives are to each other based on the actions they share, i.e. that they both need to be satisfied. The less beneficial objectives can then be put aside.

These methods have been criticized. A recent research note has showed that soft objectives can be compiled away by translating a PSP problem into a classical planning problem [9]. Empirical results also suggest that this translation based approach gives better results. This compilation is done by adding to the goal a dummy hard objective for every soft one and introduce the new actions *forgo* and *collect*. These actions can only be applied when a new *end-mode* variable is true, which corresponds to the end of classical planning. Once in this state, *collect* is applied with no cost to objectives that are satisfied while *forgo* is applied to unsatisfied objectives with costs corresponding to the utilities of the objectives. This method therefore indirectly introduces a way to abort objectives. However, this method does not produce information on when objectives should be aborted, because actions have deterministic effects and durations.

While planning under action duration uncertainty, some objectives, considered profitable at first, might need to be aborted during execution. For example, it might be because the confidence level on the deadline satisfaction of an objective has gone under the acceptable threshold. This is why a way to know when objectives should be aborted is needed, in order to restrain as much as possible the time wasted. A Mars rover is an example of a situation where one wants to know as soon as possible when an objective should be aborted. Once scientists are informed that an objective won't be satisfied, they can start planning the next mission sooner and with more information.

Some work has already been done to solve PSP under uncertainty on resource consumption [6]. Extending ideas from an earlier version of ActuPlan [2], they first compute a pessimistic and linear plan, i.e. for all actions it is assumed that everything goes wrong. Then, each action in the linear plan is a candidate branching point. Opportunistic branching plans are added when the total expected utility can be improved. Therefore, if actions end up consuming less resources then expected, more objectives can be satisfied.

3   Using Abort Actions with Deadlines

We present an approach that extends ideas from [2,3,6,9] in order to create a conditional plan with branching points that can be actions with deadlines that abort objectives.

Planning with actions having uncertain durations means that an objective can or can not be profitable depending on the actual execution time. Actions can indeed take longer than expected and, at some point, it becomes more profitable to abort some objectives and work on satisfying others that are still potentially profitable. In most current approaches presented in Section 2.2, once execution starts, the subset of objectives is static: it is not possible to remove or add elements dynamically. Therefore, in order to fully address the problem, a method is needed where that subset is dynamically created like in [6]. Some control might however be needed on the subset creation.

This is made possible by the construction of a conditional plan having different branching points depending on the actual time the actions took so far during execution. These branching points are given by intervals where, during a certain time, it is more profitable to apply a certain action while, once the interval is over, it is better to apply another one. Therefore, the conditional plan offers, at branching points, different action choices based on execution time. This behaviour leads to the creation of actions that can abort objectives, giving the possibility to modify $\mathcal{G}$ dynamically. In order to enable some control, these actions have deadlines. One could say that we adopt a positivist point of view, as the starting plan is the one that satisfies, or tries to satisfy, all objectives. Then, only if they cannot all be satisfied, they start being aborted.

3.1   ActuPlan

We extend the ActuPlan planner [2,3] to support partial satisfaction planning. ActuPlan is a planner that handles action contingency and actions with uncertain durations. Effective planning is done using a forward-search guided by the H_{max} heuristic [8]. The planner uses Bayesian networks to represent time continuously. It allows management of time variables independently of state representation, greatly reducing state space complexity compared to a planner that uses a discrete time representation.

Since action durations are uncertain, having objectives with deadlines means that the planner returns a plan with a probability of success ≤ 1. ActuPlan searches for a non-conditional plan that satisfies the goal under a threshold on the probability of success (confidence level) [3].

The planner also computes conditional plans based on execution time. Multiple plans are generated and merged into a conditional one with different branching points where a choice has to be made depending on the actual time [2].

With our method, choices include the possibility of aborting an objective dynamically if the cost to satisfy it becomes greater than its abortion cost. As previously stated, this is impossible with most current approaches to PSP since the satisfied subset is static during execution time.

3.2   Abort Actions

Based on the problem definition previously given, the abort actions are introduced as follows. To the actions set $\mathcal{A}$ is added a finite set $\{Abort(g_1), Abort(g_2)...$ $Abort(g_{|\mathcal{G}|})\}$ of size $|\mathcal{G}|$ in which each $Abort(g_i)$ is an action with precondition that g_i is in $\mathcal{G}$ and with effect of removing g_i from $\mathcal{G}$. To each $Abort(g_i)$ is assigned an abortion cost, representing how costly it is to not satisfy objective i. The cost to abort an objective is simply its utility. A deadline is also assigned to each $Abort(g_i)$, representing the maximum time when objective i be aborted. States are now modelled with an addition of $|\mathcal{G}|$ state variables, each indicating if objective g_i was aborted or not. Therefore, as in classical planning, a final state s_f is a state where every g_i from $\mathcal{G}$ is satisfied and a plan is a partially ordered sequence of actions leading from $\mathcal{I}$ to s_f. The best plan is the one with the lowest total expected cost.

This approach also naturally handles problems where there are both hard and soft objectives. An objective with an abortion cost (utility) of ∞ cannot be aborted and therefore is considered hard. Any objective with abortion cost $< \infty$ is soft and can be aborted. Once the deadline of an abort action is past, the corresponding objective becomes hard and must be satisfied. This approach also compiles away soft objectives, meaning that classical planners could solve deterministic PSP problemes using this model.

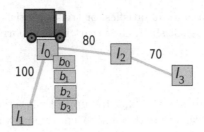

Fig. 1. Simple transport domain example

3.3 Example

Let us consider as an example this simple Transport domain problem in Fig. 1. The Transport domain has been used in the International Planning Competition (IPC) of the International Conference on Automated Planning and Scheduling (ICAPS) [7].

A truck t, initially at l_0, has to deliver 4 boxes to either locations l_1, l_2 and l_3. Here, $\mathcal{G} = \{g_0 = BoxAt(b_0, l_1), g_1 = BoxAt(b_1, l_2), g_2 = BoxAt(b_2, l_3), g_3 = BoxAt(b_4, l_3)\}$ and $\mathcal{A} = \{Drive(t, l_o, l_d), Load(t, l, b), Unload(t, l, b)\}$. In this example, $|\mathcal{A}| = 38$. $Drive$ has a cost according to a normal distribution relative to 1.2 times the distance between two locations, while $Load$ and $Unload$ have a uniform cost between 30 and 60 time units. The truck can only carry one box at a time. Let us add to $\mathcal{A}$ the 4 actions $Abort(g_1)$, $Abort(g_2)$, $Abort(g_3)$ and $Abort(g_4)$ with respective costs of 350, ∞, 600 and 500. Namely, objectives g_0, g_2 and g_3 are soft, while g_2 is hard because its abortion cost is ∞. Two problems will be compared, one where there is no deadline on abort actions and one where $abort(g_0)$ must be applied before 500 time units. Objectives also have respective satisfaction deadlines of ∞, 600, 900 and 1000.

The resulting conditional plans are shown in Fig. 2. s_0 is the initial state and intervals under actions represent branching conditions. At execution, actions are chosen based on these intervals [2].

First, the conditional plan in Fig. 2(a) with no deadline on objective abortion is analysed in order to show how the abort actions approach works. Since g_1 is a hard objective, it has to be achieved. The plan suggests to achieve it first in order to meet its deadline (states s0 to s2). From l_2, the truck returns to l_0 after unloading his box (state s3). State s4 is a branching point : a choice has to be made depending on the execution time. If g_2 was satisfied fast enough (under 263 time units), it is better to satisfy g_0 (states s4 to s9), then g_2 (states s12 to s16) and finally abort g_3. In practice, an objective i is considered aborted when branching is done on a branch that has all its sub-branches leading to the abortion of i. Here, there is no sub-branch and the main one leads to $Abort(3)$, therefore as soon as $Load(r0, l0, box)$ is applied, $Abort(3)$ is also implicitly applied.

On the other hand, if the truck arrives at state s4 past time 263, it is better to take the second branch. At this point, it is possible to know that an objective will be aborted, but not which one as there is a branching point at state 18 where one

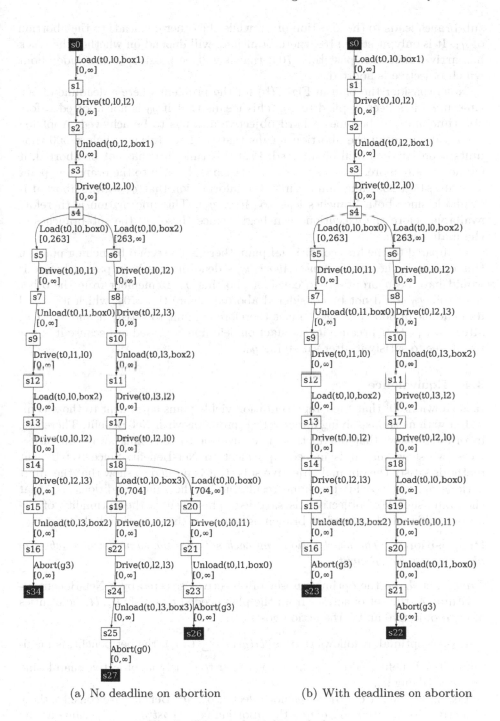

(a) No deadline on abortion   (b) With deadlines on abortion

Fig. 2. Generated conditional plans

sub-branch leads to the abortion of g_0, while the other one leads to the abortion of g_3. It is only at state s18, where branching will depend on whether the truck has arrived before or past time 704, that it will be possible to conclude about which objective is aborted.

Now, consider the plan in Fig. 2(b) for the problem where a deadline of 500 time units has been applied to g_0. This means that if g_0 is not aborted before 500 time units, it becomes a hard objective and has to be achieved. Applying such a deadline on the abortion means that we have to know before 500 time units if objective g_0 will be aborted. Past this time, if it has not be aborted, it cannot be any more. States s0 to s3 are the same, leading to the branching point at state s4 on 263 time units. First, the information that g_3 will be aborted is available since both branches lead to $Abort(g_3)$. This information is therefore available before the execution even begins since these are the sub-branches of the main branch.

Compared to the first conditional plan, there is no second branching point in this one. It is the case because $Abort(g_0)$'s deadline has been passed before it would have been profitable to consider aborting g_0. Remember from the other plan that g_0 could not be considered aborted before time 704, which it passed its deadline of 500. The agent must therefore commit to satisfy it and abort g_3 afterwards, either because its satisfaction deadline is passed or because it is not profitable to satisfy it after satisfying g_0.

3.4   Equivalence

It is shown in [9] that their representation yields plans equivalent to those computed with methods solving directly PSP problems with Net-Benefit. Therefore, in order to show that our method is equivalent to compiling away soft objectives, we show that it is indeed equivalent to Net-Benefit. Since Net-Benefit methods satisfy the optimal objective subset, it is needed to show that the abort actions approach yields the same results under the same conditions, i.e. that the same subset of objectives is satisfied. This ensures the optimality of the non-conditional plans used at branching points in the conditional plan.

Proposition 1. *The abort action approach satisfies the same optimal subset of objectives as Net-Benefit.*

Proof. Let $\mathcal{G}_s$ be the optimal subset of objectives selected by a Net-Benefit algorithm, $\mathcal{A}_p$ the set of actions from the plan needed to satisfy $\mathcal{G}_s$, $\mathcal{U}$ the utilities of each objective and $\mathcal{C}$ the action costs.

If $\mathcal{G}_s$ is optimal, it follows that $\sum_{i=1}^{|\mathcal{G}_s|} \mathcal{U}(g_i) - \sum_{i=1}^{|\mathcal{A}_p|} \mathcal{C}(a_i)$, the net benefit, is maximum. The benefit is also ≥ 0 since it is better to satisfy no objective than losing resources doing so.

Let $\mathcal{G}_a$ be the set of objectives not selected by Net-Benefit. We consider these objectives as aborted and define the abortion costs $cost(g_i)$ as the same as the objective utilities $\mathcal{U}$. $\mathcal{G}_a$ is given by $\mathcal{G} \setminus \{\mathcal{G}_s\}$. The total cost of a plan returned by our method is:

$$\sum_{i=1}^{|\mathcal{G}_a|} cost(g_i) + \sum_{i=1}^{|\mathcal{A}_p|} \mathcal{C}(a_i)$$

$$= \sum_{i=1}^{|\mathcal{G}\backslash\{\mathcal{G}_s\}|} \mathcal{U}(g_i) + \sum_{i=1}^{|\mathcal{A}_p|} \mathcal{C}(a_i)$$

$$= \sum_{i=1}^{|\mathcal{G}|} \mathcal{U}(g_i) - \sum_{i=1}^{|\mathcal{G}_s|} \mathcal{U}(g_i) + \sum_{i=1}^{|\mathcal{A}_p|} \mathcal{C}(a_i)$$

$$= \sum_{i=1}^{|\mathcal{G}|} \mathcal{U}(g_i) - (\sum_{i=1}^{|\mathcal{G}_s|} \mathcal{U}(g_i) - \sum_{i=1}^{|\mathcal{A}_p|} \mathcal{C}(a_i)).$$

Since $\sum_{i=1}^{|\mathcal{G}_s|} \mathcal{U}(g_i) - \sum_{i=1}^{|\mathcal{A}_p|} \mathcal{C}(a_i) \geq 0$ and is maximum, it follows that the total cost is minimum. Given that our method returns the plan with the minimum total cost, the same objectives as in Net-Benefit are aborted and therefore the same optimal subset is satisfied. □

4 Empirical Evaluation

Empirical tests were made with three different approaches implemented in Actuplan. All three use abort actions. First is the non-conditional planner for which a linear plan with no branching point is generated ($ActuPlan^{NC}$ [2]). Then, we test conditional planners, one for which there is no deadline on objective abortion ($ActuPlan^{C}$ [3]) and one for which there is ($ActuPlan^{C+}$). For all versions of $ActuPlan$, we set the threshold on the probability of success at $\alpha = 0.9$

The Transport and Elevators domains are adapted from the IPC of ICAPS [7]. We added abortion costs as well as abortion deadlines. The Transport domain consists of trucks, objects and locations. Trucks can move to locations, load and unload objects and may carry any number of objects at a time. An objective is to have the right object at the right location. In the Elevators domain, an objective is to have a person at the desired floor using elevators. This is done by moving elevators up and down and by people entering and leaving them.

Empirical tests were performed on a machine with a Intel QuadCore i5-3470 CPU @ 3.20GHz. Each problem has been run 100 times. Tables report the average CPU times and expected costs of plans. Planning time was limited to 360 seconds. Memory resources were limited to 2 GB.

4.1 Results

Table 1 contains results for the transport domain. $|T|$ is the number of trucks and $|\mathcal{G}|$ is the number of objects to deliver. The numbers of locations and links were kept constant.

Table 2 contains results for the elevators domain. $|E|$ is the number of elevators and $|\mathcal{G}|$ is the number of persons to bring to the correct floor. The number of floors was kept constant.

Table 1. Empirical results for the Transport domain

Size		Results					
$|T|$	$|\mathcal{G}|$	$ActuPlan^{NC}$		$ActuPlan^{C}$		$ActuPlan^{C+}$	
		CPU(s)	Cost	CPU(s)	Cost	CPU(s)	Cost
1	1	0.09	1525.0	0.08	1506.8	0.08	1506.6
1	2	0.74	1975.0	0.68	1914.6	0.86	1914.3
1	3	22.80	2175.0	23.79	2114.5	24.22	2114.3
2	1	0.40	1525.0	0.40	1506.9	0.41	1506.3
2	2	6.76	1600.0	6.59	1506.3	8.00	1507.0
2	3	48.76	1510.3	50.13	1506.6	56.33	1506.5

Table 2. Empirical results for the Elevators domain

Size		Results					
$|E|$	$|\mathcal{G}|$	$ActuPlan^{NC}$		$ActuPlan^{C}$		$ActuPlan^{C+}$	
		CPU(s)	Cost	CPU(s)	Cost	CPU(s)	Cost
1	1	0.60	83.0	0.06	81.5	0.06	81.5
1	2	0.14	92.0	0.13	90.5	0.16	90.5
1	3	99.25	192.0	118.60	190.5	154.57	190.5
2	1	0.20	68.0	0.13	65.5	0.13	65.5
2	2	3.64	83.0	4.30	76.1	5.56	76.1
2	3	> 360	-	> 360	-	> 360	-

4.2 Discussion

It was expected that non-conditional plans would have a greater expected cost than conditional ones. With branching points in a conditional plan, the agent can adapt if things go well, while it can't in the other case. This behaviour is confirmed by the empirical results, as the expected cost of non-conditional plans is always greater.

However, computing conditional a plan should take more time than computing a strictly linear one. This is expected since, in order to get a conditional plan, multiple non-conditional plans must be found. Depending on the desired confidence level, the number of deadlines and the size of the goal, a lot of non-conditional plans may be computed before merging in a conditional plan is done. This is not confirmed for the smallest problems, as the execution times are roughly the same for both conditional and non-conditional plans. Since the problems are small, there might be no branching point at all in the conditional plans and therefore there is only one plan to compute, explaining the similar times. Indeed, in both domains, it seems that as the problem gets more complex it takes significantly more time to compute a conditional plan than a non-conditional one, indicating that branching points are generated.

Another expected behaviour is that CPU times to compute conditional plans with deadlines on objective abortion would be higher than for other conditional plans. Adding deadlines on abort actions makes the problem more complex, but more information can be extracted from the plans returned. For smaller problems in both domains it doesn't seem to make a difference, but results for bigger problems show that it is more expansive to compute a conditional plan when there are deadlines on objective abortion.

In some cases, the conditional plan with deadlines is expected to return a higher plan cost as a soft objective can become hard if it is not aborted soon enough. Therefore, additional resources could be invested to satisfy it while otherwise it might have been more beneficial to satisfy other objectives. However, it is important to note that this is the desired behaviour. By adding deadlines on objective abortion, one wants to be assured that an objective will be completed if it has not been aborted before its abortion deadline. Empirical results here tend to show that conditional plans with and without abortion deadlines have the same expected cost. This is normal and means that no abortion deadline triggered an objective to become hard, thus forcing its satisfaction as in our previous example. These results are interesting because they show that it is possible to have the information on the maximum time when objectives will be aborted without necessarily incurring additional execution costs.

In general, planning times quickly get quite high compared to problems of similar sizes in a deterministic world. This is why we have only tested with small problems. The abort actions approach in itself makes the state space quite larger. As stated before, every state is modelled with $|\mathcal{G}|$ new predicates to keep trace of which objectives were aborted. It also increases the branching factor by $|\mathcal{G}|$, as in every state every abort action is applied. With deadlines on abortion, this branching factor becomes smaller as deadlines expire. However these expanded states with aborted objectives are only considered late during planning. Abort actions will generally have costs significantly greater than normal actions, given they represent the abortion of an objective. Hence, since heuristic forward planning considers the lowest total expected costs first, states where objectives have been aborted will usually be considered late.

Future work will include investigation for scalability to larger problems. In order to do so, heuristics need to be developed as to when abort actions should be applied in order to reduce the state space expansion caused by our approach.

5   Conclusion

In this paper, we presented an extension of PSP planning to deal with domains having actions with uncertain durations. We introduced new actions to explicitly abort objectives, allowing a dynamic satisfied objectives subset creation. These actions are used at branching points in a conditional plan in order to make the best choice based on the actual execution time. The advantage of our approach is that it supports deadlines on when abortion can be done. These deadlines successfully allow control on the maximum time when an objective can be aborted. The approach works and returns a conditional plan under the given threshold of probability of success.

178 S. Labranche and É. Beaudry

Acknowledgements. The authors gratefully acknowledge support from the Natural Sciences and Engineering Research Council of Canada (NSERC).

References

1. Amir, E., Engelhardt, B.: Factored planning. In: Proceedings of the Eighteenth International Joint Conference on Artificial Intelligence, pp. 929–935. Morgan Kaufmann (2003)
2. Beaudry, E., Kabanza, F., Michaud, F.: Using a classical forward search to solve temporal planning problems under uncertainty. In: Proceedings of the Association for the Advancement of Artificial Intelligence Workshops, pp. 2–8. AAAI Press (2012)
3. Beaudry, E., Kabanza, F., Michaud, F.: Planning for concurrent action executions under action duration uncertainty using dynamically generated bayesian networks. In: Proceedings of the International Conference on Automated Planning and Scheduling, pp. 10–17. AAAI Press (2010)
4. Benton, J., Do, M., Kambhampati, S.: Anytime heuristic search for partial satisfaction planning. Journal of Artificial Intelligence 173(5-6), 562–592 (2009)
5. Bresina, J., Dearden, R., Meuleau, N., Smith, D., Washington, R.: Planning under continuous time and resource uncertainty: A challenge for AI. In: Proceedings of the Conference on Uncertainty in AI, pp. 77–84 (2002)
6. Coles, A.J.: Opportunistic branched plans to maximise utility in the presence of resource uncertainty. In: Proceedings of the Twentieth European Conference on Artificial Intelligence, pp. 252–257. IOS Press (2012)
7. Gerevini, A.E., Haslum, P., Long, D., Saetti, A., Dimopoulos, Y.: Deterministic planning in the fifth international planning competition: Pddl3 and experimental evaluation of the planners. Artificial Intelligence 173(5-6), 619–668 (2009)
8. Haslum, P., Geffner, H.: Admissible heuristics for optimal planning. In: Proceedings of the Artificial Intelligence Planning Systems, pp. 140–149. AAAI Press (2000)
9. Keyder, E., Geffner, H.: Soft goals can be compiled away. Journal of Artificial Intelligence Research 36, 547–556 (2009)
10. Meuleau, N., Brafman, R.I., Benazera, E.: Stochastic over-subscription planning using hierarchies of MDPs. In: Proceedings of the International Conference on Automated Planning and Scheduling, pp. 121–130. AAAI Press (2006)
11. Nigenda, R.S., Kambhampati, S.: Planning graph heuristics for selecting objectives in over-subscription planning problems. In: Proceedings of the Conference of the International Conference on Automated Planning and Scheduling, pp. 192–201. AAAI Press (2005)
12. Smith, D.E.: Choosing objectives in over-subscription planning. In: Proceedings of the International Conference on Automated Planning and Scheduling, pp. 393–401. AAAI Press (2004)

Active Learning Strategies for Semi-Supervised DBSCAN

Jundong Li[1], Jörg Sander[1], Ricardo Campello[2], and Arthur Zimek[3]

[1] Department of Computing Science, University of Alberta, Edmonton, Canada
[2] Department of Computer Sciences, University of São Paulo, São Carlos, Brazil
[3] Institute for Informatics, Ludwig-Maximilians-Universität, Munich, Germany
{jundong1,jsander}@ualberta.ca,
campello@icmc.usp.br,
zimek@dbs.ifi.lmu.de

Abstract. The semi-supervised, density-based clustering algorithm SS-DBSCAN extracts clusters of a given dataset from different density levels by using a small set of labeled objects. A critical assumption of SS-DBSCAN is, however, that at least one labeled object for each natural cluster in the dataset is provided. This assumption may be unrealistic when only a very few labeled objects can be provided, for instance due to the cost associated with determining the class label of an object. In this paper, we introduce a novel active learning strategy to select "most representative" objects whose class label should be determined as input for SSDBSCAN. By incorporating a Laplacian Graph Regularizer into a Local Linear Reconstruction method, our proposed algorithm selects objects that can represent the whole data space well. Experiments on synthetic and real datasets show that using the proposed active learning strategy, SSDBSCAN is able to extract more meaningful clusters even when only very few labeled objects are provided.

Keywords: Active learning, Semi-supervised clustering, Density-based clustering.

1 Introduction

Clustering is the task of grouping a set of unlabeled objects based on their similarity, which is a primary task for data mining, and a common technique for further analysis of data in many fields. Among the prominent clustering paradigms is density-based clustering, based on the idea that clusters are dense regions in the data space, separated by regions of lower density. The well-known algorithm DBSCAN [1] defines local density at a point p in a data set D by two parameters: ε and MinPts. These two parameters are sometimes difficult to determine in real datasets, and often, with only one global density threshold ε, it is impossible to detect clusters of different densities using DBSCAN [2].

As a possible solution to the problem of finding clusters corresponding to locally different density thresholds, the semi-supervised clustering algorithm SS-DBSCAN has been proposed recently [3]. Semi-supervised clustering in general

M. Sokolova and P. van Beek (Eds.): Canadian AI 2014, LNAI 8436, pp. 179–190, 2014.

incorporates a small amount of prior knowledge about the data, for instance in form of labels for a small set of objects, which then is used to influence the otherwise unsupervised clustering method.

SSDBSCAN [3] uses a small amount of labeled data to find density parameters for clusters of different densities. However, the requirement that at least one object from each natural cluster is included in the subset of labeled data, is a major limitation. When objects are selected randomly for determining their labels, the set of labeled points does not necessarily achieve a good coverage of the overall clustering structure, and hence clusters may be missed when only a small number of labeled object can be provided (e.g., due to the cost associated with determining the class label of an object). To address this limitation, we propose to use active learning strategies to select the objects to be labeled, in order to obtain good clustering results with a smaller number of labeled objects.

Active learning has increasingly received attention in the machine learning and data mining community since labeled data (for both supervised and semi-supervised learning tasks) are typically hard to obtain, requiring both time and effort [4]. Instead of labeling data objects randomly, without any prior knowledge about the learning algorithm, the main task of active learning is to automatically select the most "meaningful" objects to label from a set of unlabeled objects. Two trade-off criteria are widely adopted to evaluate what is the most "meaningful" object: choosing the most informative objects that can reduce the overall uncertainty of the whole data distribution, or choosing the most representative objects that can achieve a good coverage of the whole data distribution.

SSDBSCAN [3] uses labeled objects in the following way: starting from each labeled object o, constructing a minimum spanning tree (in a transformed space of reachability distances), until an object with a different label is encountered. Then, a cluster is formed from the minimum spanning tree, whose root is the labeled object o. To form a minimum spanning tree for each cluster, the set of labeled objects should achieve a good coverage of the intrinsic clustering structure, i.e., at least one object from each cluster should be included in the subset of labeled objects. Therefore, in this paper, we focus on how to select the most representative objects by proposing a novel active learning strategy extending a Local Linear Reconstruction [5] model that can be used in combination with SSDBSCAN.

The rest of the paper is organized as follows: Section 2 reviews the related work on semi-supervised clustering and active learning; Section 3 introduces a novel active learning method LapLLR that can improve the performance of SSDBSCAN; Section 4 reports the experimental results on synthetic and real datasets; Section 5 concludes the paper and some possible future work.

2   Related Work

In this section, we first review work on semi-supervised clustering with a focus on density based approaches; then we review related work on active learning and active learning for semi-supervised clustering.

2.1   Semi-Supervised Clustering

In recent years, many semi-supervised clustering algorithms have been proposed. Wagstaff *et al.* [6] introduced must-link and cannot-link constraints to the clustering methods, showing experimentally that with these constraints, it is possible to bias the process of clustering in order to get meaningful clusters. Basu *et al.* [7] used labeled objects to generate initial seed clusters to guide the K-means clustering process. An increasing interest in the area then took place [8].

Most of the efforts in semi-supervised clustering focused on more traditional clustering approaches (e.g. K-means) until Böhm *et al.* [9] proposed HISSCLU, a hierarchical density-based clustering based on OPTICS [2]. As an improvement over HISSCLU, Lelis *et al.* [3] proposed SSDBSCAN, a semi-supervised DBSCAN method that uses labeled objects to steer DBSCAN to automatically find clusters of different densities.

In this paper, we focus on SSDBSCAN, which is therefore briefly reviewed here. Let D denote the whole dataset and let D_L denote a small subset of labeled objects in D. SSDBSCAN assumes that the class labels of objects in D_L are consistent with the density-based clustering structure of the data in the sense that if two objects have different labels, they must belong to different clusters, which are "density separable". This means that for two object p and q with different labels, there must be two different density-based clusters C_1 and C_2 (w.r.t. some density levels ε_1 and ε_2) so that $p \in C_1$ and $q \in C_2$. Assuming this label consistency, Lelis *et al.* [3] showed that one can separate two clusters C_1 and C_2 by a "cut value" that is determined by the largest edge on the density-connection paths between the two clusters. Given this result, a density cluster is then constructed in SSDBSCAN by first adding a labeled object into a cluster, and then incrementally constructing a minimum spanning tree (in a conceptual, complete graph with the objects as nodes and the edge weights as the reachability distances between the pair of nodes/objects). When a differently labeled object has to be added to the minimum spanning tree, the algorithm detects that is has crossed the border between two clusters, and the algorithm backtracks to the the largest edge in the minimum spanning tree and removes all objects that were added after this largest edge, and returns all the objects before the largest edge as a cluster. In this way, SSDBSCAN can detect clusters of different densities from each labeled object. However, for this approach to work, at least one labeled object must be available for each cluster in order to guarantee the discovery of all clusters.

2.2   Active Learning

Active learning is different from supervised learning and semi-supervised learning where all the labels of objects are obtained without considering the learning models. In active learning the learner can ask for the labeling of specific objects. Two criteria for active learning strategies have been mainly studied [4,10], namely, selecting the most informative objects, or selecting the most representative objects among the unlabeled objects.

For the active learning strategy that favors the most informative objects, uncertainty sampling [11] and committee-based sampling [12,13] as well as optimal experimental design (OED) [14] are the most widely used methods. Uncertainty sampling operates incrementally by constructing a classifier and selecting the object that is of least confidence according to the current classifier. Committee-based sampling is an extension of uncertainty sampling. The idea is to build a set of classifiers and to select those objects with the highest disagreements among these classifiers. OED aims at minimizing the variances of a parameterized model. All methods favoring informativeness select an object of highest uncertainty for determining the next label, since knowing its label is most likely to decrease the overall uncertainty.

Addressing the criterion of representativeness, Mallapragada *et al.* [15] proposed an active learning strategy based on the Min-Max criterion, where the basic idea is to build a set of labeled objects from the unlabeled dataset such that the objects in the labeled dataset are far from each other and ensure a good coverage of the dataset. Zhang *et al.* [5] proposed an active learning method named Local Linear Reconstruction (LLR) to find the most representative objects. The general idea of LLR is that the most representative objects are defined as those whose coordinates can be used to best reconstruct the whole dataset. Some other approaches like [16] tried to exploit the clustering structure of the data distribution to select the objects to be labeled.

Most of the existing semi-supervised clustering algorithms assume that the set of labeled objects have been selected in advance without considering the data distribution, typically corresponding to a random selection. Given that determining the labels of objects may be expensive and time consuming, labeling a large number of objects is often not feasible. This motivates active learning strategies to select the most meaningful objects to be labeled to optimally steer semi-supervised clustering algorithms. Vu *et al.* [17] proposed an active learning strategy for semi-supervised K-means clustering, a new efficient algorithm for active seeds selection based on a Min-Max approach that favored the coverage of the whole dataset. Experiments demonstrated that by using this active learning method, each cluster contained at least one seed after a very small number of object selections and thus the number of iterations until convergence was reduced. Basu *et al.* [18] adopted the idea of must-link and cannot-link constraints and introduced an active learning method to select the most meaningful pairwise constraints in order to improve semi-supervised clustering. Xiong *et al.* [19] proposed an iterative way to select pairwise must-link and cannot-link constraints based on a current clustering result and the existing constraints. The proposed framework is based on the concept of neighborhood to select the most informative constraints.

3 Active Learning Strategies for SSDBSCAN

In this section, we investigate how to use active learning to enable the extraction of more meaningful clusters even with very few labeled objects. In SSDBSCAN, the clustering performance is at its best when the selected objects achieve a good coverage of the clustering structure, suggesting active learning strategies favoring the most representative objects. Thus, we propose a novel active learning method called Laplacian Regularized Local Linear Reconstruction (LapLLR). The proposed method is similar to LLR [5] in that both ensure that each object can be reconstructed by only some of its nearest neighbors, and the active selected objects are those that can best represent the whole data space. However, different from LLR, we incorporate a Laplacian Graph Regularizer into the cost function of reconstructed objects to better take into account the discriminating and geometrical structure of the whole data space.

Following LLR [5], we first assume that each object in the data space can be linearly represented by some of its nearest neighbors. In many applications, the data is considered to exist on a low dimensional manifold embedded in a high dimensional space. Data in high dimensional spaces has the property of neighborhood preservation, that is, each object can be represented as a linear combination of some of its nearest neighbors [20].

Let $X = \{x_1, ..., x_k\}$ denote a dataset with k objects. Then, for each object x_i, we assume that it can be represented by a linear combination of its p-nearest neighbors in the dataset X, in this way, x_i can be formulated as:

$$x_i = S_{i1}x_1 + S_{i2}x_2 + ... S_{ik}x_k. \tag{1}$$

$$\text{s.t.} \sum_{j=1}^{k} S_{ij} = 1(S_{ij} \geq 0) \text{ and}$$

$$S_{ij} = 0, \text{ if } x_j \text{ is not among the } p \text{ nearest neighbors of } x_i$$

The constraint in Equation (1) ensures that each x_i can only be represented by its p (p is a parameter) nearest neighbors and the sum of the corresponding coefficients equals 1; coefficient S_{ij} measures how much point x_j contributes to the representation of x_i.

The optimal coefficient matrix $S = \{S_1, S_2, ... S_k\}$ for each object in X can be obtained by minimizing the following cost function:

$$\sum_{i=1}^{k} \left\| x_i - \sum_{j=1}^{k} S_{ij}x_j \right\|^2 \text{ s.t.} \sum_{j=1}^{k} S_{ij} = 1(S_{ij} \geq 0) \tag{2}$$

and $S_{ij} = 0$, if x_j is not among the p nearest neighbors of x_i

Now suppose that we already have a subset of m objects $Z = \{x_{l_1}, ..., x_{l_m}\}$ selected by some mechanisms, and suppose we want to measure the representativeness of the selected objects to see how well they can represent the data space. To measure the representativeness of these selected objects, let $Q = \{q_1, ..., q_k\}$ denote the reconstructed objects that correspond to the objects in the original dataset $X = \{x_1, ..., x_k\}$, and let $\{q_{l_1}, ..., q_{l_m}\}$ denote the subset of these reconstructed objects that correspond to the set of selected objects $\{x_{l_1}, ..., x_{l_m}\}$. Then, the "optimal" coordinates of the reconstructed objects $Q = \{q_1, ..., q_k\}$

with respect to the selected objects, represented by the indices $\{l_1, \ldots, l_m\}$, can be obtained by minimizing the following loss function:

$$\sum_{i=1}^{m} \|q_{l_i} - x_{l_i}\|^2 + \frac{\lambda}{2} \sum_{i=1}^{k} \sum_{j=1}^{k} (q_i - q_j)^2 S_{ij} \tag{3}$$

where $Q = \{q_1, \ldots, q_k\}$ is variable, λ is a regularization parameter, and the coefficient matrix S is from Equation (2). The first term of Equation (3) tries to measure the reconstruction error for the selected objects, i.e. the coordinates differences between each selected object and its corresponding reconstructed point; the second term is a Laplacian Graph Regularizer on the reconstructed objects Q by minimizing the pairwise differences between reconstructed objects, using the coefficient matrix S obtained from the original data set X. Using the same coefficient matrix for reconstructed objects, it assumes that the local neighborhood preserving property should be maintained by the reconstructed objects, which means that if an object x in the original dataset can be reconstructed by a linear combination of its p nearest neighbors, $x = S_1 x_1 +, \ldots, + S_p x_p$, then in the reconstructed data space, x's corresponding reconstructed point q, should be represented as the linear combination of $\{x_1, \ldots, x_p\}$'s corresponding reconstructed coordinates $\{q_1, \ldots, q_p\}$, with the same coefficients as in the original data set, i.e., $q = S_1 q_1 +, \ldots, + S_p q_p$.

Belkin *et al.* [21] defined the concept of a Graph Laplacian as $L = D - S$, where D is a diagonal matrix and $D_{ii} = \sum_j S_{ij}$. The weight matrix S can be determined by many methods. In this paper, we use the optimal coefficient matrix obtained by minimizing Equation (2), as described above, where S_{ij} is the coefficient of x_j in the linear combination that represents x_i.

Using a Graph Laplacian, we can rewrite the loss function in Equation (3) in the following matrix form:

$$\mathrm{Tr}((Q - X)^T \Lambda (Q - X)) + \lambda \mathrm{Tr}(Q^T L Q) \tag{4}$$

where L is the Laplacian Matrix, $\Lambda \in \mathbb{R}^{k \times k}$ is a diagonal matrix such that if $i \in \{l_1, \ldots, l_m\}$, $\Lambda_{ii} = 1$ otherwise $\Lambda_{ii} = 0$. Note that the loss function of our proposed LapLLR has a similar form as the loss function in LLR [5]. However, in our proposed method, we incorporate a Laplacian Graph Regularizer, while in LLR, the regularizer is based on Local Linear Embedding (LLE) [20], which has been designed for finding representations that are of lower dimensionality than the original data.

To obtain the minimum value of the loss function in Equation (4), we set its gradient to 0 to get the optimal solution for the reconstructed objects Q:

$$Q = (\lambda L + \Lambda)^{-1} \Lambda X \tag{5}$$

Previously, we assumed that a set of selected objects $Z = \{x_{l_1}, \ldots, x_{l_m}\}$ and therefore matrix Λ was given. For our method, we propose to select the set of objects Z so that the following loss function is minimized, in order to obtain most representative objects [5]:

$$\sum_{i=1}^{k} \|x_i - q_i\|^2 = \mathrm{Tr}((X - Q)^T(X - Q)) \tag{6}$$

$$= \mathrm{Tr}(((\lambda L + \Lambda)^{-1}\lambda L X)^T((\lambda L + \Lambda)^{-1}\lambda L X))$$

The loss function in Equation (6) makes sure that the difference between the original data and the reconstructed data should be as small as possible.

We take a sequential greedy approach as [5] to optimize Equation (6), in order to obtain the diagonal matrix Λ (the nonzero entries in the matrix Λ correspond to the objects that we want to select). For example, if we want to select the 10 most representative objects, we select the first representative object by minimizing the Equation (6) analytically (exhaustively search all possible entries) with the constraint $\sum_{i=1} \Lambda_{ii} = 1$; thus, we will obtain only one non-zero entry (denoted as l_1) in Λ, which corresponds to the first selected object. Then, we minimize Equation (6) analytically again, now setting $\sum_{i=1} \Lambda_{ii} = 2$ and keeping the previously selected non-zero entry l_1 fixed. This way the second non-zero entry can only be selected from the remaining entries. This procedure is repeated until 10 non-zero entries in Λ are obtained, corresponding to the 10 most representative objects.

4   Experimental Results

4.1   Toy Examples

For visualization, we use two synthetic two-dimensional datasets to compare our proposed active learning method with the state-of-the-art active learning method LLR [5]. These two datasets are:

- Two-circles: 32 points in a big circle, enclosing a small circle of 16 points.
- Ten clusters: 3073 points in ten clusters with different densities.

We apply LLR and the proposed LapLLR to select the most representative points in these two datasets. The experimental result on the two-circle dataset is shown in Fig. 1. Comparing the pairs of Fig. 1(a) with Fig. 1(d), Fig. 1(b) with Fig. 1(e), Fig. 1(c) with Fig. 1(f), it can be seen that the proposed LapLLR can represent the data distribution better than LLR. In Fig. 1(a), the 9th selected point is close to the 6th selected point, in Fig. 1(b), the 8th selected is also close to the 14th; the 8th, 14th, 19th and the 6th, 9th, 20th are also adjacent to each other in Fig. 1(c). In Fig. 1(d), 1(e), 1(f), there are fewer points close to each other, providing a better coverage of the whole data distribution.

In Fig. 2, we report the visualization results on the ten-clusters dataset. Comparing the pairs of Fig. 2(a) with Fig. 2(c) and Fig. 2(b) with Fig. 2(d), we see similar behavior. In Fig. 2(a), the first 50 selected points only cover 7 clusters, while the first 50 points cover 9 clusters in Fig. 2(c)). Using 75 labeled objects, the proposed active learning strategy covers all 10 clusters (Fig. 2(d)) while active learning based on LLR only covers 9 clusters (Fig. 2(b)).

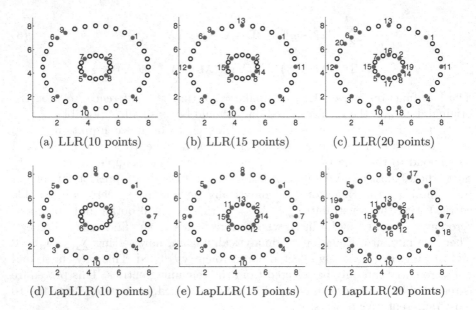

(a) LLR(10 points)   (b) LLR(15 points)   (c) LLR(20 points)

(d) LapLLR(10 points)   (e) LapLLR(15 points)   (f) LapLLR(20 points)

Fig. 1. Visualization results of the LLR active learning method and the proposed LapLLR active learning method on the two-circle datasets. (a), (b) and(c) show the results of 10,15 and 20 points, respectively, selected by the LLR method; (d), (e) and (f) show the results of 10, 15, and 20 points, respectively, selected by the proposed LapLLR method. Selected points are indicated in red color, and the number beside the red points indicate the order in which they are selected.

As mentioned before, a limitation of SSDBSCAN is that for each cluster, a labeled object is required to allow the detection of the complete clustering structure. From the previous two experiments, we can see that the proposed active learning method based on LapLLR can better cover all clusters than the active learning method based on LLR, and it achieves a good coverage of the whole data distribution already with very few labeled objects. Therefore, the proposed active learning method is more likely to help SSDBSCAN to extract clusters of different densities even with a small number of labeled objects.

4.2   Real Datasets

Next we evaluate the performance of the proposed active learning strategy on SSDBSCAN to validate if it can help SSDBSCAN to extract more meaningful clustering structures. As a measure of performance, we report the Adjusted Rand Index [22], which is commonly used for external cluster validation.

Four methods are applied to see if they can help SSDBSCAN to extract clusters of different densities:

- Random selection: randomly select objects to label, which was used in the original evaluation of SSDBSCAN.
- Min-Max [17]: select the objects to label by the Min-Max strategy to make sure the labeled objects are far away from each other.
- LLR [5]: select the objects according to LLR.
- LapLLR: the proposed active learning method which incorporates the laplacian graph regularizer.

The following four UCI datasets are used to validate the effectiveness of the proposed active learning algorithm LapLLR on SSDBSCAN: Ecoli (336 instances, 8 attributes, 8 classes), Glass (214 instances, 9 attributes, 6 classes), Iris (150 instances, 4 attributes, 3 classes) and Seeds (210 instances, 7 attributes, 3 classes). We also report the clustering performance on two datasets created from the Amsterdam Library of Object Images (ALOI) [23] by Horta and Campello [24]. From one dataset containing 5 clusters and 125 images, we use two

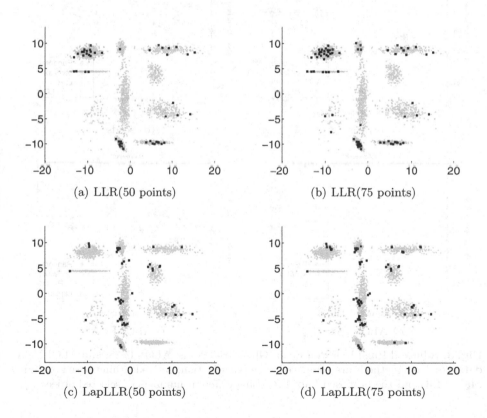

(a) LLR(50 points)

(b) LLR(75 points)

(c) LapLLR(50 points)

(d) LapLLR(75 points)

Fig. 2. Results of LLR and our LapLLR on the ten clusters dataset. (a) and (b) show the results of 50 and 75 points, respectively, selected by the LLR method; (c) and (d) show the results of 50 and 75 points, respectively, selected by the proposed LapLLR method. Selected points are indicated in darker color.

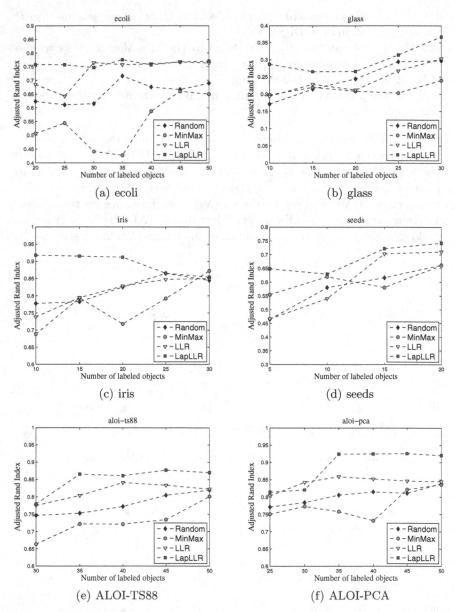

Fig. 3. Adjusted Rand Index on Ecoli, Glass, Iris, Seeds, ALOI-TS88, and ALOI-PCA datasets. Four methods are compared: random selection, active learning based on Min-Max, LLR, and the proposed LapLLR, using different numbers of selected objects.

representations, namely, the descriptor of texture statistics from the gray-level co-occurrence matrix which is denoted as "ALOI-TS88" and a 6-dimensional PCA representation of the 6 descriptors, denoted as "ALOI-PCA".

For all these 6 datasets, we set the same parameter MinPts equal to 3. The parameters p and λ are set to be equal to 5 and 0.01, respectively, in all experiments, following the suggestion from LLR [5]. There is not much difference when the parameter p is in the range of 3 to 10, but when p is set to be too large, the clustering accuracy will decrease. Both our algorithm and LLR are not very sensitive to the regularizer parameter λ. Due to the page limit, we do not discuss the effect of the parameters in detail.

As can be observed in Fig. 3, for the glass, seeds and ALOI-TS88 datasets, the proposed method LapLLR always outperforms the random selection as well as other two active learning methods. For the ecoli dataset, the difference between LLR and our LapLLR method is small, except for the smallest number of labeled objects (20 and 25), when our proposed method outperforms LLR; and both methods are better than the random selection and the Min-Max method. For the iris and ALOI-PCA datasets, when the number of labeled objects is 30, LLR is slightly better than LapLLR, but in all other cases, our method is better than all the other compared selection methods.

In summary, SSDBSCAN using the proposed LapLLR method outperforms SSDBSCAN using any of the other comparison methods, in almost all cases. In only a few cases LLR and LapLLR perform comparably with only a small difference in performance. Most importantly, LapLLR clearly outperforms the other methods when only a very small number of objects is selected for labeling. This demonstrates that the objects selected by our proposed active learning method LapLLR are indeed more representative for the cluster structure.

5   Conclusion

A major limitation of SSDBSCAN is that it needs at least one labeled object in each cluster to uncover the density-based clustering structure of a dataset, which typically requires many labeled objects, if objects are selected randomly for labeling. In this paper, we propose an active learning strategy for SSDBSCAN to select objects for labeling. Our learning strategy incorporates a Laplacian Graph Regularizer into the method of LLR, and the experiments on synthetic and real datasets show the effectiveness of this strategy when compared to random selection and other active learning strategies. The objects selected by our method result in a better performance of SSDBSCAN, compared to other common selection methods, particularly when only a very small number of objects can be labeled. This indicates that the proposed method LapLLR can achieve a better coverage of the whole data distribution.

Possible future work can be focused on two aspects: firstly, we expect to explore the effectiveness of LapLLR on other semi-supervised learning tasks; secondly, we only focus on low dimensional (under 10 dimensions) datasets in this paper, whether our method is applicable to high dimensional datasets, such as text data is another question that may be discussed in the future.

References

1. Ester, M., Kriegel, H.P., Sander, J., Xu, X.: A density-based algorithm for discovering clusters in large spatial databases with noise. In: Proc. ACM SIGKDD, pp. 226–231 (1996)
2. Ankerst, M., Breunig, M.M., Kriegel, H.P., Sander, J.: OPTICS: Ordering points to identify the clustering structure. In: Proc. ACM SIGMOD, pp. 49–60 (1999)
3. Lelis, L., Sander, J.: Semi-supervised density-based clustering. In: Proc. IEEE ICDM, pp. 842–847 (2009)
4. Settles, B.: Active learning literature survey. University of Wisconsin, Madison (2010)
5. Zhang, L., Chen, C., Bu, J., Cai, D., He, X., Huang, T.: Active learning based on locally linear reconstruction. IEEE TPAMI 33(10), 2026–2038 (2011)
6. Wagstaff, K., Cardie, C., Rogers, S., Schrödl, S.: Constrained k-means clustering with background knowledge. In: Proc. ICML, pp. 577–584 (2001)
7. Basu, S., Banerjee, A., Mooney, R.: Semi-supervised clustering by seeding. In: Proc. ICML, pp. 19–26 (2002)
8. Basu, S., Davidson, I., Wagstaff, K.: Constrained clustering: Advances in algorithms, theory, and applications. CRC Press (2008)
9. Böhm, C., Plant, C.: Hissclu: a hierarchical density-based method for semi-supervised clustering. In: Proc. EDBT, pp. 440–451 (2008)
10. Huang, S.J., Jin, R., Zhou, Z.H.: Active learning by querying informative and representative examples. In: Proc. NIPS, pp. 892–900 (2010)
11. Lewis, D.D., Gale, W.A.: A sequential algorithm for training text classifiers. In: Proc. ACM SIGIR, pp. 3–12 (1994)
12. McCallum, A., Nigam, K.: et al.: Employing EM in pool-based active learning for text classification. In: Proc. ICML, pp. 350–358 (1998)
13. Seung, H.S., Opper, M., Sompolinsky, H.: Query by committee. In: Proc. COLT Workshop, pp. 287–294 (1992)
14. Atkinson, A.C., Donev, A.N., Tobias, R.D.: Optimum experimental designs, with SAS, vol. 34. Oxford University Press, Oxford (2007)
15. Mallapragada, P.K., Jin, R., Jain, A.K.: Active query selection for semi-supervised clustering. In: Proc. ICPR, pp. 1–4 (2008)
16. Nguyen, H.T., Smeulders, A.: Active learning using pre-clustering. In: Proc. ICML, pp. 623–630 (2004)
17. Vu, V.V., Labroche, N., Bouchon-Meunier, B.: Active learning for semi-supervised k-means clustering. In: Proc. IEEE ICTAI, pp. 12–15 (2010)
18. Basu, S., Banerjee, A., Mooney, R.J.: Active semi-supervision for pairwise constrained clustering. In: Proc. SAIM SDM, pp. 333–344 (2004)
19. Xiong, S., Azimi, J., Fern, X.Z.: Active learning of constraints for semi-supervised clustering. IEEE TKDE 26(1), 43–54 (2014)
20. Roweis, S.T., Saul, L.K.: Nonlinear dimensionality reduction by locally linear embedding. Science 290(5500), 2323–2326 (2000)
21. Belkin, M., Niyogi, P., Sindhwani, V.: Manifold regularization: A geometric framework for learning from labeled and unlabeled examples. JMLR 7, 2399–2434 (2006)
22. Hubert, L., Arabie, P.: Comparing partitions. Journal of Classification 2(1), 193–218 (1985)
23. Geusebroek, J.M., Burghouts, G.J., Smeulders, A.W.M.: The Amsterdam Library of Object Images. Int. Journal of Computer Vision 61(1), 103–112 (2005)
24. Horta, D., Campello, R.J.G.B.: Automatic aspect discrimination in data clustering. Pattern Recognition 45(12), 4370–4388 (2012)

Learning How Productive and Unproductive Meetings Differ

Gabriel Murray

University of the Fraser Valley, Abbotsford, BC, Canada
gabriel.murray@ufv.ca
http://www.ufv.ca/cis/gabriel-murray/

Abstract. In this work, we analyze the *productivity* of meetings and predict productivity levels using linguistic and structural features. This task relates to the task of automatic extractive summarization, as we define productivity in terms of the number (or percentage) of sentences from a meeting that are considered summary-worthy. We describe the traits that differentiate productive and unproductive meetings. We additionally explore how meetings begin and end, and why many meetings are slow to get going and last longer than necessary.

Keywords: productivity, automatic summarization, extractive summarization, meetings.

1 Introduction

How can we quantify the intuition that some meetings are less productive than others? We can begin by defining *productivity* within the context of an automatic summarization task. If we employ *extractive* techniques to summarize a meeting by labeling a subset of dialogue act segments (sentence-like speech units) from the meeting as important, then productive meetings would seem to be ones that have a high percentage of important, summary-worthy dialogue acts, while unproductive meetings would have a low percentage of such important dialogue acts.

Given that simple definition of productivity, we can see that productivity (or lack of it) is indeed a critical issue in meetings, and that meetings differ in how productive they are. Using gold-standard extractive summaries generated by human judges on the AMI and ICSI corpora (to be described later), we can index the extracted dialogue acts by their position in the meeting and see from Figure 1 that important dialogue acts are more likely to occur at the beginning of meetings and are less likely at the end of meetings. This suggests that many meetings decrease in productivity as they go on, and may be longer than necessary.

We can also see from Figure 2 that meetings overall have a low percentage of summary-worthy dialogue acts, with an average of only 9% of dialogue acts being marked as summary-worthy in the combined AMI and ICSI corpora. Meetings also greatly differ from each other, with some having 20-25% of dialogue acts

M. Sokolova and P. van Beek (Eds.): Canadian AI 2014, LNAI 8436, pp. 191–202, 2014.

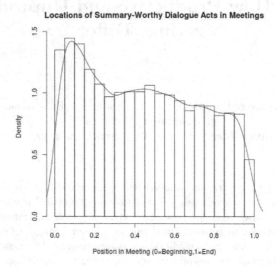

Fig. 1. Histogram/KDE of Extractive Locations

extracted, and others having only 3-4% extracted. We also see that there is a relationship between the length of a meeting and the percentage of summary-worthy dialogue acts, with longer meetings tending to have a smaller percentage of summary dialogue acts.

Another intuition is that meetings often are slow to get going (e.g. featuring idle chit-chat at first) and last longer than necessary (e.g. continuing long after the last decision items and action items have been made). Again viewing this issue from the vantage of extractive summarization, we are interested in predicting the number of dialogue acts that occur *before* the first summary-worthy dialogue act, and the number that occur *after* the last summary-worthy dialogue act. We call these *buffer* dialogue acts because they occur at the beginnings and ends of meetings.

These observations motivate us to explore meeting productivity further. Specifically, in this paper we will carry out two tasks:

- Predict the overall productivity levels of meetings, using linguistic and structural features of meetings.
- Predict the number of *buffer* dialogue acts in meetings, using the same linguistic and structural features.

We use generalized linear models (GLM's) for both tasks. Specifically, we fit a Logistic regression model for the first task and a Poisson regression for the second.

The structure of this paper is as follows. In Section 2, we discuss related work, particularly in the area of extractive meeting summarization. In Section 3, we address the first task above, while in Section 4 we address the second task.

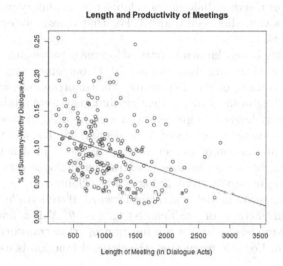

Fig. 2. Length and Productivity of Meetings

In Section 5 we describe the experimental setup, including the corpora and evaluation metrics used. Section 6 gives the experimental results, further discussed in Section 7. Finally, we summarize and conclude in Section 8.

2 Related Work

The most closely related work to ours is on meeting summarization, an area that has seen increased attention in the past ten years, particularly as automatic speech recognition (ASR) technology has improved. These range from *extractive* (cut-and-paste) approaches [1,2,3,4] where the goal is to classify dialogue acts as important or not important, to *abstractive* systems [5,6,7] that include natural language generation (NLG) components intended to describe the meeting from a high-level perspective. Carenini et al [8] provide a survey of techniques for summarizing conversational data.

This work also relates to the task of identifying action items in meetings [9,10] and detecting decision points [11,12,13]. Renals et al [14] provide a survey of various work that has been done analyzing meeting interactions. We are not aware of other work that has specifically looked in-depth at meeting productivity as we have in this paper.

3 Predicting the Overall Productivity Levels of Meetings: Task 1

In Task 1, the goal is to predict the overall productivity of a meeting, given some linguistic and structural features of the meeting. The productivity is measured as the percentage of meeting dialogue acts labeled as summary-worthy. That is, we are predicting a value between 0 and 1. For that reason, we employ Logistic regression for this task.

Logistic regression is well-known in natural language processing, but is usually used in cases where there are dichotomous (0/1) outcomes, e.g. in classifying dialogue acts as extractive or non-extractive [15]. Unfortunately, we do not have gold-standard labeling of meetings indicating that they were productive or non-productive. However, Logistic regression can also be used in cases where each record has some associated numbers of successes and failures, and the dependent variable is then a proportion or percentage of successes. That is our case here, where each meeting has some number of extractive dialogue acts ("successes") and some remaining non-extractive dialogue acts ("failures").

The Logistic regression model is straight-forward. If we have features (or *predictors*) X and parameters (or *coefficients*) θ, then $\theta^T X$ is a linear predictor. Generalized linear models include some function $g()$ that transforms the predictions. In the case of Logistic regression, the sigmoid function is used:

$$g = \frac{1}{1 + e^{-\theta^T X}}$$

Thus, the predictions are constrained to fall between 0 and 1.

For this task, the meeting-level features we use are described below, with abbreviations for later reference. We group them into feature categories, beginning with **term-weight (tf.idf)** features:

- **tfidfSum** The sum of $tf.idf$ term scores in the meeting.
- **tfidfAve** The average of $tf.idf$ term scores in the meeting.
- **conCoh** The conversation cohesion, as measured by calculating the cosine similarity between all adjacent pairs of dialogue acts, and averaging. Each dialogue act is represented as a vector of $tf.idf$ scores.

Next are the structural features relating to meeting and dialogue act **length**:

- **aveDALength** The average length of dialogue acts in the meeting.
- **shortDAs** The number of dialogue acts in the meeting shorter than 6 words.
- **longDAs** The number of dialogue acts in the meeting longer than 15 words.
- **countDA** The number of dialogue acts in the meeting.
- **wordTypes** The number of unique word types in the meeting (as opposed to word tokens).

There are several **entropy** features. If s is a string of words, and N is the number of words types in s, M is the number of word tokens in s, and x_i is a word type in s, then the word entropy *went* of s is:

$$went(s) = \frac{\sum_{i=1}^{N} p(x_i) \cdot -\log(p(x_i))}{(\frac{1}{N} \cdot -\log(\frac{1}{N})) \cdot M}$$

where $p(x_i)$ is the probability of the word based on its normalized frequency in the string. Note that word entropy essentially captures information about type-token ratios. For example, if each word token in the string was a unique type then the word entropy score would be 1. Given that definition of entropy, the derived **entropy** features are:

- **docEnt** The word entropy of the entire meeting.
- **speakEnt** This is the speaker entropy, essentially using speaker ID's instead of words. The speaker entropy would be 1 if every dialogue act were uttered by a unique speaker. It would be close to 0 if one speaker were very dominant.
- **speakEntF100** The speaker entropy for the first 100 dialogue acts of the meeting, measuring whether one person was dominant at the start of the meeting.
- **speakEntL100** The speaker entropy for the last 100 dialogue acts of the meeting, measuring whether one person was dominant at the end of the meeting.
- **domSpeak** Another measure of speaker dominance, this is calculated as the percentage of total meeting DA's uttered by the most dominant speaker.

We have one feature relating to **disfluencies**:

- **filledPauses** The number of filled pauses in the meeting, as a percentage of the total word tokens. A filled pause is a word such as *um*, *uh*, *erm* or *mm − hmm*.

Finally, we use two features relating to **subjectivity / sentiment**. These features rely on a sentiment lexicon provided by the SO-Cal sentiment tool [16].

- **posWords** The number of positive words in the meeting.
- **negWords** The number of negative words in the meeting.

4   Predicting the Number of *Buffer* Dialogue Acts in Meetings: Task 2

In Section 1, we introduced the term *buffer* dialogue acts to describe the dialogue acts that occur *before* the first summary-worthy dialogue act and *after* the last summary-worthy dialogue act in the meeting. Intuitively, a high total number of buffer dialogue acts can indicate an unproductive meeting, e.g. a meeting that was either slow to get going or continued longer than necessary, or both. In Task 2, we want to predict the number of buffer dialogue acts for each meeting. For this experiment, we predict the *total* number of buffer dialogue acts, combined from both the beginning and end of the meeting, though we could alternatively predict those separately.

Since our task now is to predict non-negative count data, we use Poisson regression. Like Logistic regression, Poisson regression is a type of generalized linear model. If we have our linear predictor $\theta^T X$, then the transformation function $g()$ for Poisson regression is:

$$g = e^{\theta^T X}$$

Thus, the output is constrained to be between 0 and ∞. For this task, we use the same features/predictors as described in Section 3.

5   Experimental Setup

In this section we briefly describe the corpora and evaluation methods used in these experiments.

5.1   Corpora

In analyzing meeting productivity, we use both the AMI [17] and ICSI [18] meeting corpora. These corpora each include audio-video records of multi-party meetings, as well as both manual and speech recognition transcripts of the meeting discussions. The main difference between the two corpora is that the AMI meetings are scenario-based, with participants who are role-playing as members of a fictitious company, while the ICSI corpora features natural meetings of real research groups.

As part of the AMI project on studying multi-modal interaction [14], both meeting corpora were annotated with extractive and abstractive summaries, including many-to-many links between abstractive sentences and extractive dialogue act segments. We use these gold-standard summary annotations in the following experiments.[1]

5.2   Evaluation

For the following regression experiments, we evaluate the fitted models primarily in terms of the *deviance*. The deviance is -2 times the log likelihood:

$$Deviance(\theta) = -2 \, log[\, p(y|\theta) \,]$$

A lower deviance indicates a better-fitting model. Adding a random noise predictor should decrease the deviance by about 1, on average, and so adding an informative predictor should decrease the deviance by more than 1. And adding k informative predictors should decrease the deviance by more than k.

For both tasks, we perform an in-depth analysis of the individual features used and report the θ parameters of the fitted models. We report a parameter

---

[1] While we utilize the dialogue act segmentation from those annotations, in this work we make no attempt to classify dialogue act *types* (e.g. *inform*, *question*, *backchannel*) [19].

estimate to be significant if it is at least two standard errors from zero. For the Logistic regression model, the θ parameters can be interpreted in terms of the *log odds*. For a given parameter value θ_n, a one-unit increase in the relevant predictor is associated with a change of θ_n in the log odds. For the Poisson model, given a parameter value θ_n, a one-unit increase in the relevant predictor is associated with the output being multiplied by e^{θ_n}.

6 Results

In this section we present the results on both tasks, first using a Logistic regression model to predict the productivity of each meeting, and then using a Poisson regression model to predict the number of buffer dialogue acts for each meeting.

6.1 Logistic Regression: Task 1 Results

For the productivity prediction task, Table 1 shows the deviance scores when using a baseline model (the "null" deviance, using just a constant intercept term), when using individual predictor models, and when using a combined predictor model. We see that the combined model has a much lower deviance (2843.7) compared with the null deviance (4029.7). Using 16 predictors, we expected a decrease of greater than 16 in the deviance, and in fact the decrease is 1186. We can see that the individual predictors with the largest decreases in deviance are *wordTypes*, *docEnt*, *domSpeak* and *negWords*.

Table 1. Logistic Regression: Deviance Using Single and Combined Predictors

Feature	Deviance
null (intercept)	4029.7
tfidfSum	3680.3
tfidfAve	3792.8
conCoh	3825.1
aveDALength	4029.7
shortDAs	3690.7
longDAs	3705.9
countDA	3637.8
wordTypes	3599.4
docEnt	3652.3
domSpeak	3575.2
speakEnt	3882.6
speakEntF100	3758.9
speakEntL100	3825.8
filledPauses	3986.9
posWords	3679.2
negWords	3612.5
COMBINED-FEAS	**2843.7**

Table 2 gives the parameter estimates for the predictors. For completeness, we report parameter estimates using each predictor in a univariate (single predictor) and multivariate (combined predictors) model. Note that the signs, values and significance of the parameter estimates can change between the univariate and multivariate models, e.g. due to correlations between predictors. We restrict most of our discussion to the univariate models. In the univariate models, all parameter estimates except for *aveDALength* are significant (at least two standard errors from zero). Of the three features giving the largest decreases in deviance, *docEnt* and *domSpeak* have positive parameter values while *speakEnt* has a negative value. That is, an increase in the word entropy of the document is associated with an increase in the log odds of productivity, as is an increase in the dominance of the most dominant speaker. This latter fact is also reflected in the negative value of the *speakEnt* parameter. The greater the dominance of the most dominant speaker, the lower the speaker entropy, and the greater the log odds of productivity.

Table 2. Logistic Regression: Parameter Estimates (significance indicated by boldface)

Feature	θ (Univariate)	θ (Multivariate)
null (intercept)	-	-4.644e+00
tfidfSum	**-3.172e-05**	**2.686e-04**
tfidfAve	**-0.062929**	**-1.139e-01**
conCoh	**8.62097**	-6.542e-02
aveDALength	0.0001451	**2.908e-01**
shortDAs	**-4.896e-04**	**7.970e-04**
longDAs	**-1.460e-03**	**-8.122e-03**
countDA	**-2.865e-04**	**-5.551e-04**
wordTypes	**-5.563e-04**	**-1.689e-03**
docEnt	**5.2647**	**1.912e+00**
domSpeak	**2.3336**	**1.391e+00**
speakEnt	**-4.3420**	-1.801e-01
speakEntF100	**-2.4884**	-4.008e-01
speakEntL100	**-2.79522**	5.711e-01
filledPauses	**3.59026**	-1.590e-01
posWords	**-0.0092129**	7.038e-04
negWords	**-0.007798**	-3.103e-04

We use the trained model to predict on 26 held-out test meetings, and the results are shown in Figure 3, which plots predicted values (the x-axis) against the observed-predicted values (the y-axis). This shows that our model tends to under-predict on the held-out meetings.

6.2 Poisson Regression: Task 2 Results

For the buffer dialogue act prediction task, Table 3 shows the null deviance baseline, the deviance scores for individual predictors, and the deviance of the

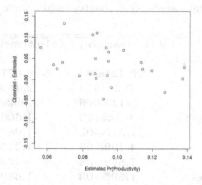

Fig. 3. Productivity Prediction on Hold-Out Test Meetings

Table 3. Poisson Regression: Deviance Using Single and Combined Predictors

Feature	Deviance
null (intercept)	12951.8
tfidfSum	8695.1
tfidfAve	9710.2
conCoh	12457.0
aveDALength	12274.0
shortDAs	9377.7
longDAs	8795.5
countDA	8660.4
wordTypes	8309.1
docEnt	8514.2
domSpeak	11873.0
speakEnt	12540.0
speakEntF100	11014.0
speakEntL100	12542.0
filledPauses	12493.0
posWords	8701.6
negWords	9339.1
COMBINED-FEAS	**6229.3**

combined model. We can see that the combined model exhibits drastically lower deviance over the null baseline. The decrease is 6722.5, where a decrease of 16 would be expected by adding random noise predictors. The best three predictors in terms of lowering the deviance are *countDA*, *wordTypes* and *docEnt*. The *wordTypes* predictor is likely an effective predictor because, as with *countDA*, it is a correlate of meeting length. Similarly, *docEnt* is likely effective because shorter meetings tend to have higher word entropy.

Table 4 shows the parameter estimates for the predictors, again using both univariate and multivariate models. Of the three best features mentioned above, *countDA* has a positive parameter estimate, meaning that longer meetings tend to have more buffer dialogue acts at the beginnings and ends of meetings. So

Table 4. Poisson Regression: Parameter Estimates (significance indicated by boldface)

Feature	θ (Univariate)	θ (Multivariate)
null (intercept)	-	4.535
tfidfSum	**9.421e-05**	**-6.071e-04**
tfidfAve	**0.201454**	**2.466e-01**
conCoh	**-11.8006**	**-3.092e+00**
aveDALength	**0.145172**	**-1.600e-01**
shortDAs	**1.319e-03**	**-1.080e-03**
longDAs	**4.499e-03**	**1.215e-02**
countDA	**7.849e-04**	**2.428e-03**
wordTypes	**1.596e-03**	**-1.689e-03**
docEnt	**-16.4528**	**-3.810e+00**
domSpeak	**-3.70093**	1.659e-01
speakEnt	**7.11037**	**-1.801e-01**
speakEntF100	**6.1627**	**1.622e+00**
speakEntL100	**4.07004**	**-3.535e+00**
filledPauses	**-10.63467**	-1.283e+00
posWords	**0.0265242**	**7.185e-03**
negWords	**0.0198312**	**-6.018e-03**

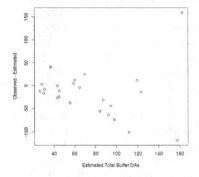

Fig. 4. Buffer DA Prediction on Held-Out Test Meetings

not only do longer meetings take up more of your time, but that time is not necessarily well-spent. Meetings with a large number of word types (the *word-Types* predictor) also tend to have more buffer dialogue acts, while higher word entropy *docEnt* tends to indicate fewer buffer dialogue acts.

Examining the parameter estimate for *speakEnt* (as well as *domSpeak*), we see that meetings with a dominant participant tend to have fewer buffer dialogue acts. And meetings that contain many sentiment words (both *posWords* and *negWords*) tend to have more buffer dialogue acts. Together these findings suggest that meetings with many active participants who are expressing opinions do not always make best use of the time.

Figure 4 shows prediction on the 26 held-out test meetings. Here we see a tendency to over-predict the number of buffer dialogue acts in the test set.

7 Discussion

If we accept that examining meeting summaries is a good proxy for examining meeting productivity, then looking at our two tasks overall it seems we can draw the following conclusions:

- Shorter meetings tend to be more productive.
- Meetings with a dominant participant, i.e. a leader, are more productive.
- Meetings with a large number of sentiment words tend to be less productive.

Though both the Logistic and Poisson models for the two tasks show great improvement in comparison with their respective null deviances, the deviances of the combined predictor models are still fairly high. We plan to do further research on other predictors that may indicate productivity or lack of it. It may also be worthwhile to do some gold-standard labeling of meeting productivity, e.g. enlisting human annotators to make judgments about how productive a meeting was, how well the participants managed the time, whether they achieved their desired decision items, etc. We would also like to make use of participant summaries, which are included in the AMI corpus.

8 Conclusion

In this work, we examined the issue of meeting productivity through the lens of extractive summarization. We carried out two tasks. First, we predicted the productivity levels of meetings using linguistic and structural features. Then we introduced the idea of *buffer* dialogue acts that occur before the first summary-worthy dialogue act and after the last summary-worthy dialogue act in the meeting, and we predicted the number of buffer dialogue acts in a meeting using the same linguistic and structural features. For both tasks, we analyzed and interpreted the individual features used, and found that combined predictor models far outperformed the baselines.

References

1. Murray, G., Renals, S., Carletta, J.: Extractive summarization of meeting recordings. In: Proc. of Interspeech 2005, Lisbon, Portugal, pp. 593–596 (2005)
2. Galley, M.: A skip-chain conditional random field for ranking meeting utterances by importance. In: Proc. of EMNLP 2006, Sydney, Australia, pp. 364–372 (2006)
3. Xie, S., Favre, B., Hakkani-Tür, D., Liu, Y.: Leveraging sentence weights in a concept-based optimization framework for extractive meeting summarization. In: Proc. of Interspeech 2009, Brighton, England (2009)

4. Gillick, D., Riedhammer, K., Favre, B., Hakkani-Tür, D.: A global optimization framework for meeting summarization. In: Proc. of ICASSP 2009, Taipei, Taiwan (2009)
5. Kleinbauer, T., Becker, S., Becker, T.: Indicative abstractive summaries of meetings. In: Proc. of MLMI 2007, Brno, Czech Republic (2007) (poster)
6. Murray, G., Carenini, G., Ng, R.: Generating and validating abstracts of meeting conversations: a user study. In: Proc. of INLG 2010, Dublin, Ireland (2010)
7. Liu, F., Liu, Y.: Towards abstractive speech summarization: Exploring unsupervised and supervised approaches for spoken utterance compression. IEEE Transactions on Audio, Speech and Language Processing 21(7), 1469–1480 (2013)
8. Carenini, G., Murray, G., Ng, R.: Methods for Mining and Summarizing Text Conversations, 1st edn. Morgan Claypool, San Rafael (2011)
9. Purver, M., Dowding, J., Niekrasz, J., Ehlen, P., Noorbaloochi, S.: Detecting and summarizing action items in multi-party dialogue. In: Proc. of the 9th SIGdial Workshop on Discourse and Dialogue, Antwerp, Belgium (2007)
10. Murray, G., Renals, S.: Detecting action items in meetings. In: Popescu-Belis, A., Stiefelhagen, R. (eds.) MLMI 2008. LNCS, vol. 5237, pp. 208–213. Springer, Heidelberg (2008)
11. Hsueh, P.-Y., Moore, J.D.: Automatic decision detection in meeting speech. In: Popescu-Belis, A., Renals, S., Bourlard, H. (eds.) MLMI 2007. LNCS, vol. 4892, pp. 168–179. Springer, Heidelberg (2008)
12. Fernández, R., Frampton, M., Ehlen, P., Purver, M., Peters, S.: Modelling and detecting decisions in multi-party dialogue. In: Proc. of the 2008 SIGdial Workshop on Discourse and Dialogue, Columbus, OH, USA (2008)
13. Bui, T., Frampton, M., Dowding, J., Peters, S.: Extracting decisions from multi-party dialogue using directed graphical models and semantic similarity. In: Proceedings of the SIGDIAL 2009, London, UK (2009)
14. Renals, S., Bourlard, H., Carletta, J., Popescu-Belis, A.: Multimodal Signal Processing: Human Interactions in Meetings, 1st edn. Cambridge University Press, New York (2012)
15. Murray, G., Carenini, G.: Summarizing spoken and written conversations. In: Proc. of EMNLP 2008, Honolulu, HI, USA (2008)
16. Taboada, M., Brooke, J., Tofiloski, M., Voll, K., Stede, M.: Lexicon-based methods for sentiment analysis. Computational Linguistics 37(2), 267–307 (2011)
17. Carletta, J.E., et al.: The AMI meeting corpus: A pre-announcement. In: Renals, S., Bengio, S. (eds.) MLMI 2005. LNCS, vol. 3869, pp. 28–39. Springer, Heidelberg (2006)
18. Janin, A., Baron, D., Edwards, J., Ellis, D., Gelbart, D., Morgan, N., Peskin, B., Pfau, T., Shriberg, E., Stolcke, A., Wooters, C.: The ICSI meeting corpus. In: Proc. of IEEE ICASSP 2003, Hong Kong, China, pp. 364–367 (2003)
19. Dielmann, A., Renals, S.: DBN based joint dialogue act recognition of multiparty meetings. In: Proc. of ICASSP 2007, Honolulu, USA, pp. 133–136 (2007)

Gene Functional Similarity Analysis by Definition-based Semantic Similarity Measurement of GO Terms

Ahmad Pesaranghader[1], Ali Pesaranghader[2], Azadeh Rezaei[1], and Danoosh Davoodi[3]

[1] Multimedia University (MMU), Jalan Multimedia, 63100 Cyberjaya, Malaysia
[2] Universiti Putra Malaysia (UPM), 43400 UPM, Serdang, Selangor, Malaysia
[3] University of Alberta, Edmonton, AB T6G 2R3. Alberta, Canada
{ahmad.pgh,ali.pgh,azadeh.rezaei}@sfmd.ir, danoosh@ualberta.ca

Abstract. The rapid growth of biomedical data annotated by Gene Ontology (GO) vocabulary demands an intelligent method of semantic similarity measurement between GO terms remarkably facilitating analysis of genes functional similarities. This paper introduces two efficient methods for measuring the semantic similarity and relatedness of GO terms. Generally, these methods by taking definitions of GO terms into consideration, address the limitations in the existing GO term similarity measurement methods. The two developed and implemented measures are, in essence, optimized and adapted versions of Gloss Vector semantic relatedness measure for semantic similarity/relatedness estimation between GO terms. After constructing optimized and similarity-adapted definition vectors (Gloss Vectors) of all the terms included in GO, the cosine of the angle between terms' definition vectors represent the degree of similarity or relatedness for two terms. Experimental studies show that this semantic definition-based approach outperforms all existing methods in terms of the correlation with gene expression data.

Keywords: Semantic Similarity, Gene Functional Similarity, Gene Ontology, Gene Expression, MEDLINE, Bioinformatics, BioNLP, Computational Biology.

1 Introduction

Gene Ontology (GO) [1] describes the attributes of genes and gene products (either RNA or protein, resulting from expression of a gene) using a structured and controlled vocabulary. GO consists of three ontologies: biological process (BP), cellular component (CC) and molecular function (MF), each of which is modeled as a directed acyclic graph. In recent years, many biomedical databases, such as Model Organism Databases (MODs) [2], UniProt [3], SwissProt [4], have been annotated by GO terms to help researchers understand the semantic meanings of biomedical entities. With such a large diverse biomedical dataset annotated by GO terms, computing functional or structural similarity of biomedical entities has become a very important research topic. Many researchers have tried to measure the functional similarity of genes or proteins based on their GO annotations. Since different biomedical researchers may annotate the same or similar gene function with different but semantically similar GO

M. Sokolova and P. van Beek (Eds.): Canadian AI 2014, LNAI 8436, pp. 203–214, 2014.
© Springer International Publishing Switzerland 2014

terms based on their research findings, a reliable measure of semantic similarity of GO terms is critical for an accurate measurement of genes functional similarities.

Generally, the output of a similarity measure is a value, ideally normalized between 0 and 1, indicating how semantically similar two given terms (concepts) are. Mainly, with respect to the semantic relatedness measurement, there are two models of computational technique available: taxonomy-based model and distributional model. In taxonomy-based model, proposed measures take advantage of lexical structures such as taxonomies. Measures in this model largely deal with semantic similarity measurement as a specific case of semantic relatedness. The notion behind distributional model comes from Firth idea (1957) [5]: "a word is characterized by the company it keeps". In these measures, words specifications are derived from their co-occurrence distributions in a corpus. These co-occurred features (words) will be represented in vector space for the subsequent computation of relatedness.

With respect to GO, existing studies propose different methods for measuring the semantic similarity of GO terms. These methods include: node-based, edge-based, and hybrid measures, which all get categorized under taxonomy-based model, and because of this characteristic they have limitations associated with this model.

Gloss Vector semantic relatedness measure is a distributional-based approach with a wide application in NLP (Natural Language Processing). Generally, this measure by constructing definitions (Glosses) of the concepts from a predefined thesaurus, estimates semantic relatedness of two concepts as the cosine of the angle between those concepts' Gloss vectors. Considering efficiency of distributional-based model in general and Gloss Vector semantic relatedness measure in specific, none of the existing studies dealing with gene products and protein families has exploited this beneficial information of GO terms' definitions for genes functionality measurement. This paper, drawing upon this aspect of GO terms, attempts for more accurate GO terms semantic similarity estimation leading to a reliable gene functionality analysis.

The remainder of the paper is organized as follows. We enumerate existing measures of semantic similarity dealing specifically with genes functional similarity in Section 2. These measures are already evaluated against protein and gene products datasets. In Section 3, we list data and resources employed for our experiments of genes functional similarity analysis. In Section 4, we introduce our measures as a new approach for genes functional similarity measurement using GO terms' definitions. In brief, our proposed semantic similarity and relatedness measures are optimized and similarity-adapted versions of Gloss Vector semantic relatedness measure for an improved estimation of similarity and relatedness between GO terms. Regarding to the experiments, in Section 5, we represent and evaluate the results of Pearson's correlation for the discussed semantic measures against an available gene expression dataset. Finally, we state conclusion and scopes of future studies in Section 6.

2   Related Works

The proposed semantic similarity measures are discussed in two ways: those measures which are already evaluated in the context of genetics, and those measures which are implemented in other areas (especially in NLP applications) rather than genetics.

2.1   Semantic Similarity Measures Applied in GO context

Semantic similarity is a technique used to measure the likeness of concepts belonging to an ontology. Most early semantic similarity measures were developed for linguistic studies in natural language processing. Recently, semantic similarity measurement methods have been applied to and further developed and tailored for biological uses. A semantic similarity function returns a numerical value describing the closeness between two concepts or terms of a given ontology.

Generally, Gene Ontology (GO), as a Directed Acyclic Graph (DAG), is a hierarchical structure of GO terms by which gene products for their characteristics of biological process (BP), cellular component (CC) and molecular function (MF) get annotated. In genetics, two gene products that have similar function are more likely to have similar expression profiles and be annotated to similar GO terms [6]. Therefore, a comparison and analysis of the similarity between gene expressions of two gene products would be more efficient knowing the similarity scores of them obtained by a semantic similarity measure. Moreover, in the context of PPI, semantic similarity can also be used as an indicator for the plausibility of an interaction because proteins that interact in the cell (in vivo) are expected to participate in similar cellular locations and processes. Thus, semantic similarity measures would be useful for scoring the confidence of a predicted PPI using the full information stored in the ontology. Considering GO and gene products annotations as information resources, the proposed semantic similarity measures employing these resources are as follows.

Resnik Measure - Resnik [7] uses the concept of "information content" (IC) to define a semantic similarity measure. The IC for a concept located in an ontology (or taxonomy) is based on the probability $p(c)$ of occurrence for that concept in a corpus.

$$p(c) = \frac{tf + if}{\sum_{i=1}^{n}(tf_i + if_i)} \tag{1}$$

tf is term frequency of concept c in the corpus, *if* is inherited frequency or frequencies of c's descendants in the ontology all summed together, and n is the total number of concepts in the ontology

In the context of genetics, by considering GO annotations as the corpus, *tf* is the frequency of gene products annotated by c, and *if* is the frequency of gene products annotated by c's descendants in GO. Generally, information content measures how informative one concept (term) is. IC of the concept c is calculable by:

$$IC(c) = -\log(p(c)) \tag{2}$$

The more information two terms share the higher is their similarity. The shared information is captured by the set of common ancestors in the graph. The amount of shared information and thus the similarity between the two terms is quantified by the IC of the least common ancestors or LCAs. This leads to the following formula for semantic similarity measurement between two concepts in an ontology:

$$sim_{Resnik}(c_1, c_2) = \max IC(LCA(c_1, c_2)) \tag{3}$$

Jiang and Conrath Measure – As Resnik measure only considers ancestors' IC and ignores input concepts' level of specificity, Jiang and Conrath [8] deal with this issue:

$$dis_{J\&C}(c_1,c_2) = \min(IC(c_1) + IC(c_2) - 2 \times IC(LCA(c_1,c_2))) \tag{4}$$

Lin Measure – As Jiang and Conrath was originally a distance measure which was not also normalized, Lin [9] proposed a new measure:

$$sim_{Lin}(c_1,c_2) = \max(\frac{2 \times IC(LCA(c_1,c_2))}{IC(c_1) + IC(c_2)}) \tag{5}$$

simGIC Measure – Basically, simGIC [10] (or Graph Information Content similarity) is a functional similarity of gene products. This measure directly employs information contents of the GO terms by which two compared gene products are annotated. For two gene products A and B with annotation sets of T_A and T_B, simGIC is given by:

$$simGIC(A,B) = \frac{\sum_{t \in T_A \cap T_B} IC(t)}{\sum_{t \in T_A \cup T_B} IC(t)} \tag{6}$$

Wang Measure - Wang [11] attempts to improve existing measures by aggregating the semantic contributions of ancestor terms in the GO graph. Formally, a GO term c can be represented as $DAG_c = (c; T_c; E_c)$ where T_c is the set including term c and all of its ancestor terms in the GO graph, and E_c is the set of edges connecting the GO terms in DAG_c (edges which connect T_c terms). By defining the semantic value (SV) of term c as the aggregate contribution of all terms in DAG_c to the semantics of term c, Wang proposes terms closer to term c in DAG_c contribute more to its semantics. The semantic value (SV) of a GO term c is:

$$SV(c) = \sum_{t \in T_c} S_c(t) \tag{7}$$

where S_c is semantic contribution of term c or its ancestors into c's meaning. The semantic contribution of term t to term c is calculable by:

$$S_c(t) = \begin{cases} 1 & \text{if } t=c \\ \max\{\omega_e \times S_c(t') | t' \in \text{children of } t\} & \text{if } t \neq c \end{cases} \tag{8}$$

where ω_e is the "semantic contribution factor" for edge $e \in E_c$ linking term t with its child term t'. Finally, the semantic similarity between two GO terms a and b is:

$$sim_{Wang}(a,b) = \frac{\sum_{t \in T_a \cap T_b}(S_a(t) + S_b(t))}{SV(a) + SV(b)} \tag{9}$$

simRel Measure - Schlicker [12] introduces a new measure of similarity between GO terms that is based on Lin. The measure simRel takes into account how close terms are to their LCA as well as how detailed the LCA is. To do this, they consider that the probability derived by (1) decreases when the depth of a term grows, i.e. more detailed terms have a lower probability than coarser terms. This gives a measure for the relevance of terms: the lower their probabilities the higher is their relevance.

$$simRel(c_1, c_2) = \max(\frac{2 \times IC(LCA(c_1, c_2))}{IC(c_1) + IC(c_2)} \times (1 - p(c))) \qquad (10)$$

Even though all of the proposed semantic similarity measures have shown their usefulness on the conducted experiments, each of them suffers from some limitations. Resnik, as mentioned, does only consider LCAs. Lin and J&C tend to overestimate the estimated levels of similarity and simRel even goes farther. Wang needs calculation of empirical value of "semantic contribution factor" based on gene classification in advance. And in general, all of the measures discussed above do not consider GO terms definitions in their approach while this asset is likely to be a dependable supply. Therefore, in this paper, we attempt to extract this rich source of information to calculate more accurate estimation of semantic similarity. To develop an enhanced definition-based similarity measure, we optimize and then adapt Gloss Vector semantic relatedness measure for similarity measurement of GO terms.

2.2   Gloss Vector Semantic Relatedness Measure

Patwardhan and Pedersen [13] introduced the Gloss Vector semantic relatedness measure by combination of terms' definitions from a thesaurus and co-occurrence data from a corpus. In their approach, every word in the definition of one term from WordNet gets replaced by its context vector from the co-occurrence data from a corpus, and then all of these context vectors summed together build that term's definition vector (Gloss Vector). After definition vector construction for all the terms in the same way, the relatedness will be the cosine of the angle between two input terms associated definition vectors. The Gloss Vector measure is highly valuable as it employs empirical knowledge implicit in a corpus and terms' definitions. In brief, Gloss Vector method gets completed through five successive steps which are:

1) Construction of first order co-occurrence matrix by scanning and counting bigrams' frequencies from the corpus (this matrix is square and symmetric),
2) Removing insignificant words using low and high frequency cut-off points (done by elimination of very low/high frequent bigrams),
3) Developing terms' (concepts) extended definitions through concatenation of directly linked terms' definitions to the target term's definition using a taxonomy or a linked thesaurus (WordNet in the original research),
4) Constructing definition matrix (all definition vectors) by employing results of step 2 (cut-off first-order matrix) and step 3 (terms' extended definitions) and mentioned technique for definition vectors computation, and finally
5) Estimation of semantic relatedness for a concept pair (pair of input terms).

Figure 1 illustrates the entire procedure.

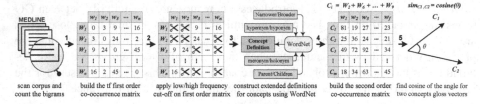

Fig. 1. 5 steps of Gloss Vector relatedness measure

As a clarifying example in the GO context, after stop-words removal and porter stemmer implementation on MEDLINE (as the employed corpus in this study) and then construction of first-order co-occurrence matrix using bigrams frequencies of the rest (changed MEDLINE), low and high frequency cut-offs for removing insignificant bigrams from this matrix get enforced. For the concept *"male germ-cell nucleus"* (GO:0001673) with the GO definition of *"The nucleus of a male germ cell, a reproductive cell in males"*, the definition vector (second order co-occurrence vector) will be constructed. For this propose, after removing stop-words and applying porter stemmer on the definition, all the available first-order co-occurrence vectors (from cut-off step) of the words seen in *"male germ-cell nucleus"* definition which are *nucleus, male, germ, cell*, and *reproductive* will be added together (considering *male* and *cell* two times). In this sample for the sake of similarity we used only definition of the concept and not its extended definition reachable by appendage of definitions of its directly linked terms in GO. For other GO terms, the procedure of definition vector construction is the same. In order to find the semantic relatedness between two concepts (GO terms) we need to find the cosine of the angle between their computed definition vectors. It is clearly justifiable that *"male germ-cell nucleus"* shows higher relatedness score to term *"female germ-cell nucleus"* (GO:0001674) with the GO definition of *"The nucleus of the female germ cell, a reproductive cell in females"* instead of term *"glutamate synthase activity"* (GO:0015930) with the GO definition of *"Catalysis of the formation of L-glutamine and 2-oxoglutarate from L-glutamate, using NADH, NADPH or ferredoxin as hydrogen acceptors"*.

Considering reliability and applicability of Gloss Vector relatedness measure it still suffers from two drawbacks:

1) In cut-off stage only bigrams frequencies are considered without taking frequencies of individual terms into account. This causes many informative words get removed as they tend to be less frequent. In our first introduced measure, Optimized Gloss Vector relatedness measure, by applying Pointwise Mutual Information (PMI) we deal with this issue statistically. Our research work in [14] conducted on MeSH demonstrates the superiority of this optimized measure over regular Gloss Vector relatedness measure.

2) This measure is basically developed for relatedness measurement rather than similarity estimation (similarity is only a specific case of relatedness). Our second measure, Adapted Gloss Vector similarity measure, by taking GO structure into account, attempts to adapt Gloss Vector measure for more accurate similarity measurement. Our experiments in [15], over an available dataset of concept-pairs, by considering MeSH structure, shows this adapted approach outperforms other measures on the task of similarity estimation.

3   Experimental Data

3.1   Gene Ontology (GO)

Gene ontology (GO), is a major bioinformatics initiative to unify the representation of gene and gene product attributes across all species. The project mainly aims at maintaining and developing its controlled vocabulary of gene and gene product attributes; annotating genes and gene products, and assimilate and disseminate annotation data; providing tools for easy access to all aspects of the data provided by the project. The designed GO, as already mentioned, comprises three sub-ontologies: cellular component (CC), molecular function (MF) and biological process (BP). For our experiment in this study the GO and all the required GO annotations were downloaded from the official gene ontology site and the SGD website.

3.2   MEDLINE Abstract

MEDLINE[1] contains over 20 million biomedical articles from 1966 to the present. The database covers journal articles from almost every field of biomedicine including medicine, nursing, pharmacy, dentistry, veterinary medicine, and healthcare. For the current study we used MEDLINE abstracts as the corpus to build a first-order term-term co-occurrence matrix for the later computation of second-order co-occurrence matrix. This second-order matrix, on which we apply optimization and adaptation, gets used for the introduced measures. We employ the 2013 MEDLINE abstract.

3.3   Gene Expression Datasets

The gene expression dataset is from the study by Jain and Bader [16] in which the gene expression dataset for *Saccharomyces cerevisiae* was downloaded from GeneMANIA and other contained data from various microarray experiments. Test datasets were prepared from 5000 *S.cerevisiae* gene pairs randomly selected from a list of all possible pairs of proteins in their gene expression dataset. This was done independently for CC, BP, and MF annotations of gene products (including IEA annotations).

4   Method and Measures

4.1   PMI and Optimized Gloss Vector Semantic Relatedness Measure

Pointwise Mutual Information (PMI) is a measure of association used in information theory. In computational linguistics, PMI for two given terms indicates the likelihood of finding one term in a text document that includes the other term. PMI gets formulated as:

$$\text{PMI}(t_1, t_2) = \log \frac{p(t_1, t_2)}{p(t_1) \times p(t_2)} \tag{11}$$

$p(t_1, t_2)$ is the probability that concepts t_1 and t_2 co-occur in a same document, and $p(t_1)$ and $p(t_2)$ for t_1 and t_2 respectively are the probabilities of their occurrence in a document

---

[1] http://mbr.nlm.nih.gov/index.shtml

The result of PMI measure represents how associated the terms t_1 and t_2 are, by considering frequencies of t_1 and t_2 individually (marginal probability) and frequency of t_1 and t_2 co-occurred (joint probability). It is highly possible t_1 and t_2 co-occur in a context with low frequency but they are associated with and descriptive of each other.

We take advantage of PMI for our first measure, Optimized Gloss Vector relatedness measure, for statistical elimination of insignificant features (terms). In order to integrate this statistical association measure into the Gloss Vector measure procedure, in our approach, we 1) ignore the low and high frequency cut-off step in Gloss Vector measure, 2) multiply terms' IDF (inverse document frequency - without *log*) in terms' TF (term frequency) first-order vectors by considering all definitions as a collection of documents 3) construct a normalized second order co-occurrence matrix from first-order co-occurrence matrix having terms' definitions, 4) build PMI-on-SOC matrix by enforcing PMI on the normalized second-order co-occurrence matrix to find relative association of concepts or terms (rows of matrix) and words or features (columns of matrix), and 5) apply low and high level of association cut-offs on PMI-on-SOC matrix. Figure 2 illustrates the entire procedure of our proposed measure of Optimized Gloss Vector semantic relatedness. The last produced matrix (optimized PMI-on-SOC matrix) is loaded in memory and gets used for the measurement of relatedness between two GO terms (concepts). In this matrix, each row stores the calculated optimized definition vectors of its associated GO terms and the cosine of the angles between them indicates the degree of semantic relatedness.

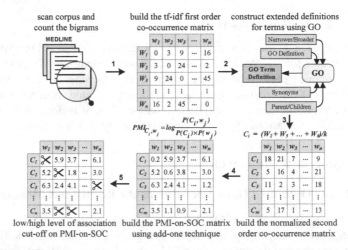

Fig. 2. 5 steps of Gloss Vector measure optimization using PMI

In Gloss Vector relatedness measure, for constructing definition vector (Gloss Vector) of a concept, all first-order co-occurrence vectors of words included in the concept' definition get summed together. For applying PMI on the second order co-occurrence matrix properly, we need to have all the vectors normalized first. To construct a normalized second-order co-occurrence vector for a concept, after adding all first order co-occurrence vectors of the constituent words in its definition, we will divide the result vector by the quantity of these added vectors. Also, as a general rule of thumb, PMI has a limitation for selecting strongly associated bigrams as it is biased towards low frequency words. In order to resolve this weakness, the simple yet helpful add-one technique is employed. In add-one technique, before applying PMI on a matrix, all the elements of the matrix get incremented by 1 unit.

4.2   GO Structure and Adapted Gloss Vector Semantic Similarity Measure

Here, by considering the concepts' taxonomy (GO structure), we would construct the enriched second-order co-occurrence matrix used for the measurement of similarity between GO terms (concepts). For this purpose, by using optimized PMI-on-SOC matrix, we can calculate enriched gloss vectors of the concepts (GO terms) in the GO taxonomy. Our formula to compute the enriched gloss vector for a concept is:

$$\text{Vector}(c_j) = \frac{ca_i(c_j) + ia_i(c_j)}{\sum_{k=1}^{m} \left(ca_i(c_k) + ia_i(c_k)\right)} \quad \forall i <= n$$

$$\text{Vector} \in R^n \text{ , n is the quantity of features}$$
$$1 \leq j \leq m \text{ , m is the quantity of concepts}$$

(12)

where $ca_i(c_j)$ (concept association) is the level of association between concept c_j and feature i (already computed by PMI in the previous stage); $ia_i(c_j)$ (inherited association) is the level of association between concept c_j and feature i inherited from c_j's descendants in the taxonomy calculable by summation of all concept's descendants' levels of association with feature i; and finally, the denominator is summation of all augmented concepts (concepts plus their descendants) levels of association with feature i. All of these levels of association' values are retrievable from optimized PMI-on-SOC matrix. The collection of enriched gloss vectors builds enriched gloss matrix which get used for the final semantic similarity measurement.

The intuition behind the second stage is that the higher concepts in an ontology tend to be more general, and so associated with lots of words (features in their vectors) and the lower concepts should be more specific and therefore associated with less number of features. By adding up the computed optimized second order vector of one concept with its descendants' vectors we attempt to enforce the forgoing notion regarding to the abstractness/concreteness of the concepts which by itself implicitly implies the idea of commonality for similarity measurement as well.

5   Experiments and Discussions

As, in most cases, gene products are annotated with more than one GO term in the same GO ontology (BP, CC or MF), there are several methods to measure the functional similarity of gene products based on the semantic similarity of these GO terms. The common methods are: MAX and AVE, BMA (best-match average) [16]. MAX and AVE respectively define functional similarity between gene products as the maximum or average semantic similarity values over the GO terms annotating the genes. BMA for two annotated gene products p and q is the average of best matches of GO terms of each gene product against the other. Each of these methods has been used and shown to be relevant in a specific circumstance. The BMA method is found to be the best from a biological perspective [10] and is shown to be more suitable for protein function prediction [11]. As such, in this study, we use BMA method given by the following formula:

$$BMA(p,q) = \frac{1}{n+m} \left(\sum_{t \in T_p^x} \max_{s \in T_q^x} S(s,t) + \sum_{t \in T_q^x} \max_{s \in T_p^x} S(s,t) \right) \tag{13}$$

where $S(s,t)$ is semantic similarity (or relatedness) score between terms s and t, T_r^X is a set of GO terms in X representing the molecular function (MF), biological process (BP) or cellular component (CC) ontology annotating a given gene products r and $n=|T_p^X|$ and $m=|T_q^X|$ are the number of GO terms in these sets.

BMAs of the available dataset of 5000 gene product pairs for all the semantic measures discussed in this study get calculated separately. This would help to evaluate which measure produces more reliable results. This evaluation is done against the available standard reference of 5000 genes expressions. As already stated, two gene products that have similar function are more likely to have similar expression profiles and be annotated to similar GO terms. Considering this fact, by comparing BMA results of simRel, Wang, Jiang and Conrath, Lin, Gloss Vector relatedness, and our Optimized Gloss Vector relatedness and Adapted Gloss Vector similarity measures, next to the result of sim-GIC measure, we can discover which similarity or relatedness measure estimates better results of functional similarity. This comparison is done through Pearson's correlation between gene expression data and simGIC and BMAs results. Pearson's correlations between genes expressions and semantic similarity and relatedness measures on *S.cerevisiae* dataset for CC, BP and MF ontologies are shown in figure 3.

Considering figure 3, we can see while Adapted Gloss Vector semantic similarity measure (AdpGlossVec) shows better correlation with gene expression in comparison with previously proposed similarity measures, Optimized Gloss Vector relatedness measure (OptGlossVec) has the highest correlation with gene expression data in cellular component (CC), molecular function (MF) and biological process (BP) ontologies. This indicates, while the majority of studies emphasize employing similarity measures in their approaches, definition-based relatedness measures can yield better results correlated with gene expression data. Table 1 highlights the calculated results of Pearson's correlation presented in figure 3. It lists the correlation results for four measures of Resnik, GlossVec, AdpGlossVec and OptGlossVec.

By comparing the results of OptGlossVec and AdpGlossVec with GlossVec (Gloss Vector relatedness measure) outputs we also notice our consideration of optimization and adaptation on Gloss Vector relatedness measure for producing better similarity and relatedness estimations have been effective. These presented results are when the optimum cut-off points of frequency (Gloss Vector) and level of association (Optimized and Adapted Gloss Vector) were applied. These cut-off thresholds are found after examination of several cut-off points separately and then comparison of their correlations against gene expression data with each other.

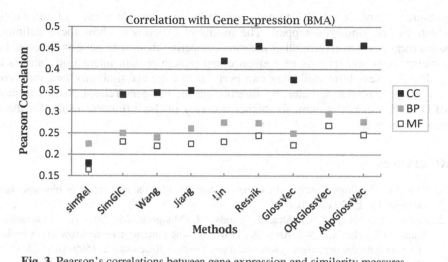

Fig. 3. Pearson's correlations between gene expression and similarity measures

Table 1. Pearson's correlation of 4 semantic measures on 3 GO ontologies

semantic measure	Pearson's correlation		
	molecular function	biological process	cellular component
Resnik	0.2442	0.2735	0.4550
GlossVec	0.2221	0.2493	0.3752
OptGlossVec	**0.2674**	**0.2968**	**0.4638**
AdpGlossVec	0.2463	0.2765	0.4563

Table 2. The best low/high cut-off points of Gloss Vector measures

ontology	GlossVec (low/high cut-off)	OptGlossVec (low/high cut-off)	AdpGlossVec (low/high cut-off)
cellular component (CC)	800 / 15400	1.4 / 4.2	1.6 / 4.0
biological process (BP)	650 / 13000	1.1 / 5.0	1.5 /4.9
molecular function (MF)	700 / 15000	1.3 / 4.7	1.7 / 4.8

Table 2 represents for any of CC, BP or MF ontologies which cut-off thresholds have been selected as the optimum points of insignificant features elimination. As the table indicates, the high and low cut-off points, enforced on PMI-on-SOC matrix, for the Adapted Gloss Vector similarity (AdpGlossVec) and Optimized Gloss Vector relatedness (OptGlossVec) measures should not be necessarily the same.

6 Conclusion

Considering genes with similar expression patterns would have higher similarity in GO based measures as they are annotated with semantically similar GO term, we introduced two semantic similarity and relatedness measures mainly dependant on GO terms' definitions. By comparing them with available semantic similarity measures we observed GO terms' definitions are valuable source of information which was previously untapped. The results of experiment show these definition-based measures outperform other measure comparing them with gene expression data; however, more evaluations on sequence and protein-protein interaction datasets are needed to assess how well they can perform to estimate similarity and interaction between gene products. Later, by filtering out the not-gene-related articles from the MEDLINE, we will attempt for higher accuracy of the estimated results for the semantic similarity.

References

1. The Gene Ontology Consortium. Gene ontology: tool for the unification of biology. Nature Genetics 25, 25–29 (2000)
2. Stein, L.D., Mungall, C., Shu, S., Caudy, M., Mangone, M., Day, A., Nickerson, E., Stajich, J.E., Harris, T.W., Arva, A., Lewis, S.: The generic genome browser: A building block for a model organism system database. Genome Research 12, 1599–1610 (2002)
3. The UniProt Consortium. The uniprot consortium: The universal protein resource (uniprot). Nucleic Acids Research, pp. 190–195 (2008)
4. Kriventseva, E.V., Fleischmann, W., Zdobnov, E.M., Apweiler, R.: Clustr: a database of clusters of swiss-prot+trembl proteins. Nucleic Acids Research 29, 33–36 (2001)
5. Firth, R.: A Synopsis of Linguistic Theory 1930-55. In Studies in Linguistic Analysis (1957)
6. Sevilla, J.L., Segura, V., Podhorski, A., Guruceaga, E., Mato, J.M., Martinez-Cruz, L.A., Corrales, F.J., Rubio, A.: Correlation between Gene Expression and GO Semantic Similarity. IEEE/ACM Trans. Comput. Biol. Bioinformatics, 330–338 (2005)
7. Resnik, P.: Using Information Content to Evaluate Semantic Similarity in a Taxonomy. In: Proceedings of the 14th International Joint Conference on Artificial Intelligence (1995)
8. Jiang, J.J., Conrath, D.W.: Semantic Similarity based on Corpus Statistics and Lexical Taxonomy. In: International Conference on Research in Computational Linguistics (1997)
9. Lin, D.: An Information-theoretic Definition of Similarity. In: 15th International Conference on Machine Learning, Madison, USA (1998)
10. Pesquita, C., Faria, D., Bastos, H., Ferreira, A.E.N., Falcao, A.O., Couto, F.M.: Metrics for GO based protein semantic similarity: a systematic evaluation (2008)
11. Wang, J.Z., Du, Z., Payattakool, R., Yu, P.S., Chen, C.F.: A new method to measure the semantic similarity of GO terms. Bioinformatics 23, 1274–1281 (2007)
12. Schlicker, A., Albrecht, M.: FunSimMat - a comprehensive functional similarity database
13. Patwardhan, S., Pedersen, T.: Using WordNet-based Context Vectors to Estimate the Semantic Relatedness of Concepts. In: Proceedings of the EACL 2006 Workshop, Making Sense of Sense: Bringing Computational Linguistics and Psycholinguistics Together, Trento, Italy, pp. 1–8 (2006)
14. Pesaranghader, A., Muthaiyah, S., Pesaranghader, A.: Improving Gloss Vector Semantic Relatedness Measure by Integrating Pointwise Mutual Information: Optimizing Second-Order Co-occurrence Vectors Computed from Biomedical Corpus and UMLS. In: International Conference on Informatics and Creative Multimedia, pp. 196–201 (2013)
15. Pesaranghader, A., Rezaei, A., Pesaranghader, A.: Adapting Gloss Vector Semantic Relatedness Measure for Semantic Similarity Estimation: An Evaluation in the Biomedical Domain. In: Proceedings of the 3rd Joint International Semantic Technology (2013)
16. Shobhit, J., Bader, G.D.: An improved method for scoring protein-protein interactions using semantic similarity within the gene ontology. BMC Bioinformatics (2010)

Text Representation Using Multi-level Latent Dirichlet Allocation

Amir H. Razavi and Diana Inkpen

School of Electrical Engineering and Computer Science
University of Ottawa
{araza082,diana}@eecs.uottawa.ca

Abstract. We introduce a novel text representation method to be applied on corpora containing short / medium length textual documents. The method applies Latent Dirichlet Allocation (LDA) on a corpus to infer its major topics, which will be used for document representation. The representation that we propose has multiple levels (granularities) by using different numbers of topics. We postulate that interpreting data in a more general space, with fewer dimensions, can improve the representation quality. Experimental results support the informative power of our multi-level representation vectors. We show that choosing the correct granularity of representation is an important aspect of text classification. We propose a multi-level representation, at different topical granularities, rather than choosing one level. The documents are represented by topical relevancy weights, in a low-dimensional vector representation. Finally, the proposed representation is applied to a text classification task using several well-known classification algorithms. We show that it leads to very good classification performance. Another advantage is that, with a small compromise on accuracy, our low-dimensional representation can be fed into many supervised or unsupervised machine learning algorithms that empirically cannot be applied on the conventional high-dimensional text representation methods.

Keywords: Latent Dirichlet Allocation (LDA), Text representation, Topic extraction, Text mining, Multilevel representation.

1   Introduction

For many years, classification of text data has been regarded as a practical and effective text mining task. In order to improve the performance of such an important task, we always need an informative and expressive method to represent the texts [18] [17]. In this regard, if we consider the words as the smallest informative units of a text, there is a variety of well-known quantitative information measures that can be used to represent a text. Such methods have been used in a variety of information extraction projects, and in many cases have even outperformed some syntax-based methods. There are a variety of Vector Space Modeling (VSM) methods which have been well explained and compared, for example in [20]. However, these kinds of representations disregard valuable knowledge that could be inferred by considering the different types of relations between the words. These major relations are actually the essential components that, at a higher level, could express concepts or explain the main topic of a

M. Sokolova and P. van Beek (Eds.): Canadian AI 2014, LNAI 8436, pp. 215–226, 2014.

text. A representation method which could add some kind of relations and dependencies to the raw information items, and illustrate the characteristics of a text in a more extensive manner, could play an important role in knowledge extraction, concept analysis and sentiment analysis tasks.

In this paper, the main focus is on how we represent the topics of the texts. Thus, we first introduce a LDA topic-based representation method as the selected approach, and in the second stage, we build a multi-level topic representation based on the first step. In the third stage, we run machine learning algorithms on a representation that combines various topical representation levels, in order to explore the most discriminative representation for the task of text classification.

2   Background and Related Work

In most text classification tasks, the text are represented as a set of independent units like unigrams / bag of words (BOW), bigrams, and/or multi-grams which construct the feature space, and the text is normally represented only by the assigned value (binary, frequency, or TF-IDF[1]), which is explicitly about the existence of the features in the text [19]. In this case, since most lexical features occur only a few times in each context, if at all, the representation vector tends to be very sparse. This method has two disadvantages. First, very similar contexts may be represented by different features in the space. Second, in short texts, we will have too many zero features for machine learning algorithms, including supervised classification methods.

Capturing the right sense of a word in its context is a critical issue in the representation methods. When we review the literature in this area, we find some useful hypotheses, such as: "You shall know a word by the company it keeps" [8], and that the meanings of words are (largely) determined by their distributional patterns; this is known as the Distributional Hypothesis [10] [11], which state that words which occur in similar contexts tend to be similar. There are many works about semantic similarity based on the Distributional Hypothesis [14].

In 2003, Blei, Ng and Jordan presented the Latent Dirichlet Allocation (LDA) model and a Variational Expectation-Maximization algorithm for training their model. These topic models are a kind of hierarchical Bayesian models of a corpus [2]. The model can unveil the main themes of a corpus, which can potentially be used to organize, search, and explore the documents of the corpus. In the LDA topic modeling, a "topic" is a distribution over the feature space of the corpus and each document can be represented by several topics with different weights. The number of topics and the proportion of vocabulary that create each topic are considered as two hidden variables of the model. The conditional distribution of these variables, given an observed set of documents, is regarded as the main challenge of the model.

Griffiths and Steyvers in 2004 applied a derivation of the Gibbs sampling algorithm for learning LDA models [9]. They showed that the extracted topics capture a meaningful structure of the data. The captured structure is consistent with the class labels assigned by the authors of the articles. The paper presents further applications

---

[1] Term frequency–inverse document frequency.

of this analysis, such as identifying "hot topics" by examining temporal dynamics and tagging some abstracts to help exploring the semantic content. Since then, the Gibbs sampling algorithm was shown as more efficient than other LDA training methods, e.g., variational EM and Expectation-Propagation [15]. This efficiency is attributed to a famous attribute of LDA namely, "the conjugacy between the Dirichlet distribution and the multinomial likelihood". This means that the conjugate prior is useful, since the posterior distribution is the same as the prior, and it makes inference feasible; therefore, when we are doing sampling, the posterior sampling becomes easier. Because of this, the Gibbs sampling algorithm was applied for inference in a variety of models which extend LDA [21], [7], [4], [3], [13].

Recently, Mimno et al. presented a hybrid algorithm for Bayesian topic modeling in which the main effort is to combine the efficiency of sparse Gibbs sampling with the scalability of online stochastic inference [16]. They used their algorithm to analyze a corpus that included 1.2 million books (33 billion words) with thousands of topics. They showed that their approach reduces the bias of variational inference and can be generalized by many Bayesian hidden-variable models.

3   Datasets

In order to have a proper evaluation on our multi-level LDA representation, we conducted experiments and evaluation on two well-known textual datasets which are publicly available and can be used and compared in the future. We needed to run experiments on topic/subject classified datasets. The main difference between the two selected datasets is the number of train / test data samples and the distribution of the topics in the data that let us to also compare the performance of the proposed method in two cases: in the first dataset we have a balanced distribution over the class labels, while in the second dataset the distribution is unbalanced over the same set of topic labels.

3.1   Reuters R8 Subset

The first dataset that we chose to run our experiments on was the well-known R8 subset of the Reuters-21578 collection (from UCI machine learning repository[2]), a typical text classification dataset benchmark. The source document collection was downloaded from the CSMining Group's datasets[3]. The data includes the 8 most frequent classes of Reuteres-21578; hence the topics that will be considered as class labels in our experiments are "acq, crude, earn, grain, interest, money, ship, and trade".

In order to follow the Sebastiani's convention [18], we also call these sets R8. In addition to R8, the R10 subset was used by some researchers and it contains 10 classes, as the name indicates). The only difference between R10 and R8 is that the classes "corn" and "wheat", which are intimately related to the class "grain" were removed. The distribution of documents per class in the R8 subset is shown in the Table 1.

---

[2] http://archive.ics.uci.edu/ml/index.html
[3] http://csmining.org/index.php/data.html

Table 1. Class distribution of training and test subsets for R8

Class	# of train documents	# of test documents	Total # of documents
Acq	1596	696	2292
Earn	2840	1083	3923
Grain	41	10	51
Interest	190	81	271
Money-fx	206	87	293
Ship	108	36	144
Trade	251	75	326
Crude	253	121	374
Total	5485	2189	7674

According to the numbers from Table 1, the baseline of any classification experiment over this dataset may be considered as 51%, for a trivial classifier that puts everything in the most frequent class, which is "earn".

3.2   Reuters Transcribed Subset

The second dataset that we chose for our experiments is the Reuters Transcribed Subset. This is a selection of 20 files from each of the 10 largest classes in the Reuters-21578 collection. The data includes 10 directories labeled by topic name, each containing 20 files of transcriptions in that class (except for the 'trade' directory, which has 21 files). The topics that will be considered as class labels in our experiments are "acq, crude, earn, grain, interest, money, ship and trade" (the "corn" and "wheat" classes were removed for consistency with our first dataset). Since the 8 class labels (topics) are distributed evenly over the Reuters Transcribed Subset data, the baseline for any classification experiment over this dataset may be considered as 12.5%.

4   Method

4.1   Preprocessing

In the preprocessing stage, initially, internet addresses and email addresses were filtered out. Then all the delimiters such as spaces, tabs or newline characters, in addition to some characters like: "\ \r : ( ) ` 1 2 3 4 5 6 7 8 9 0 \' , ; = \ [ ] ; / < > { } | ~ @ # $ \ % ^ & * _ + " have been removed from documents, whereas the expressive characters like: " - . ' ' ! ? " were kept. Punctuations (such as quotes, " ") could be useful for determining the scope of speaker's message. This step considerably reduces the size of feature space and prevents the system from dealing with a large number of unreal tokens[4] as features for our classifiers and LDA estimation/inferences.

---

[4] For example: "aaaaaaa", "buzzzzzzzzzzzzzzz" and "--------a".

Two types of stop-words removal were performed: static stop words removal and corpus based dynamically stop words removal. For the first one, we tokenized the documents individually to be passed to the static stop-word removal step that is based on an extensive list of stop-words which has been already collected specifically for the Reuter corpus. In the second one, additional stop words were determined based on their frequency, distribution and the tokenization strategy over the corpus (i.e., unigrams, bigrams, 3 or 4 grams). We removed tokens with very high frequency relative to the corpus size where those appear in every topical class (i.e., those are almost useless for the topic identification task).

The output of this stage passed to the stemming process through the Snowball stemming algorithm. The output of this stage was formatted for two different purposes; first, an ".arff" file to be used as our training/testing datasets for the classification task; second, a standard text file format ".txt" to be fed to the LDA topical estimation / inference modeling[5].

4.2   LDA Multi-level Topic Modeling

For our goal of topic extraction from the two Reuters subsets, we developed a method based on the original version of LDA presented in [2]. LDA is a generative probabilistic model of a corpus. The basic idea is that the documents are represented as a weighted relevancy vector over latent topics, where a topic is characterized by a distribution over words. We applied and modified the code originally written by Heinrich [12] based on the theoretical description of Gibbs Sampling. A remarkable attribute of the chosen method is that lets a word to participate in more than one topical subset, based on its different senses / usages in its context.

The R8 subset that we used for the LDA topical representation was already passed through the preparation and filtration processes (the pre-processing). In this way, each document is represented by a number of topics in which each topic contains a small number of words inside (i.e., each topic consists in a cluster of words); and each word can be assigned to more than one topic across the entire input data (e.g., polysemous words can be in more than one topic). Therefore, the LDA method assigns some clusters of words as topics, with different weights, for each document.

For example; the following is one topical cluster: {"investment", "success", "plan", "company", "organization", "rate", "market", "sale", "contract", "profit"} extracted by the LDA model estimation process. The number of topics and the number of words inside each topic are two parameters of the method that can be adjusted as needed. In this research, the number of words in each topic has been set to maximum 10 words in each cluster. We observed that increasing the number of words inside each topic decreases the consistency of the topical clusters and make them noisy. These topical clusters will be regarded as dimensions / features of a new vector space, to represent the corpus in a lower-dimensional space. We may have more than one cluster, in which each feature (word) in the feature space belongs to, with some degree of

---

[5] The details are included to help the research becomes replicable.

membership. Since the number of clusters is another parameter of the LDA algorithm, in the first level, we initially choose N = 256 as the number of topical clusters. Then, in order to avoid aggressive topical cluster merging, which may cause the loss of meaningful topics, we set the number of clusters to N = N / 2 in order to obtain more general clusters / topics, and we continued this process at the next levels, for more generalization, until the number of topic clusters reaches the number of expected classes.

By running the LDA topic estimation algorithm, we have a topical cluster membership distribution vector for each document in the corpus. This can be considered as a new representation of our documents in a space with N dimensions at each level. We applied exactly the same process to add another N / 2 dimensions (the number of topical clusters) for each document, and keep adding dimensions for the next levels,   as long as N is greater than or equal the number of corpus classes (8 in our case). According to this procedure, the total number of topic-based representation levels is equal to six on our dataset (six levels).

The final step is to integrate all the extracted features in the six levels as one integrated topical representation of the corpus. This representation then will be compared with the initial BOW representation, and we will also combine the two representations (BOW and multi-level LDA features), in order to increase the discrimination power of the features for text classification task.

4.3   Text Topic Classification

As mentioned in Section 3, the first classification dataset consists of 5485 training and 2189 testing short / medium length documents listed in 8 categorical topics. For the two datasets, we initially applied the TF-IDF method which is a classic method that gives higher weights to terms that are frequent in a document, but rare in the whole corpus. For this representation, we also applied the Snowball stemming algorithm (in order to reduce the feature space). After removing stop-words and stemming, we obtained 17387 words as the BOW feature set of the R8 subset for the general topic classification task. For evaluation of our representation over small number of train/ test data (versus a large number of training and testing sets of the R8 dataset), a set of stop-words removed form of the TF-IDF based representation of Reuters Transcribed Subset was selected. For the BOW representation of the second subset we also applied Snowball stemming algorithm on the feature space which includes 5480 words.

As the second and complementary representation of our two datasets, we used the integrated topical representation vector of the documents calculated using the LDA technique, which produced 504 topical features for the 6 levels (256 + 128 + 64 + 32 + 16 + 8 = 504).

To conduct our empirical performance evaluation of a supervised machine learning algorithm, it is good to have two disjoint subsets: training and testing. Partitioned training and testing datasets can provide reliable results only when we have enough samples to split into large enough subsets for the training and testing processes. This was the case for the R8 subset, as shown in Table 1; and it was not the case for the Reuters transcribed subset. When we do not have enough instances to split into large enough training and testing sets, we evaluate our classification process based on stratified 10-fold cross-validations. This means that we split the entire dataset into 10 almost equal size and class distribution folds, then train a classifier 10 times on a different 9 fold integration of the entire 10-folds, and test it on the 10th one. We did this for both datasets.

Before we integrated the 6 levels of topical representations, we used them individually for our text classification task and noticed that the discriminative power of the individual levels are about 10 to 30 percentage points less than the corresponding BOW representation. In other words, replacing any level of topical representation decreases the classification accuracy, comparing to the BOW representation. We found that the level with 32 topical dimensions was the most discriminative level, but still about 10 percentage points lower than the BOW representation.

We will see in section 5 that the LDA multi-level topical representation solely in many cases is able to outperform that BOW representation. Note that the topical representation of the corpus is a relatively low-dimensional representation of the corpus compared to the BOW high-dimensional representation. This allows more machine learning algorithms to be used in real-world settings with about the same performance. However, the integration of our low-dimensional representation vectors with the conventional BOW representation can boost the classification accuracy in the high dimensional space.

We evaluated the BOW representation and the multi-level LDA representation separately, and then we integrated the two representations. When we integrated them into one representation, we obtain 17891 features (17387 words plus 504 topics) for the first subset and 5984 features (5480 words plus 504 topics) for the second subset.

As part of the supervised machine learning core of the system, we trained a variety of classifiers, in order to evaluate the benefits of the text representation models. As classifiers for our experiments, we chose Support Vector Machines (SVM) because of the usually high performance, Multinomial Naïve Bayes (NB), because of the good performance on text data, and Decision Trees (DT), since the learning model is in a human- comprehensible form.

5   Results and Discussion

We initially ran our selected classification algorithms on the three representations (BOW, topical based and the integrated one) over the R8 dataset. We found the Precision, Recall, F-measure, Accuracy, True Positive (TP) and False Positive (FP) rates (the most common and declarative evaluation measures recently used in most machine learning papers), and calculated their weighted average (Wtd. Avg.) value for our experiments. For example, the weighted average value of "Recall" is calculated by averaging the recall of each class value, weighted by the percentage of that value in the test-set. We conducted the classification evaluation by training on the training set and testing on the separate test set.

Table 2. Comparison of the classification evaluation measures for different representation methods on the split R8 data (5485 training and 2189 testing documents)

Evaluation measure → / Representation/ Classifier used ↓	TP Rate Avg.	FP Rate Avg.	Precision Avg.	Recall Avg.	F1-Measure Avg.	Accuracy %
BOW / SVM	0.933	0.028	0.93	0.933	0.929	93.33
LDA Topics / SVM	0.959	0.016	0.96	0.959	0.959	95.89
LDA+BOW / SVM	0.970	0.011	0.970	0.970	0.970	**97.03**
BOW / NB	0.952	0.013	0.956	0.952	0.952	95.20
LDA Topics / NB	0.946	0.017	0.944	0.946	0.944	94.61
LDA+BOW / NB	0.955	0.01	0.957	0.955	0.956	**95.52**
BOW / DT	0.915	0.037	0.914	0.915	0.915	91.54
LDA Topics / DT	0.918	0.031	0.92	0.918	0.918	91.78
LDA+BOW / DT	0.921	0.032	0.921	0.921	0.92	**92.10**

As a second scenario, we also trained and test the same set of classifiers using 10-fold cross-validation on the whole dataset, to check the stability of the results when training and testing sets are rotationally changed by stratified 10-fold cross-validation (this means that the classifier is trained on nine parts of the data and tested on the remaining part, then this is repeated 10 times for different splits, and the results are averaged over the 10 folds; this is repeated 10 times).

We performed experiments with the three classification algorithms (SVM, Multinomial NB, and DT), for each of the three representations, to check the stability the results. We changed the "Seed", which is a random parameter of the 10-fold cross-validation in order to avoid the accidental "over-fitting".

The evaluation measures calculated over the three representations for the R8 data set are shown in Tables 2 and 3. We report the rate of true positives, the rate of false positives, the precisions, recalls and F-measure averaged over the 8 classes, and the accuracy of the classification task.

According to the best of our knowledge, the accuracy of our integrated representation method on the Reuters R8 dataset is higher than any simple and combinatory representation method from related work, which reports accuracies of 88%-95% [6], [1], [22], while 96% was reached with SVM on a complex representation method based on kernel functions and Latent Semantic Indexing [5].

For the second subset, since the dataset only consisted of 161 short / medium length documents labeled with the 8 classes, we performed our evaluation process using the 10-fold cross-validation method. The calculated values for each of the evaluation measures are shown in the table 4.

We recall that in this dataset the class values are almost evenly distributed in all training and testing subsets (12.5% baseline). Similarly to the results in tables 1 and 2, each of the evaluation measures that appear in table 4 is a macro average of 8 class label values (one class label vs. the other) multiplied by the 10 folds of the 10-fold-cross-validation. For example, the Recall measure is the average of the Recalls values

Table 3. Comparison of the classification evaluation measures for different representation methods on entire R8 data, using 10 fold cross-validation

Evaluation measure → Representation/ Classifier used ↓	TP Rate Avg.	FP Rate Avg.	Precision Avg.	Recall Avg.	F1- Measure Avg.	Accuracy %
BOW / SVM	0.947	0.021	0.947	0.947	0.946	94.67
LDA Topics / SVM	0.959	0.016	0.960	0.959	0.959	95.89
LDA+BOW / SVM	0.973	0.01	0.973	0.973	0.973	**97.29**
BOW / NB	0.949	0.015	0.951	0.949	0.950	94.91
LDA Topics / NB	0.926	0.035	0.926	0.926	0.922	92.57
LDA+BOW / NB	0.946	0.015	0.947	0.946	0.965	**94.59**
BOW / DT	0.904	0.04	0.904	0.904	0.947	90.40
LDA Topics / DT	0.917	0.034	0.918	0.917	0.917	91.73
LDA+BOW / DT	0.919	0.032	0.919	0.919	0.919	**91.88**

Table 4. Comparison of the classification evaluation measures for different representation methods on the Reuters Transcribed Subset, using 10 fold cross-validation

Evaluation measure → Representation/ Classifier used ↓	TP Rate Avg.	FP Rate Avg.	Precision Avg.	Recall Avg.	F1- Measure Avg.	Accuracy %
BOW / SVM	0.580	0.043	0.572	0.582	0.562	58.11
LDA Topics / SVM	0.577	0.045	0.598	0.567	0.579	57.72
LDA+BOW / SVM	0.647	0.037	0.644	0.647	0.643	**64.65**
BOW / NB	0.562	0.049	0.552	0.562	0.542	56.21
LDA Topics / NB	0.538	0.051	0.568	0.537	0.559	54.31
LDA+BOW / NB	0.627	0.041	0.624	0.627	0.623	**62.68**
BOW / DT	0.516	0.054	0.536	0.514	0.517	51.74
LDA Topics / DT	0.565	0.042	0.558	0.551	0.545	56.72
LDA+BOW / DT	0.617	0.041	0.614	0.617	0.613	**61.68**

of the 8 classes, while for each class the value is the macro average of 10 Recall values calculated for the 10 runs of cross-validation. Since the distribution is balanced, averaging over 80 runs, the numbers tend to stay at the mean value of their range.

The best results were obtained with the SVM classifier for the integrated representation BOW and LDA Topics, achieving an accuracy of 97% (51% baseline) for the first data set and 65% (12.5% baseline) for the second subset. Although the small number of documents in each fold may increase the variance of the results from fold to fold, our results on the second subset also confirm the applicability of the presented method even for corpora with a small number of documents.

The improvement over the BOW representation is statistically significant, according to a paired t-test. It is known that the BOW representation is difficult to outperform in topic classification tasks. The fact that our integrated representation succeeded shows that the features that we added bring valuable semantic information.

6 Conclusions and Future Work

We designed and implemented a multi-level text representation method that we tested on the Reuters R8 dataset and on the Reuters Transcribed dataset. Our system applied LDA topical modeling estimation / inference for the topic-based representation purpose. The method was evaluated using Multinomial Naïve Base, SVM and Decision Tree classification algorithms.

We showed that the proposed method is not only useful for dimensionality reduction of the usual high-dimensional representations of the textual datasets (e.g., BOW) without compromising the performance, but also the performance of the classifiers reveals that integrating the proposed representation with the conventional BOW representation can improve the overall discrimination power of the classifiers. However, the quality of the topic-based representation potentially can even be boosted by using a larger textual background resource collected in the same domain in order to build the LDA models.

Our text classification method has the several advantages. In the LDA representation each document is represented by the LDA weighted membership distribution of the topical word clusters, with a classification performance almost similar to that of the BOW representation; hence any other high dimensional vector representation of any collection of documents can be also replaced by its LDA weighted membership distribution, in order to reduce the dimensionality and consequently to deal with the curse of dimensionality without compromising the classification performance. The lower-dimensional representation can be used for any supervised / unsupervised machine learning algorithm that cannot be applied on high-dimensional data.

The performance of the topical-based representation method via the LDA algorithm can simply be improved by adding a source of background data in the same domain.

One limitation of our method is that the current design is based on case insensitive text. The method could be developed based on case sensitive texts for more precise presentation treatment of named entities.

Other directions of future work are to use some resources such as "Wordnet Domain" in our method in order to improve the quality of topical groups extracted via LDA, and to compare the performance of the proposed classification method in informal / unstructured and formal / structured corpora.

References

1. Aggarwal, C.C., Zhao, P.: Towards graphical models for text processing. Knowledge Information Systems (2012), doi:10.1007/ s10115-012-0552-3
2. Blei, D.M., Griffiths, T.L., Jordan, M.I., Tenenbaum, J.B.: Hierarchical topic models and the nested Chinese restaurant process. In: Proceedings of the Conference on Neutral Processing Information Systems, NIPS 2003 (2003)

3. Blei, D.M., McAulie, J.: Supervised topic models. In: Proceedings of the Conference on Neutral Processing Information Systems, NIPS 2007 (2007)
4. Blei, D.M., Ng, A., Jordan, M.: Latent Dirichlet Allocation. Journal of Machine Learning Research 3, 993–1022 (2003)
5. Cardoso-Cachopo, A., Oliveira, A.L.: Combining LSI with other Classifiers to Improve Accuracy of Single-label Text Categorization. In: Proceedings of the First European Workshop on Latent Semantic Analysis in Technology Enhanced Learning, EWLSATEL 2007 (2007)
6. Chen, Y.-L., Yu, T.-L.: News Classification based on experts' work knowledge. In: Proceedings of the 2nd International Conference on Networking and Information Technology IPCSIT 2011, vol. 17. IACSIT Press, Singapore (2011)
7. McCallum, A., Wang, X.: Topic and role discovery in social networks. In: Proceedings of IJCAI 2005 (2005)
8. Firth, J.R., et al.: Studies in Linguistic Analysis. A synopsis of linguistic theory, 1930-1955. Special volume of the Philological Society. Blackwell, Oxford (1957)
9. Griffiths, T.L., Steyvers, M.: Finding scientific topics. Proceedings of the National Academy of Sciences 101(suppl. 1), 5228–5235 (2004)
10. Harris, Z.: Distributional structure. In: Katz, J.J., Fodor, J.A. (eds.) The Philosophy of Linguistics. Oxford University Press, New York (1964)
11. Harris, Z.: Distributional structure. In: Katz, J.J. (ed.) The Philosophy of Linguistics, pp. 26–47. Oxford University Press (1985)
12. Heinrich, G.: Parameter estimation for text analysis. Technical Report (2004), For further information please refer to JGibbLDA at the following link: http://jgibblda.sourceforge.net/
13. Li, W., McCallum, A.: Pachinko allocation: Dag-structured mixture models of topic correlations. In: Proceedings of ICML 2006 (2006)
14. McDonald, S., Ramscar, M.: Testing the distributional hypothesis: The influence of context on judgements of semantic similarity. In: Proceedings of the 23rd Annual Conference of the Cognitive Science Society 2001 (2001)
15. Minka, T., Lafferty, J.: Expectation propagation for the generative aspect model. In: Proceedings of the 18th Annual Conference on Uncertainty in Artificial Intelligence, UAI 2002 (2002), https://research.microsoft.com/minka/papers/aspect/minka-aspect.pdf
16. Mimno, D., Hoffman, M., Blei, D.M.: Sparse stochastic inference for latent Dirichlet allocation. In: Proceedings of International Conference on Machine Learning, ICML 2012 (2012)
17. Pan, X., Assal, H.: Providing context for free text interpretation. In: Proceedings of Natural Language Processing and Knowledge Engineering, pp. 704–709 (2003)
18. Sebastiani, F.: Classification of text, automatic. In: Brown, K. (ed.) The Encyclopedia of Language and Linguistics, 2nd edn., vol. 14, pp. 457–462. Elsevier Science Publishers, Amsterdam (2006)
19. Jones, K.S.: A statistical interpretation of term specificity and its application in retrieval. Journal of Documentation 28(1), 11–21 (1972)
20. Turney, P.D., Pantel, P.: From frequency to meaning: Vector space models of semantics. Journal of Artificial Intelligence Research (JAIR) 37, 141–188 (2010)

21. Wang, X., McCallum, A.: Topics over time: A non-Markov continuous-time model of topical trends. In: Proceedings of ACM SIGKDD conference on Knowledge Discovery and Data Mining, KDD 2006 (2006)
22. Yuan, M., Ouyang, Y.X., Xiong, Z.: A Text Categorization Method using Extended Vector Space Model by Frequent Term Sets. Journal of Information Science and Engineering 29, 99–114 (2013)

A Comparison of h^2 and MMM for Mutex Pair Detection Applied to Pattern Databases

Mehdi Sadeqi[1], Robert C. Holte[2], and Sandra Zilles[1]

[1] Department of Computer Science, University of Regina
Regina, SK, Canada S4S 0A2
{sadeqi2m|zilles}@cs.uregina.ca
[2] Department of Computing Science, University of Alberta
Edmonton, AB, Canada T6G 2E8
holte@cs.ualberta.ca

Abstract. In state space search or planning, a pair of variable-value assignments that does not occur in any reachable state is considered a mutually exclusive (mutex) pair. To improve the efficiency of planners, the problem of detecting such pairs has been addressed frequently in the planning literature. No known efficient method for detecting mutex pairs is able to find all such pairs. Hence, the number and type of mutex constraints detected by various algorithms are different from one another.

The purpose of this paper is to study the effects on search performance when errors are made by the mutex detection method that is informing the construction of a pattern database (PDB). PDBs are deployed for creating heuristic functions that are then used to guide search. We consider two mutex detection methods, h^2, which can fail to recognize a mutex pair but never regards a reachable pair as mutex, and the sampling-based method MMM, which makes the opposite type of error. Both methods are very often perfect, i.e. they exactly identify which pairs are mutex and which are reachable. In the cases that they err that we examine in this paper, h^2's errors cause search to be moderately slower (7%−24%) whereas MMM's errors have very little effect on search speed or suboptimality, even when its sample size is quite small.

1 Introduction

In heuristic state space search and planning, the goal is to find a (preferably optimal, i.e., least-cost) path from a start state to a goal. The state space is usually determined implicitly using a set of operators defining the problem domain. The states in this state space are specified using a set of variables with every variable having a set of possible values. A pair of variable-value assignments that do not co-occur in any reachable state is considered to be a "mutually exclusive pair" or mutex pair for short. For example, consider a representation of the well-known 8-puzzle (3 × 3-sliding-tile puzzle) where each position of the puzzle corresponds to one variable representing the tile in that location (e.g. variable UL might correspond to the upper-left corner position representing the numbered tile or "no

M. Sokolova and P. van Beek (Eds.): Canadian AI 2014, LNAI 8436, pp. 227–238, 2014.
© Springer International Publishing Switzerland 2014

tile" in that position). An example of mutex pair here would be (UL="no tile" and BR="no tile"), where BR is the variable refers to the bottom-right position indicating what is in that position. Since there is only one empty position in this puzzle, these two variables cannot simultaneously have the value "no tile".

Mutex pair detection can be used for improving the performance of planning systems as was first shown in Graphplan [2]. In this planner, if there is no valid plan that allows two actions or two facts at the same level of reasoning, the latter are considered mutex. This motivated the adoption or development of mutex detection methods in other planners [3, 4, 6–10, 13, 14, 17–20, 23, 24].

h^2 is a state-of-the-art method for mutex pair detection in planning [12]. It is a conservative approach to mutex detection, i.e., it might consider mutex pairs as reachable but it will never consider a reachable pair mutex. h^2 is very effective for a large number of domains, for example, it perfectly detects all mutex pairs for Scanalyzer, the Blocks World with Table Positions, and almost all sizes of the Sliding-Tile Puzzle [22]. Despite its effectiveness, this method may fail to detect all mutex pairs. This happens, for example, for a special kind of mutex pair in the 2×2-Sliding-Tile Puzzle and in the so-called stack representation of the Blocks World with n blocks and p table positions. In the 2×2-Sliding-Tile Puzzle, some of the mutex pairs that state that two specific distinct tiles reside in two specific distinct locations are missed by h^2. This only happens in this very small version of the puzzle since such pairs are never mutex in larger versions of the puzzle [22]. In the stack representation of the n-Blocks World with p table positions, h^2 fails to detect all mutex pairs that state, for some $i < n$ and some $j > i$, that $n - i$ blocks are stacked on table position a while a specific block is at height j on table position $b \neq a$. This kind of variable-value assignment pairs is mutex since it entails the existence of more than n blocks [21].

A recently introduced mutex detection method is MMM [21]. Unlike all other approaches, MMM is not conservative, it errs "on the other side." It never considers a mutex pair reachable, but might consider a reachable pair mutex.

Mutex detection has several applications, most notably search space pruning. As an example, consider the search space of regression planning where nodes are partial assignments of state variables and edges are actions. Backward search starts from the goal and stops when we reach an assignment of state variables consistent with the start state. One way to reduce search effort is to remove nodes that contain mutex pairs. Another application of mutex detection is explicating some of the implicit constraints of propositional planning tasks. This is done by translating propositional planning domains into representations with multi-valued state variables and revealing dependencies between variables.

Mutex detection can also be used for improving abstraction-derived heuristic functions. These are functions that estimate the distance-to-goal for any state and can be used for directing search. Abstraction is a popular method for deriving admissible heuristics, i.e., heuristics that never overestimate the true distances and that guarantee A* and IDA* to find optimal solutions. One creates an abstract version of the original state space and uses the true distances in the abstract state space as heuristic values, which are then stored in an efficient data structure called a pattern database (PDB) [5]. The PDB is essentially a table listing abstract states along with the corresponding heuristic values, and is built by moving backwards starting from the abstract goal applying the abstract versions of the operators in a breadth-first manner. Unfortunately, PDBs may include abstract states to which no reachable original state is mapped by the abstraction. Such abstract states are called spurious; they may create short-cuts in the abstract space and thus lower heuristic values [25]. In many cases, spurious states contain mutex pairs. Hence, by removing some of the shortcuts created by spurious abstract states, mutex detection can help to improve the quality of heuristics, and thus to speed up search.

The purpose of this paper is to study the effects on search performance when errors are made by the mutex detection method (h^2 or MMM) that is informing the construction of a PDB. We will see that both methods are often perfect, i.e. they exactly identify which pairs are mutex and which are reachable. We show several cases in which one or the other is not perfect. When h^2 errs (fails to identify some of the mutexes), a moderately negative effect on search speed is observed, but the heuristics remain admissible. MMM's errors may introduce inadmissibility, but in our experiments, suboptimal solutions were rarely produced and, when they did occur, the suboptimality was always very small.

2 Background

In this section we introduce the methods behind h^2 and MMM. We further explain how PDB creation is modified when taking mutex pairs into account.

2.1 Mutex Detection Methods

Most existing mutex detection methods use invariant synthesis in the process of mutex detection. The state-of-the-art mutex detection method h^2 discovers mutex pairs as a special case of "at-most-one" invariants consisting of only two atoms.[1] The h^2 invariant synthesis process can be summarized as follows:

- The (pairs of) atoms of the initial state are reachable.
- An operator is considered applicable if all single atoms and pairs of atoms in its preconditions are reachable.
- An applicable operator turns reachable all its single add effects and all pairs made in one of the following ways:
  - from the add effects of the operators,
  - any add effect combined with any previous reachable atom which is not deleted by the operator and is not mutex with one of its preconditions.

MMM is a sampling-based method that can be summarized as follows [21]:

---

[1] For more background on invariants the reader is referred to [12].

1. Fix a number N of (not necessarily distinct) pairs of variable-value assignments to be sampled.
2. To sample at least N pairs of variable-value assignments, sample N_s states, where N_s is the smallest integer such that $N_s \cdot \binom{m}{2} \geq N$. All pairs of variable-value assignments are extracted from the N_s sampled reachable states.
3. If a pair of variable-value assignments is not seen in this process, it is considered to be a mutex pair.

The number N of pairs to be sampled is determined by Berend and Kontorovich's [1] upper bound on the expected probability mass of the elements not seen after taking N i.i.d. samples from any fixed distribution. A good sampling method of reachable states is also an essential part for the success of this approach. For details of the MMM approach see [21].

A property distinguishing MMM from existing mutex detection methods is that it errs "on the other side." While existing methods never consider a reachable pair mutex, MMM never considers a mutex pair reachable, but might consider a reachable pair mutex. To minimize the risk of missing reachable pairs, it is desirable to use a near-uniform sampling process, so that no pairs of variable-value assignments have too small a probability of being sampled. Unfortunately, there is no known method for sampling states in a way that creates a near-uniform sample of the contained pairs of variable-value assignments, and further one does in general not sample pairs i.i.d. when sampling states. Furthermore, even with uniform sampling there would be no *guarantee* that MMM finds all reachable pairs; however, we would have a guaranteed minimum probability of finding all reachable pairs. The potential for MMM to miss some reachable pairs must be taken into consideration when building PDBs.

2.2 Pattern Databases

A well-known approach for directing search is by using heuristic functions. These are functions that estimate the distance from any given state s to a goal state. A heuristic function is *admissible* if it never overestimate the actual distances. An admissible heuristic function guarantees that heuristic search algorithms like A* and IDA* find optimal solutions when using this function. By creating an abstract version of the original state space and using the true distances in this abstract state space, one can generate an admissible heuristic function. The speed up in search achieved by A* and IDA* depends on the quality of the heuristic function: the closer the heuristic values are to the actual distances, the more efficient A* and IDA* will be.

Creating the h^2-modified PDB is quite straightforward. The only difference to the original PDB creation is that while moving backwards from the abstract goal, an abstract state containing a mutex pair is not added to the open list.

Because MMM errs on the other side, the MMM-modified PDB creation needs more work. The corresponding steps can be summarized as follows:

1. The PDB is built as usual.
2. MMM is used forward from the start state in the original state space to find mutex pairs.
3. An auxiliary PDB is built finding paths to abstract states that do not pass through abstract states containing pairs that are considered mutex by MMM. This means that, in the process of creating this PDB, abstract states containing a mutex pair are not added to the open list.
4. Distances in the original PDB will be replaced by those from the auxiliary PDB as long as the latter are not infinite.

The above explanation is given this way for the ease of understanding the MMM-modified PDB creation process. The actual implementation of the described method is more efficient in the sense that instead of building a separate PDB, the states considered spurious are flagged in the original PDB. As usual, the final PDB will be used for guiding the search. Since MMM might flag some reachable pairs as mutex, the resulting heuristic might be inadmissible. Though this can cause A* or IDA* to find suboptimal solutions, the suboptimality is bounded (additively) by the maximum amount that MMM increases a value in the original PDB (p.219, [11]). It should be noted that a suboptimal path will be found only if *all* the optimal paths are blocked by an MMM mistake.

3   Experimental Setup

The purpose of our experiment is to evaluate the effects on search of h^2 failing to find all mutex pairs and MMM failing to find all reachable pairs. The choice of domains, representations, and abstractions were driven by this goal, and the results reported below are only for those combinations in which one or both of h^2 and MMM are "imperfect" in the sense just described. Such combinations are rather rare; for the majority of the combinations we considered both methods were perfect (this is consistent with the findings described in [21]). All domains were represented using PSVN [16]. The representations used for the domains were intentionally chosen so that many mutex pairs exist, and we deliberately chose versions of each domain that were small enough that (1) problem instances could be solved reasonably quickly with a single PDB; and (2) if we were not able to analytically compute the exact number of reachable pairs, the state space was small enough that we could determine the exact number of reachable pairs by enumerating all reachable states.

3.1   Domains and Representations

The experimented domains and representations are as follows:

Towers of Hanoi. In the n-Disks Towers of Hanoi with p Pegs, a state describes the constellation of n disks stacked on p named pegs. In every move, a disk can be transferred from one peg to another provided that all disks on the destination peg are larger than the moving disk. The goal is to stack up all disks in decreasing order, from bottom to top, on the goal peg from a given start state using the legal moves.

We encode a state of the n-Disk Towers of Hanoi with p pegs as a vector of length pn, where for every peg n binary variables encode whether a disk is on this peg or not. h^2 is perfect on the size of this domain that we tested (14 disks, 4 pegs), but MMM is not. This is different than the encoding used in [21].

Blocks World with Table Positions. In the n-Blocks World with p Table Positions, a state describes the constellation of n blocks stacked on a table with p named positions, where at most one block can be located in a "hand." In every move, either the empty hand picks up the top block off one of the stacks on the table, or the hand holding a block places that block onto an empty table position or on top of a stack of blocks. The goal is to stack up all numbered blocks in increasing order, from bottom to top, on the goal position from a given start state using the legal moves.

We consider two PSVN representations of the n-Blocks World with p distinct table positions. In the first one, called the *top representation* [21], a state vector has $1 + p + n$ components, each containing either the value 0 or one of n block names: (i) the first component is the name of the block in the hand, (ii) the next p components are the names of the blocks immediately on table positions 1 through p, (iii) the last n components are the names of the blocks immediately on top of blocks $a, b, c, \ldots$. In each case, the value 0 means "no block." In the versions of the Blocks World we considered (3 table positions, 9 and 12 blocks), h^2 and MMM are both perfect with this representation.

In the *height* representation, a state is a vector of length $1 + 3n + p$, where (i) the first component is the name of the block in the hand or 0 if the hand is free, (ii) for every block, 3 components encode its table position, its height relative to the table and whether there is any block on top of this block and (iii) the last p components encode whether there is any block on a table position. In the versions of the Blocks World we considered, MMM is perfect with this representation but h^2 is not.

Sliding-Tile Puzzle. In the $n \times m$-Sliding-Tile Puzzle ($n \times m$-puzzle for short), representing an $n \times m$ grid, in which tiles numbered 1 through $n \cdot m - 1$ each fill one grid position and the remaining grid position is blank. A move consists of swapping the blank with an adjacent tile. The goal is to have the numbered tiles in increasing order from top left corner to bottom right corner with the blank tile in the bottom right position.

In the *standard representation* of this puzzle, states are vectors of length $n \cdot \ell$, where each component corresponds to a grid position and represents the number of the tile in this position (B, if the position is blank). In the *dual representation*, a vector component corresponds to either the blank or one of the tiles. The value of a vector component represents the grid position at which the corresponding tile is located. In the version of the Sliding-Tile puzzle we considered (3×4), h^2 and MMM are both perfect with both of these representations.

Scanalyzer. In the n-Belt Scanalyzer, a state describes the placement of n plant batches on n conveyor belts along with information indicating which batches have been "analyzed." (For a detailed description of this domain, see [15].) In a *rotate* move, a batch can be switched from one conveyor belt in the upper half and vice versa. In a *rotate-and-analyze* move, a batch can simultaneously be transferred and analyzed from the topmost conveyor belt to the bottommost one while the batch at the bottommost conveyor belt is moved to the topmost one without any change to its "analyze" state. Once a batch is analyzed, it will remain analyzed henceforward.

In the PSVN representation of the n-Belt Scanalyzer [15] (for even n), a state is a vector of length $2n$ in which each belt corresponds to two components: the name of the batch on that belt and a flag indicating whether that batch is analyzed. h^2 and MMM are both perfect on the size of this domain that we considered (12 belts), although prior research [21] shows that MMM is not perfect on sufficiently large sizes of this domain.

Barman. We use the multi-valued representation derived by Fast Downward's preprocessing algorithm from `barman-opt11-strips_prob01-003.pddl`. Neither h^2 nor MMM is perfect on this domain.

Transport. We use the multi-valued representation derived by Fast Downward's preprocessing algorithm from `transport-opt08-strips_prob01.pddl`. MMM is perfect on this domain but h^2 is not, although it is perfect on the larger versions we tried.

3.2 Mutex Calculation

In our experiments, MMM and h^2 each calculate their set of reachable pairs once for each domain, not once for each start state used for testing. All the domains except Barman and Transport have invertible operators. For these, the calculation of reachable pairs began at the goal state. For Transport and Barman this calculation began at the start state in the PDDL problem definition file.

3.3 MMM's Sampling Method

The sampling method used for MMM is a simple random walk modified to cope with non-invertible operators. While doing a random walk, when we generate a child state, if there is no operator that generates its parent when applied to the child, we add this child state, along with the parent, to a list of states that do not return to their parents. When generating a node in the random walk process, we check if a state belongs to this list. If it does, the parent of this state is also considered as a child in generating the next state of the random walk.[2] The length of the random walk was equal to the sample size N_s, as determined by the method described in [21] with $p = 0.00001$.

4 Experimental Results

Our experiments involve building a PDB based on the mutexes discovered by one of the methods (h^2 or MMM) and then solving a set of problem instances with

---

[2] Although this is not the most efficient approach for this purpose, it is enough for the purpose of this study.

IDA* using that PDB. For h^2, optimal solutions will be found so we measure search performance in terms of the number of nodes expanded compared to the number that would have been expanded if all mutex pairs had been known when the PDB was built. These results are presented in the first subsection below. For MMM, we measure the suboptimality of the solutions found in addition to the search performance. These results are presented in the second subsection below.

The problem instances (start states) used for evaluation are generated as follows. For Barman and Transport, the start states were chosen uniformly at random from the set of states reachable from the start state in the PDDL problem definition file (the state spaces were small enough that this set of reachable states could be enumerated). For the other domains, we generated start states uniformly at random using domain-specific knowledge.

4.1   Effect of Mutexes Missed by h^2

The smallest size of the Transport domain (the only size we tried on which h^2 is imperfect) has one state variable that can take on 5 values (0...4). We defined a domain abstraction that maps two of these values (0 and 4) to the same value (0). The average number of nodes expanded over 100 start states is 52 using the h^2-based PDB, compared to 46 for a PDB based on all mutex pairs. This difference represents a 13% reduction in search speed.

In the Barman representation there are 62 state variables, of which we projected out all except for variables 1, 4, 7, 10, 11, 15, 18, 19, 23, 27, 32, 36, 40, 41, 42, 43, 47, 50, 54, 58, and 62. If IDA* is run with the PDB based on this abstraction on our 100 start states, it expands $179,963,835$ nodes on average. The PDB based on the mutex pairs that h^2 finds reduces this only slightly, to $179,863,552$ nodes. When knowing all the mutex pairs, only $145,221,715$ nodes would be expanded, so the mutexes h^2 misses cause IDA* to be 24% slower.

In the height representation of the Blocks World with 9 blocks and 3 table positions we evaluate ten abstractions that project out 17-19 variables from the state representation. Table 1 compares the number of nodes expanded using the h^2-derived PDB with the PDB based on all mutex pairs. The numbers shown are an average over 1,000 problem instances with an average solution length of 37.073. The "Ratio" column shows that IDA* is between 7% and 17% slower because of the mutex pairs missed by h^2.

4.2   Effect of the Reachable Pairs Missed by MMM

In the Towers of Hanoi with 14 disks and 4 pegs there are $6,188$ reachable pairs and MMM found $5,783$ of them with the sample size it computed ($147,151$ states). The abstraction we used projected out 4 variables, the ones indicating whether or not disks 3, 7, 10, and 12 are on the goal peg (peg 4). Using the PDB based on this abstraction (which contained no spurious states) IDA* expanded $407,939,036$ nodes on average over 100 problem instances with an average solution length of 84.93. Using the MMM-based PDB gives slightly better results—always optimal solutions are found and $405,102,108$ nodes are expanded.

Table 1. Comparison of the h^2-based PDB with the PDB based on all mutex pairs for the Blocks World with 9 blocks and 3 table positions, in terms of the average number of nodes expanded by IDA*

#	h^2 mutexes	all mutexes	Ratio
1	49,713,497	46,330,261	1.07
2	46,383,586	43,370,655	1.07
3	24,517,691	22,781,118	1.08
4	45,702,201	41,787,185	1.09
5	41,955,106	39,756,764	1.10
6	32,572,769	29,487,744	1.10
7	17,508,997	15,725,339	1.11
8	53,238,425	48,043,270	1.11
9	16,599,782	14,641,797	1.13
10	28,087,835	24,006,009	1.17

Table 2. Results on MMM-based PDBs for the Sliding-Tile puzzle, over various sample sizes used by MMM

#Samples	# Pairs Found	Avg. Solution Length	# Nodes Expanded
5,000	7,743	34.987	8,260,410
7,500	8,306	35.381	4,034,902
10,000	8,679	34.871	4,721,158
12,500	8,798	34.743	4,900,564
15,000	8,829	34.737	1,873,844
20,000	8,851	34.737	4,876,682

In the Barman domain there are $8,546$ reachable pairs of which MMM found $7,834$ with the sample size it computed ($171,566$ states). The abstraction we used is the same as that used for h^2 (see above). Using the PDB based on this abstraction, IDA* expanded $179,963,835$ nodes on average over the same problem instances used for h^2. Using the MMM-based PDB gives exactly the same results—all solutions found are optimal and the same number of nodes is expanded. Using a PDB based on all mutex pairs only $145,221,715$ nodes would have been expanded.

To get additional data on the effect of MMM's mistakes we severely reduced the sample size it was given on two domains where the sample size it computed was sufficient to find all reachable pairs. The first such experiment was with the dual representation of the 3×4 Sliding-Tile Puzzle. The abstraction we used projects out tiles 1, 2, 6, 7, 8, 10 and 11. The optimum average solution length over 1,000 problem instances is 34.737. With the PDB for this abstraction IDA* expands $12,891,609$ nodes on average, and with the PDB based on all mutex pairs it expands $4,876,682$ nodes. There are $8,856$ reachable pairs in this domain. Table 2 shows, for various sample sizes, the number of reachable pairs MMM found in a sample of that size, and the average solution length and the number of nodes expanded using the MMM-based PDB for that sample size. For all sample sizes, the solutions are within 2% of optimal and the number of nodes expanded is close to the one using the PDB based on all mutex pairs.

The second such experiment was with the top representation of the Blocks World top with 12 blocks and 3 table positions. The abstraction we used mapped constant 0 and 1 to 0, constants 2, 3, and 4 to 1, constants 5 and 6 to 2, constants

7, 8, and 9 to 3, and constants 10, 11, and 12 to 4. The optimum average solution length over 100 problem instances is 50.02. With the PDB for this abstraction IDA* expands $338,837,610$ nodes on average, and with the PDB based on knowing all mutex pairs it expands $219,554,160$ nodes. There are $16,744$ reachable pairs in this domain. Table 3 shows, for a variety of sample sizes, the number of reachable pairs MMM found in a sample of that size, and the average solution length and the number of nodes expanded using the MMM-derived PDB for the given sample size. Even for small sample sizes, the solutions found are optimal and the number of nodes expanded is close to the number expanded using the PDB based on all mutex pairs. The average heuristic values over $1,000$ problem instances are also shown in this table. Although the inadmissibility of the heuristic due to missing reachable pairs can cause A* or IDA* to find suboptimal solutions (see the second row of the table), the suboptimality will be bounded (p.219, [11]).

Table 3. Results on MMM-based PDBs for the Blocks World with 12 blocks and 3 positions in top representation, over various sample sizes used by MMM. The average heuristic value of the original PDB over $1,000$ problem instances is 23.214.

#Samples	# Pairs Found	Avg. Solution Length	# Nodes Expanded	Average Heuristic
2,500	6,769	50.02	334,008,819	23.246
5,000	10,504	50.26	292,895,093	23.744
7,500	12,137	50.02	245,575,442	23.770
10,000	13,212	50.02	232,661,738	23.896
15,000	14,870	50.02	230,423,751	23.912
20,000	15,860	50.02	223,020,627	23.956
25,000	16,414	50.02	221,228,345	23.966
30,000	16,493	50.02	219,554,160	23.968
35,000	16,539	50.02	220,180,915	23.956
40,000	16,654	50.02	219,554,160	23.968
45,000	16,700	50.02	219,554,160	23.968
50,000	16,726	50.02	219,554,160	23.968

As a final example, consider a second abstraction of the Blocks World with 12 blocks and 3 table positions in top representation, one that contains no mutexes. The abstraction we used mapped constants 0, 1 and 2 to 0, constants 3, 4, and 5 to 1, constants 6, 7 and 8 to 2, constants 9, 10, 11, and 12 to 3. With the PDB for this abstraction IDA* expands $11,315,347,373$ nodes on average. There are $16,744$ reachable pairs in this domain. Table 4 shows, for a variety of sample sizes, the number of reachable pairs MMM found in a sample of that size, and the average solution length and the number of nodes expanded using the MMM-derived PDB for the given sample size. The average heuristic values over $1,000$ problem instances are also shown in this table. With every sample size, the solutions found are optimal. In some cases when the sample size is very small, the number of nodes expanded is less than the number expanded using the original PDB. This means that missing reachable pairs by MMM does not necessarily yield suboptimality; we can even gain a speed-up in search without sacrificing solution quality. As mentioned before, the inadmissibility caused by missing reachable pairs can make A* or IDA* find suboptimal solutions. However, the missing pairs do not cause any suboptimality in solution length in the 100 problem instances experimented here.

Table 4. Solution length, number of nodes expanded, averaged over 100 problem instances and average heuristic value of the MMM-based PDB over 1,000 problem instances of the Blocks World with 12 blocks and 3 table positions in top representation. The average heuristic value of the original PDB over 1,000 problem instances is 18.058.

#Samples	# Pairs Found	Avg. Solution Length	# Nodes Expanded	Average Heuristic
2,500	6,769	50.02	10,089,469,726	18.222
5,000	10,504	50.02	10,934,356,821	18.120
7,500	12,137	50.02	10,996,998,475	18.136
10,000	13,212	50.02	11,315,347,373	18.058
15,000	14,870	50.02	11,267,254,232	18.058
20,000	15,860	50.02	11,315,347,373	18.058
25,000	16,414	50.02	11,315,347,373	18.058
30,000	16,493	50.02	11,315,347,373	18.058
35,000	16,539	50.02	11,315,347,373	18.058
40,000	16,654	50.02	11,315,347,373	18.058
45,000	16,700	50.02	11,315,347,373	18.058
50,000	16,726	50.02	11,315,347,373	18.058

5 Conclusions

The purpose of this paper was to study the effects on search performance when errors are made by the mutex detection method that is informing the construction of a pattern database (PDB). We have considered two mutex detection methods, h^2, which can fail to recognize a mutex pair but never regards a reachable pair as mutex, and MMM, which makes the opposite type of error. Both methods are very often perfect, i.e., they exactly identify which pairs are mutex and which are reachable. In the cases that they err that we have examined in this paper, h^2's errors cause search to be moderately slower (7%−24%) than it otherwise would be, but its PDB is guaranteed to be an admissible heuristic. MMM's errors can cause its PDB to be an inadmissible heuristic but in our experiments its errors rarely caused a non-optimal solution to be found and when they did the solutions were extremely close to optimal. MMM's errors were expected to speed up search but this was not observed; they had little effect on search speed. Our experiments also showed that MMM can perform very well with quite small sample sizes.

Acknowledgements. This work was supported by the Natural Sciences and Engineering Research Council of Canada (NSERC).

References

1. Berend, D., Kontorovich, A.: The missing mass problem. Stat. and Prob. Lett. 82, 1102–1110 (2012)
2. Blum, A.L., Furst, M.L.: Fast planning through planning graph analysis. Artif. Intell. 90(1), 1636–1642 (1995)

3. Bonet, B., Geffner, H.: Planning as heuristic search: New results. In: Biundo, S., Fox, M. (eds.) ECP 1999. LNCS (LNAI), vol. 1809, pp. 360–372. Springer, Heidelberg (2000)
4. Chen, Y., Huang, R., Xing, Z., Zhang, W.: Long-distance mutual exclusion for planning. Artif. Intell. 173(2), 365–391 (2009)
5. Culberson, J., Schaeffer, J.: Pattern databases. Comput. Intell. 14(3), 318–334 (1998)
6. Edelkamp, S., Helmert, M.: MIPS: The Model-Checking Integrated Planning System. AI Magazine 22(3), 67–72 (2001)
7. Fox, M., Long, D.: The automatic inference of state invariants in TIM. J. Artif. Intell. Res. 9, 367–421 (1998)
8. Gerevini, A., Saetti, A., Serina, I.: Planning through stochastic local search and temporal action graphs in lpg. J. Artif. Int. Res. 20, 239–290 (2003)
9. Gerevini, A., Schubert, L.K.: Discovering state constraints in DISCOPLAN: Some new results. In: AAAI/IAAI, pp. 761–767 (2000)
10. Gerevini, A., Schubert, L.: Inferring state constraints for domain-independent planning. In: AAAI/IAAI, pp. 905–912 (1998)
11. Harris, L.R.: The heuristic search under conditions of error. Artificial Intelligence 5(3), 217–234 (1974)
12. Haslum, P.: Admissible Heuristics for Automated Planning. Linköping Studies in Science and Technology: Dissertations. Dept. of Computer and Information Science, Linköping Univ. (2006)
13. Haslum, P., Bonet, B., Geffner, H.: New admissible heuristics for domain-independent planning. In: AAAI, pp. 1163–1168 (2005)
14. Helmert, M.: The Fast Downward planning system. J. Artif. Intell. Res. 26, 191–246 (2006)
15. Helmert, M., Lasinger, H.: The Scanalyzer domain: Greenhouse logistics as a planning problem. In: ICAPS, pp. 234–237 (2010)
16. Hernádvölgyi, I., Holte, R.: PSVN: A vector representation for production systems. Technical Report TR-99-04, Dept. of Computer Science, Univ. of Ottawa (1999)
17. Kautz, H.: SATPLAN04: Planning as satisfiability. In: 4th International Planning Competition Booklet (2004)
18. Kautz, H., Selman, B.: Pushing the envelope: Planning, propositional logic, and stochastic search, pp. 1194–1201. AAAI Press (1996)
19. Penberthy, J., Weld, D.: Temporal planning with continuous change. In: AAAI, pp. 1010–1015 (1994)
20. Rintanen, J.: An iterative algorithm for synthesizing invariants. In: AAAI/IAAI, pp. 806–811 (2000)
21. Sadeqi, M., Holte, R.C., Zilles, S.: Detecting mutex pairs in state spaces by sampling. In: Cranefield, S., Nayak, A. (eds.) AI 2013. LNCS (LNAI), vol. 8272, pp. 490–501. Springer, Heidelberg (2013)
22. Sadeqi, M., Holte, R.C., Zilles, S.: Using coarse state space abstractions to detect mutex pairs. In: SARA, pp. 104–111 (2013)
23. Scholz, U.: Extracting state constraints from PDDL-like planning domains. In: AIPS Workshop on Analyzing and Exploiting Domain Knowledge for Efficient Planning, pp. 43–48 (2000)
24. Vidal, V., Geffner, H.: Branching and pruning: An optimal temporal pocl planner based on constraint programming. Artif. Intell. 170, 298–335 (2006)
25. Zilles, S., Holte, R.C.: The computational complexity of avoiding spurious states in state space abstraction. Artif. Intell. 174, 1072–1092 (2010)

Ensemble of Multiple Kernel SVM Classifiers

Xiaoguang Wang[1], Xuan Liu[1], Nathalie Japkowicz[1], and Stan Matwin[2,3]

[1]School of Electrical Engineering and Computer Science, University of Ottawa, Canada
{Bwang009,Nat}@eecs.uottawa.ca,
Xliu107@.uottawa.ca
[2]Faculty of Computer Science, Dalhousie University, Canada
[3]Institute for Computer Science, Polish Academy of Sciences, Poland
Stan@cs.dal.ca

Abstract. Multiple kernel learning (MKL) allows the practitioner to optimize over linear combinations of kernels and shows good performance in many applications. However, many MKL algorithms require very high computational costs in real world applications. In this study, we present a framework which uses multiple kernel SVM classifiers as the base learners for stacked generalization, a general method of using a high-level model to combine lower-level models, to achieve greater computational efficiency. The experimental results show that our MKL-based stacked generalization algorithm combines advantages from both MKL and stacked generalization. Compared to other general ensemble methods tested in this paper, this method achieves greater performance on predictive accuracy.

Keywords: multiple kernel learning, stacked generalization, ensemble learning.

1   Introduction

Ensembles of models are sets of models whose outputs are combined into a single output or prediction. Stacked generalization [1] is a heterogeneous ensemble method for combining multiple classifiers (base models) by learning a meta-level classifier based on the output of the base-level classifiers, estimated via cross-validation. When we choose the base learners for ensemble methods such as stacked generalization, Kernel based methods [5] [6] such as Support Vector Machines (SVMs) [4] could be one of the choices. Joachims et al. [24] show that combining two kernels is beneficial if both of them use different data instances as support vectors and achieve approximately the same performance. Recent developments on SVMs and other kernel methods have shown the need to consider multiple kernels. This provides flexibility and reflects the fact that typical learning problems often involve multiple, heterogeneous data sources.

The reasoning is similar to combining different classifiers: instead of choosing a single kernel function, it is better to have a set and let an algorithm do the picking or combination step. MKL can be useful in two aspects:

M. Sokolova and P. van Beek (Eds.): Canadian AI 2014, LNAI 8436, pp. 239–250, 2014.
© Springer International Publishing Switzerland 2014

- Since a kernel plays the role of defining the similarity between instances, different kernels correspond to different notions of similarity, and using a specific kernel may be a source of bias. To avoid this we can import a learning method to pick the best kernel or use a combination of a kernel set. In allowing a learner to choose from a set of kernels, a better solution can be found.
- Different kernels may use inputs from different representations. Since different kernels may have different measures of similarity, combining kernels can be done to combine multiple information sources.

There are many outstanding advantages of MKL. However, most MKL algorithms have some limitations in application. For instance, most MKL methods do not consider the group structure between the combined kernels. In multiple kernels learning (MKL), increasing the number of candidate kernels leads to better accuracy, but also increases the training time significantly [3]. Our idea is: if we separate the kernel set into subsets, each subset of kernels can lead to a different combination of kernels. Using an ensemble method such as majority voting or stacked generalization, the outputs of each MKL model can be combined into a single prediction. In this case, fewer kernels are required to be handled by each base MKL learner of this ensemble model than the number of kernels need to be handled by using a single MKL model. Since stacked generalization is an ideal method for parallel computation, using MKL as the base learner, this stacking MKL method can combine more kernels and process more instances in a fixed time than using a single MKL model. MKL performs better than the single kernel method but has a high cost. By enforcing sparse coefficients, MKL also generalizes feature selection to kernel selection. Stacked generalization is an efficient algorithm to combine heterogeneous classifiers and it can also benefit from diversity of data distribution. Our target is to combine advantages from both methods. The idea of combining two ensemble methods is frequently used. For instance, Wolpert et al. [15] present several ways that stacking can be used in agreement with the bootstrap procedure to achieve a further improvement on the performance of bagging for some regression problems. Kai et al. [16] also present ways of combining bagging and stacking for classification.

Contributions:

- In this paper, we combine the multiple kernel learning algorithms with stacked generalization (denoted as SMKL). Experimental results show that this algorithm can benefit from both methodologies.
- Results also show that even without using parallel computation, SMKL can process more instances with the same number of kernels or combine more kernels with the same number of instances than MKL. This makes SMKL adaptable to real world applications.
- We analyze this algorithm and compare it to other ensemble methods. A statistical explanation of how this method works is also given.

Section 2 is about the background and related work. In section 3, we present the algorithm SMKL and present some discussion and related work about it. After these developments, we present an experimental section (section 4) that illustrates the efficiency of our algorithm and some concluding remarks. In section 5, we give a detailed analysis about ensemble methods using SVMs as a base learner. Section 6 is the conclusion followed by references.

2   Related Theory and Work

Stacked generalization [1] is a way of combining multiple models that have been learned for a classification task. This method has also been used for regression and unsupervised learning [14] [15].

In the most common form of stacked generalization, the first step is to collect the output of each model into a new set of data. For each instance in the original training set, this data set includes all models' predictions of that instance for every class along with its true classification. In the second step, this new data is treated as the data for another learning problem. A learning algorithm is employed to solve this problem.

The key idea of MKL is to learn a linear combination of a given set of base kernels by maximizing the margin between the two classes or by maximizing kernel alignment. We can think of kernel combination as a weighted average of kernels and consider the weight $\beta \in \mathbb{R}_+^P$ and $\sum_{m=1}^P \beta_m = 1$, where P denotes the number of weights. Suppose one is given n $m \times m$ symmetric kernel matrices $K_j, j = 1, \ldots, n$, and m class class labels $y_i \in \{1, -1\}, i = 1, \cdots, m$. A linear combination of the n kernels under an ℓ_1 norm constraint is considered:

$$K = \sum_{j=1}^n \beta_j K_j, \beta \geq \emptyset, \|\beta\|_1 = 1 \tag{1}$$

where $\beta = (\beta_1, \ldots, \beta_n)^T \in \mathbb{R}^n$, and $\emptyset$ is the n dimensional   vector of zeros. Geometrically, different scaling of the feature spaces leads to different embeddings of the data in the composite feature space. The goal of MKL is to learn the optimal scaling of the feature spaces and maximize the so-called "separability" of the two classes in the composite feature space.

In kernel methods, the choice of a kernel function is critical, since it completely determines the embedding of the data in the feature space. Ideally, this embedding should be learnt from the training data. In practice, a simplified version of this very challenging problem is often considered: given multiple kernels capturing different "views" of the problem, an "optimal" combination of them must be learned.

Lanckriet et al. [9] have proposed to use the soft margin of SVM as a measure of separability, that is, to learn the weight β by maximising the soft margin between the two classes. Bach et al. [10] have reformulated the problem and then proposed a SMO algorithm for medium-scale problems. Cortes et al. [22] discuss the suitability of the 2-norm for MKL. In their paper they conclude that using the 1-norm improves the performance for a small number of kernels, but not for a large number of kernels. Meanwhile, the 2-norm increases the performance significantly for larger sets of candidate kernels and never decreases it.

The performance improvement of MKL comes at a price. Learning the entire set of models and then combining their predictions is computationally more expensive than learning just one simple model. The computational complexity of MKL is very high for two major reasons: 1). Similar to normal kernel based methods, MKL needs to compute kernel functions for each sample-pair over the training set; 2). MKL needs to optimize the classifier parameters and kernel weights in an alternative manner, thus learning global optimal parameters would incur intensive computation. More specifically, MKL that use optimization approaches to learn combination parameters have high computational complexity, since they are generally modeled as a semi definite programming (SDP) problem, a quadratically constrained quadratic programming (QCQP) problem, or a second-order cone programming (SOCP) problem. MKL can also be modeled as a semi-infinite linear programming (SILP) problem [10], which uses a generic linear programming (LP) solver and a canonical SVM solver in the inner loop. This method is more efficient than previous methods [10], but the computational complexity is still very high.

In recent years, there has been an effort made to reduce the computational complexity of SVM algorithm [25] [26]. For MKL, Chen et al. [27] have proposed a method by dividing the global problem with multiple kernels into multiple local problems, each of which is optimized in a local processor with a single kernel. In this paper, we present an alternative method by combining stacked generalization with MKL. The following section will give the details of this framework.

3   Stacked Generalization on Multiple Kernel SVMs

Here we gave the detailed steps of the stacked generalization base on MKLs method as below:

- Step1: Given a data set $S = \{(y_n, x_n), n = 1, ..., N\}$, where y_n is the class value and and x_n is a vector representing the attribute values of the nth instance, randomly split the data into J almost equal parts $S_1, ..., S_J$. Define S_j and $S^{-j} = S - S_j$ to be the the test and training sets for the jth fold of a J-fold cross-validation.

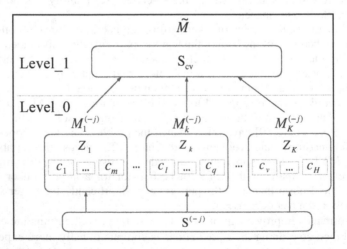

Fig. 1. This figure illustrates the j-fold cross-validation process in level-0; the level-1data set S_{cv} at the end is used to produce level-1 model $\tilde{M}$

- Step 2: Instead of choosing K learning algorithms $\{Z_1, Z_2, \ldots, Z_K\}$ directly as the level-0 generalizers, here we choose H number of kernels $C = \{c_1, c_2, \ldots, c_H\}$. Divide this set into K groups. Call the Multi-kernel algorithm as introduced in section II.b with each group of kernels to build the base learning algorithms $\{Z_1, Z_2, \ldots, Z_K\}$. Invoke the kth algorithm on the data in the training set S^{-j} to induce a model M_k^{-j}, for k=1,...,K. These are called level-0 models. For each instance x_n in S_j, the test set for the jth cross-validation fold, let Z_{kn} denote the prediction of the model M_k^{-j} on x_n . At the end of the entire cross-validation process, the data set assembled from the outputs of the K models is: $S_{CV} = \{(y_n, Z_{1n}, \ldots, Z_{Kn}), n = 1, \ldots, N\}$. These are the level-1 data.
- Step3: Use the same Multi-kernel algorithm as in step 2 with all the kernels of $C = \{c_1, c_2, \ldots, c_H\}$ to be the learning algorithm V^m as the level-1 generalizer. A model $\tilde{M}$ can be derived for y as a function of $(Z_1, \ldots, Z_K)$. This is the level-1 model. To complete the training process, the final level-0 models M_k, k=1,...,K, are are derived using all the data in S.Now let us consider the classification process, which uses the models M_k, k=1,...,K, in conjunction with $\tilde{M}$. Let z_k denotes the prediction output vector of function Z_k. Given a new instance, models M_k produce a vector $(z_1, \ldots, z_K)$. This vector is input to the level-1 model $\tilde{M}$, whose output is the final classification result for that instance.

This algorithm is presented in Fig.2 and Fig.1 gives its workflow.

In this SMKL algorithm, we choose $MKL(C_k)$ as the level-0 generalizers instead of of using common classifiers and we use $MKL(C)$ as the level-1 generalizer. In the next sections we will discuss the relationship between the attributes used in the Meta learning set and the Meta learning algorithm used for learning the Meta model.

Input: Learning set S; Number of folds for meta-data generation J; Meta learning algorithms V^m; Base learning algorithms $\{Z_1, Z_2, \ldots, Z_K\}$; Kernel set $C = \{c_1, c_2, \ldots, c_H\}$; And Multi-kernel Function $MKL()$.
Output: Ensemble E
$E = \emptyset$
$\{S_1, S_2, \ldots, S_J\} = SplitData(S, J)$
$L^m = \emptyset$
$\{C_1, C_2, \ldots, C_K\} = SplitSet(C, K)$
for k= 1 to K **do**
  $Z_k = MKL(C_k)$
end for
$V^m = MKL(C)$
for j= 1 to J **do**
  for k= 1 to K **do**
  $M_k^j = Z_k(S - S_j)$
  end for

$$L_j^m = \bigcup_{x_i \in S_j} \{(M_1^j(x_i), M_2^j(x_i), \dots M_K^j(x_i), y_i)\}$$

```
end for
```

$$L^m = \bigcup_{j=1}^{J} L_j^m$$
$$M^m = V^m(L^m)$$
$$\{M_1, M_2, \dots M_K\} = \{Z_1(S), Z_2(S), \dots, Z_K(S)\}$$
$$E = (\{M_1, M_2, \dots M_K\}, M^m)$$

```
return E
```

Fig. 2. Algorithm of SMKL

4   Experimental Result

In this section, we validate the usefulness of the proposed stacked generation on MKL (SMKL) with experimental evidence on datasets.

Table 1. Datasets details

Datasets	Table Column Head			
	dim	*n_pts*	*n_negative*	*n_positive*
Abalone(6,12)	8	526	259	267
Abalone(9,10)	8	1323	689	634
Balance-scale(1,3)	4	576	288	288
CMC(1,3)	9	1140	629	511
Glass(1,2)	9	146	70	76
Heart-statlog	13	270	150	120
Ionosphere	34	351	126	225
Liver-disorders	6	345	145	200
Monk3	6	122	62	60
Sonar	60	208	97	111
Tae(1,2)	5	99	49	50
Vehicle(1,2)	18	429	212	217

Table 2. Kernels details

1)Gaussian(2.1,1)	2)Polynomial(1)	3)Sigmoid(0.1)
4)Exponential(10,10)	5)Spherical(10)	6)Gaussian(20,1)
7)Circular(1)	8)Gaussian(100,10)	9)InverseMultiQuadric(1)
10)Gaussian(10,10)	11)T-Student(1)	12)Gaussian(1,1)
13)Linear	14)Spline	15)Chi-square(1)
16)RationalQuadratic(10)	17)Polynomial(10)	18)HistogramIntersection
19)ANOVA(1)	20)Distance(1)	21)Spherical(1)
22)Wavelet(1,1)	23)Sigmoid(0.01)	24)Polynomial(0.1)
25)Polynomial(2)	26)Gaussian(10,1)	27)Cauchy(1)
28)RationalQuadratic(1)	29)T-Student(0.1)	30)InverseMultiQuadric(10)

We use twelve real-world datasets from the UCI Repository of machine learning databases [7]. Details of these datasets are given in Table 1 while "dim" denotes feature number and "n_pts" denotes instance number. The numbers in the parentheses beside the name of the datasets are the classes' numbers which are chosen from the original datasets which is multi-class. For example, Abalone (6, 12) denotes the class No. 6 and No. 12 which are chosen from the original Abalone dataset.

For the MKL method, Sonnenburg et al.'s algorithm [11] is used in our experiment so the multi-kernel method is modeled as a semi-infinite linear programming (SILP) problem for large scale MKL problem. To get robust performance on all kinds of datasets, 2-norm MKL algorithm [22] is used in our experiment. All experiments use 10 folds cross validation.

For stacked generalization and SMKL, we choose j=10 for the level_0 j-fold inner cross-validation process. 30 different kernels are chosen for MKL, SMKL and ensemble methods using single kernel SVMs. Table 2 gives the details of these kernels. The format of these kernels is kernel_name (gamma, cost). For all kernels the epsilon is 1e-5, and the coef0 is 0. For SMKL and Voting+MKL, we randomly divide the kernel set into a different number of groups and repeat this procession ten times for every cross validation to get the experimental result.

4.1 Comparing SMKL with Other Ensemble Methods

In this experiment, we first compare SMKL with all its base learners. Among the results, kernel number 18 (HistogramIntersection kernel) in the kernel list (TABLE II.) is the best of all base kernels using Nemenyi's post-hoc test [23] method. Then we compare SMKL with other ensemble methods. Table 3 gives the result. SVM(*best*) denotes using svm with the best kernel in our kernel list. Boosting_best denotes Adaboost [13] using SVM [3] with the kernel number 18 in our kernel list. Bagging_best denotes Bagging [12] using SVM [5] with kernel number 18 in our kernel list; Voting (30 kernels) denotes Majority voting on the 30 single kernel SVMs; SSVM (30 kernels) denotes stacked generalization on the 30 single kernel SVMs, MKL (30 kernels) denotes MKL using 30 kernels. Voting+MKL denotes dividing the kernel sets into different groups randomly and then using MKL (30 kernels) on each group to generate the base models; finally majority voting is used to combine the generated base models.

As Nemenyi's post-hoc test [23] is a non-parametric statistical test for multiple classifiers and multiple domains, we performed this test on the results in Table 3. We rank the accuracies for each domain with different classifiers using the following formula to calculate the q value between different classifiers [23].

$$q_{i,j} = \frac{\overline{R_i} - \overline{R_j}}{\sqrt{\frac{k(k+1)}{6n}}} \tag{2}$$

Where k is the number of classifiers and n is the number of datasets. The sums of the ranks of all tested classifiers are shown in Table 4. Therefore, we conclude that the difference between SMKL and other classifiers are significant and SMKL has better performance than all the other classifiers.

4.2 Experiments about Computational Complexity

In this experiment, we compare the running time of SMKL with MKL. m denotes the number of kernels and n denotes the number of instances. We report running time

Table 3. Experiment result of comparing SMKL with other methods

Datasets	methods name (accuracy is in percentage)							
	SVM (best)	Adaboost _best	Bagging _best	Voting (30kernels)	MKL (30kernels)	SSVM (30kernels)	Voting +MKL	SMKL (30kernels)
Abalone(6,12)	92.21±0.5	49.23±0.0	91.64±1.1	92.02±0.7	92.21±0.9	92.21±0.4	92.21±0.2	92.97±0.9
Abalone(9,10)	60.31±0.3	52.08±0.0	60.39±0.8	59.03±1.1	61.00±2.0	60.85±0.7	61.53±1.2	60.84±0.8
Balancescale	99.83±0.1	54.86±5.1	99.48±0.2	98.26±0.3	99.65±0.1	96.88±0.2	98.44±0.4	99.83±0.1
CMC	68.25±0.3	55.18±0.0	67.72±1.7	67.28±1.0	67.54±0.4	64.30±2.1	66.14±0.7	66.75±1.4
Glass	82.25±1.2	47.94±0.0	82.19±3.7	80.82±0.6	80.82±1.5	78.06±3.4	81.51±2.0	82.19±1.2
Heartstatlog	76.30±4.9	55.56±0.0	79.63±1.8	55.56±0.0	84.44±1.1	83.33±0.3	69.63±3.2	83.33±0.5
Ionosphere	92.59±1.3	76.64±6.9	92.88±0.9	94.87±0.7	94.30±0.6	94.30±1.4	94.87±0.5	96.01±0.7
Liver-disorders	69.86±1.2	42.03±0.0	67.83±4.8	59.71±3.0	72.17±1.0	73.62±0.9	70.43±2.3	71.30±0.4
Monk3	93.44±0.6	50.82±0.0	93.44±0.4	89.34±0.6	89.34±0.5	90.98±0.2	90.16±0.7	91.80±0.3
Sonar	87.02±1.4	87.02±1.1	87.02±0.5	83.17±0.9	87.50±0.1	87.02±0.2	86.06±0.9	87.50±0.3
Tae	72.73±0.4	49.49±0.0	74.75±1.8	74.75±2.7	71.72±3.2	74.75±2.2	76.74±1.8	77.78±0.3
Vehicle	61.07±1.1	61.07±0.5	64.59±0.2	50.58±5.7	64.57±0.4	63.00±1.0	49.65±1.9	65.73±0.4

Table 4. The rank of accuracies of all the datasets and classifiers (lower rank sum score is better)

Datasets	1 SVM (best)	2 Adaboost _best	3 Bagging _best	4 Voting (30kernels)	5 MKL (30kernels)	6 SSVM (30kernels)	7 Voting +MKL	8 SMKL (30kernels)
Abalone(6,12)	4	10	9	8	6	6	6	2.5
Abalone(9,10)	9	10	5	7	2	3	1	4
Balancescale	8	10	3	5	2	6	4	1
CMC	2.5	10	1	4	2.5	7	6	5
Glass	6	10	2.5	6	6	8	4	2.5
Heartstatlog	5	9.5	7	9.5	1	2.5	8	2.5
Ionosphere	8.5	10	6	2.5	4.5	4.5	2.5	1
Liver-disorders	5	10	6.5	9	2	1	4	3
Monk3	1.5	10	1.5	8	8	5	6	4
Sonar	9	4	4	7	1.5	4	6	1.5
Tae	9	10	4	4	6.5	4	2	1
Vehicle	6	5	2	9	3	4	10	1
Rank sums	71	107.5	48	75.5	40	51.5	58	27.5

results (Athlon ™ II X2 240 2.81G processor, 2.75G RAM) in Table 5. SMKL_p denotes SMKL using parallel computing on level_0.

Table 5 shows that even without using parallel computation on level_0, SMKL can process more instances with the same number of kernels or combine more kernels with the same number of instances than MKL. If we apply parallel computation on level_0, SMKL can get more efficiency on learning.

Table 5. Running times in seconds for SMKL and MKL. (Left) Ionosphere data with fixed number of data points n and varying number of kernels m; (Right) Ionosphere data with fixed number of kernels m and varying number of data points n.

Ionosphere, n=351				Ionosphere, m=12			
m	SMKL	SMKL_p	MKL	n	SMKL	SMKL_p	MKL
6	62	50	33	351	170	142	147.5
12	170	142	147	702	432	291	459
24	562	423	738	1404	1497	988	1779
48	1135	997	3393	2808	9927	6519	19488
96	3568	2781	*	5616	36253	27436	*
192	*	*	*	11232	*	*	*

5   Discussions

5.1   Ensemble Methods Using SVMs as Base Learners

Although Evgeniou et al. [20] experimentally found that with accurate parameter tuning, single SVMs and ensembles of SVMs perform similarly, using ensemble methods or multiple kernel methods is always a more stable and robust methodology than using a single model. The statistical reason is that when we learn a model on the learning data, the resulting model can have more or less good predictive performance on these learning data. However, even if this performance is good, this does not guarantee good performance on the unseen data. Therefore, when learning single models, although there are evaluation techniques that minimize this risk, we can easily end up with a bad model. By taking into account several models and averaging their predictions, we can reduce the risk of selecting a very bad model.

In several real-world problems, SVM ensembles are reported to give improvements over single SVMs [17] [18], but a few works also showed negative experimental results about ensembles of SVMs [19].

The same situation occurs in our experiments. As shown in the experiment in section 4.1, stacking multiple SVMs, MKL and SMKL all show better predictive performance than all single SVMs while Voting+SVMs does not improve. Using the relative best single SVMs which is the kernel number 18 (in 30 kernels), bagging improves the predictive performance a little but boosting totally fails on this model.

We can give fundamental explanations to the above mentioned differences using the bias-variance analysis framework: the error of a learning algorithm can be divided into a part due to the functional form used by the algorithm (bias) and a part that is due to the instability of the algorithms (variance). Bagging, stacking and random forests reduce the variance part.

Boosting mainly reduces the bias part, but also reduces the variance portion. It can be viewed as an incremental forward stagewise regression procedure with regularization (Lasso penalty), which maximizes the margin between the two classes, much like the approach of support vector machines [21]. In boosting, when we increase the weights of examples that have not been correctly predicted by previous base models (or decrease the weights of the correctly predicted examples) and learn a new base model, the weight we increase may easily be above what the base learner needs because SVM is not a weak learner. Due to this, the complementary base models we try to generate by learning subsequent models may make the process hardly converge. In this situation, using SVM as the base learner for boosting may lead to worse performance, such as the result shown in our experiment. (Table 3 and Table 4)

Because of its simplicity, Voting allows for some theoretical analysis of its efficiency. However, since it requires neither cross-validation nor level-learning, unless level-0 generalizers perform comparably to one another, Voting cannot be comparable to other ensemble methods as shown in our experiments (Table 3 and Table 4)

Bagging reduces only the variance under the bootstrap assumption. On the other hand, SVMs can be strong, low-biased learners, but this property depends on the proper selection of the kernel and its parameters. If we can identify low-biased base learners with a relatively high unbiased variance, bagging SVM can lower the error [17]. As shown in Table 3 and Table 4, kernel number 18 is selected as the best single SVM from the given kernel set, which means that we identify it as a low-biased base learner using cross-validation. Bagging SVM (kernel number 18) does improve the prediction performance as shown in Table 3 and Table 4.

5.2   The Reason Why SMKL Works

Although the idea of multiple kernel methods is similar to combining different classifiers, to be strict, MKL is not the same as traditional ensemble methods since it combines kernels instead of combining classifiers.

Giorgio et al. [17] analyze the bias-variance decomposition of error in SVMs. The analysis shows that the minimum of the overall error, bias, net-variance, unbiased and biased variance occurs often in different SVM models. These different behaviors of different SVM models could be in principle exploited to produce diversity in multiple kernel learning. As we have already mentioned, different kernels correspond to different notions of similarity of instances and may be using inputs coming from different representations. Thus by combining kernels, it is possible to combine multiple information sources and decrease the bias created by specific kernels. Moreover, by enforcing sparse coefficients, MKL also generalizes feature selection to kernel selection. These are the reasons why MKL is always more robust and has better predictive performance than single kernel classifier. By using MKL models as base learners for heterogeneous ensemble methods such as stacked generalization or voting, we can get the benefit of combining kernels. In addition, the diversity of different MKL models still exists. Since stacked generalization chooses level-1 generalizer to get the benefit of this diversity, using stacked generalization is a better choice than voting as the experiment result shows in Table 3 and Table 4 (compare Voting +MKL with SMKL).

The principle of how SMKL works is similar to Random forests [13], one of the most successful ensemble approaches. Random forests combine two sources of diversity of the base models: Variations in the learning data set (achieved through different bootstrap samples, as in bagging) and a randomized base-level learning algorithm. SMKL also combines multiple sources of diversity of the base models.

5.3   About the Computational Complexity

SMKL is ideal for parallel computation. The construction of each level-0 model proceeds independently; no communication with the other modeling processes is necessary. This feature makes SMKL applicable in the real world. Suppose the average computation time required for a MKL learning algorithm is t, SMKL requires g models and each model employs J-fold cross-validation. Assuming that time t is needed to derive each of the h level-0 models and the level-1 model, the learning time for SMKL is $T = (h(J + 1) + 1)t$. From Table 5 we find that the increase of running time of MKL is much higher than that of SMKL when we increase the number of kernels or instances. Since the experiment in Table 5 is set on one computer, SMKL does not benefit greatly from parallel computation. Higher efficiency can be expected if we implement SMKL on parallel computation with more processors.

6   Conclusions and Future Work

In this paper, we have introduced an approach which uses multiple kernel methods as base learners for stacked generalization process. The experimental results show that this method obtains the advantages from both the multiple kernel learning method and the stacked generalization method. Moreover, we have analyzed and compared many widely used ensemble methods which use kernel methods such as SVMs as base learners. Our future work includes implementation of this method for parallel computation with a large dataset and adopting more kernel functions, and testing the effect of kernel subsets arrangement in our method. Moreover, we will use further bias-variance decomposition analysis on this method.

References

1. Wolpert, D.H.: Stacked Generalization. Neural Networks 5, 241–259 (1992)
2. Kai, M.T., Ian Witten, H.: Issues in Stacked Generalization. Journal of Artificial Intelligence Research 10, 271–289 (1999)
3. Bach, F.R., Lanckriet, G.R.G., Jordan, M.I.: Multiple kernel learning, conic duality, and the SMO algorithm. In: Proceedings of the 21st International Conference on Machine Learning (2004)
4. Vapnik, V.: The Nature of Statistical Learning Theory. Springer (1999)
5. Scholkopf, B., Smola, A., Muller, K.: Kernel principal component analysis. Advances in Kernel Methods: Support Vector Learning, 327–352 (1999)
6. Shawe-Taylor, J., Cristianini, N.: Kernel Methods for Pattern Analysis. Cambridge University Press (2004)
7. Frank, A., Asuncion, A.: UCI Machine Learning Repository. University of California, School of Information and Computer Science, Irvine, CA (2010),
   http://archive.ics.uci.edu/ml
8. Gehler, P.V., Nowozin, S.: Infinite kernel learning. Technical report, Max Planck Institute for Biological Cybernetics (2008)
9. Lanckriet, G., Cristianini, N., Bartlett, P., Ghaoui, L.E., Jordan, M.: Learning the kernel matrix with semidefinite programming. Journal of Machine Learning Research 5, 27–72 (2004)
10. Bach, F., Lanckriet, G., Jordan, M.: Multiple kernel learning, conic duality, and the smo algorithm. In: Proceedings of the 21st International Conference on Machine Learning, pp. 41–48 (2004)
11. Sonnenburg, S., Raetsch, G., Schaefer, C., Scholkopf, B.: Large scale multiple kernel learning. Journal of Machine Learning Research 7, 1531–1565 (2006)
12. Breiman, L.: Bagging predictors. Machine Learning 24(2), 123–140 (1996)
13. Ho, T.K.: The random subspace method for constructing decision forests. IEEE Transactions on Pattern Analysis and Machine Intelligence 20(8), 832–844 (1998)
14. Schapire, R.E.: The strength of weak learnability. Machine Learning 5, 197–227 (1990)
15. Wolpert, D.H., Macready, W.G.: Combining stacking with bagging to improve a learning algorithm. Technical Report SFI-TR-96-03-123, Santa Fe Institute, Santa Fe, New Mexico (1996)
16. Kai, M.T., Witten, H.I.: Stacking Bagged and Dagged Models. In: ICML, pp. 367–375 (1997)

17. Valentini, G., Dietterich, T.G.: Bias-Variance Analysis of Support Vector Machines for the Development of SVM-Based Ensemble Methods. Journal of Machine Learning Research 5, 725–775 (2004)
18. Kim, H.C., Pang, S., Je, H.M., Kim, D., Bang, S.Y.: Pattern Classification Using Support Vector Machine Ensemble. In: Proceedings of the International Conference on Pattern Recognition, vol. 2, pp. 20160–20163. IEEE (2002)
19. Valentini, G., Dietterich, T.G.: Low Bias Bagged Support Vector Machines. In: Fawcett, T., Mishra, N. (eds.) Proceedings of the Twentieth International Conference, Machine Learning, pp. 752–759. AAAI Press, Washington (2003)
20. Evgeniou, T., Perez-Breva, L., Pontil, M., Poggio, T.: Bounds on the Generalization Performance of Kernel Machine Ensembles. In: Langley, P. (ed.) Proc. of the Seventeenth International Conference on Machine Learning, pp. 271–278. Morgan Kaufmann (2000)
21. Friedman, J.H., Hastie, T., Tibshirani, R.J.: Additive logistic regression: a statistical view of boosting. Technical report, Stanford University, Department of Statistics (1998)
22. Cortes, C., Mohri, M., Rostamizadeh, A.: L2 regularization for learning kernels. In: Proceedings of the 25th Conference on Uncertainty in Artificial Intelligence (2009)
23. Japkowicz, N., Shah, M.: Evaluating Learning Algorithms: A Classification Perspective. Cambridge University Press (2011)
24. Joachims, T., Cristianini, N., Shawe-Taylor, J.: Composite kernels for hypertext categorisation. In: Proceedings of the 18th International Conference on Machine Learning (2001)
25. Alham, N.K., Li, M., Liu, Y.: A distributed SVM ensemble for image: Classification and annotation. In: Proceedings of the 9th International Conference on Fuzzy Systems and Knowledge Discovery, pp. 1581–1584. IEEE, Piscataway (2012)
26. Chang, E.Y., Zhu, K., Wang, H.: PSVM: Parallelizing support vector machines on distributed computers. Adv. Neural Inf. Process Syst. 20, 1–8 (2007)
27. Chen, Z.Y., Fan, Z.P.: Parallel multiple kernel learning: A hybrid alternating direction method of multipliers. Knowledge and Information Systems (2013)

Combining Textual Pre-game Reports and Statistical Data for Predicting Success in the National Hockey League

Josh Weissbock and Diana Inkpen

University of Ottawa, Ottawa, Canada
{jweis035,diana.inkpen}@uottawa.ca

Abstract. In this paper, we create meta-classifiers to forecast success in the National Hockey League. We combine three classifiers that use various types of information. The first one uses as features numerical data and statistics collected during previous games. The last two classifiers use pre-game textual reports: one classifier uses words as features (unigrams, bigrams and trigrams) in order to detect the main ideas expressed in the texts and the second one uses features based on counts of positive and negative words in order to detect the opinions of the pre-game report writers. Our results show that meta classifiers that use the two data sources combined in various ways obtain better prediction accuracies than classifiers that use only numerical data or only textual data.

Keywords: machine learning, natural language processing, sentiment analysis, monte carlo method, ice hockey, NHL, national hockey league.

1 Introduction

Sports prediction, especially in ice hockey, is an application in which automatic classifiers cannot achieve high accuracy [1]. We believe this is due to the existence of an upper bound as a result of parity, the difference in skill between the best and worst teams, and the large role of random chance within the sport.

We expand upon our previous work in machine learning to forecast success in a single hockey game [1]. Similar to our previous model, we train a classifier on statistical data for each team participating. In addition, we use pre-game textual reports and we apply sentiment analysis techniques on them. We train a classifier on word-based features and another classifier on the counts of positive and negative words. We then use these individual classifiers and create a meta-classifier, feeding the outputs of the individual classifiers into the second level classifier, in a cascade. We compare several meta-classifiers, one uses a cascade-classifier and the other two use majority voting and the highest confidence of the first level classifiers.

This method returns an accuracy that improves upon our previous results and achieves an accuracy that is higher than "the crowd" (gambling odds) and expert statistical models by the hockey prediction website http://puckprediction.com/.

M. Sokolova and P. van Beek (Eds.): Canadian AI 2014, LNAI 8436, pp. 251–262, 2014.
© Springer International Publishing Switzerland 2014

This application is of interest to those who use meta-classifiers as it is successfully being used in an area that has little academic research and exposure, as well as improves upon the results from traditional approaches of a single classifier.

2   Background

There is little previous work in academic on sports predictions and machine learning for hockey. This is likely because the sport itself is difficult to predict due to the low number of events (goals) a match, and the level of international popularity for ice hockey is mcuh lower than other sports. Those who have explored machine learning for sports predictions have mainly looked at American Football, Basketball and Soccer.

Within hockey, machine learning techniques have been used to explore the attacker-defender interactions to predict the outcomes with an accuracy over 64.3% [2]. Data mining techniques have been used to analyze ice hockey and create a model to score each individual players contributions to the team [3]. Ridge regression to estimate an individual players contributions to his team's expected goals per 60 minutes has been analyzed [4]. Poisson process have been used to estimate rates at which National Hockey League (NHL) teams score and yield goals [5]. Statistical analysis of teams in the NHL when scoring and being scored against on the first goal of the game [6]. Due to the low number of events (goals) they found that the response to conceding the first goal plays a large role in which team wins. Using betting line data and a regression model and it has been found that teams in a desperate situation (e.g., facing elimination) play better than when not playing under such pressures [7].

Other sports have used machine learning to predict the outcome of games and of tournaments. In soccer, Neural Networks have achieved a 76.9% accuracy [8] in predicting the 2006 World Cup by training on each stage of the tournament (a total of 64 games). Neural Networks have also predicted the winners of games in the 2006 Soccer World Cup and achieved a 75% accuracy [9].

Machine learning has been used in American football with success. Neural networks have been employed to predict the outcome of National Football League (NFL) games using simple features such as total yardage, rushing yardage, time of possession, and turnover differentials [10]. Training on the first 13 weeks and testing on the 14th and 15th week of games they achieved 75% accuracy. Neural networks were able to predict individual games [11], at a similar accuracy of 78.6%, using four statistical categories of yards gained, rushing yards gained, turnover margin and time of possession.

Basketball has had plenty of coverage in game and playoff prediction with the use of machine learning. Basketball games can easily have over 100 events a night and this is reflected in the higher accuracies. In prediction of single games, neural networks have predicted at 74.33% [12], naive bayes predicts at 67% [13], multivariate linear regression predicts at 67% [14], and Support Vector Machines predict at 86.75% [15]. In terms of predicting playoff tournaments, Support Vector Machines trained on 2400 games over 10 years and predicted

30 playoff games with an accuracy of 55% (despite his higher accuracy over 240 regular games) [15]. Naive Bayes have been trained on 6 seasons of data to predict the 2011 NBA playoffs [16]. The prediction were that the Chicago Bulls will win the championship, but they were ultimate eliminated in the semi-finals.

In our previous work [1] we explored predicting the outcome of a single game in hockey. We used 14 different statistical data for each team as features. These features included both traditional statistics that are published by the league (e.g., Goals For, Goals Against, Wins, Location etc) and Performance Metrics which are used by hockey analysts (e.g. Offensive Zone Time Estimates, Estimations on the effects of Random Chance, Goals For/Against Rates). After trying a number of machine learning algorithms, our best results came from using a tuned SVM that acheived an accuracy of 59.3%. Further work showed that by using a voting meta-classifier with SVM, NaiveBayes and NeuralNetworks we could increase that accuracy to 59.8%. Using the Correlation-based Feature Subset Selection from Weka [17] we found the most important features to predicting a single game were: Goals For, Goals Against and Location. Traditional statistics outperformed the Performance Metrics in machine learning despite the fact that performance metrics have been shown to be better predictors in the long term.

3   Upper Bounds

We found in our previous experiments that no matter what we tried we were not successful in predicting the NHL with an accuracy higher than 60%. We decided to explore this further and it is our assumption that there is an upper bound that exists in sport predictions that makes it improbable to predict at 100%.

We used a method similar to Burke [18] who looked at prediction within the NFL by comparing observed, theoretical and a mixed-variation win/loss records. His findings conclude that the NFL has an upper bounds of approximately 76%. This seems to hold with the NFL-related research, as the authors have not been able to achieve higher results.

Rather than look at win/loss records we compared the observed win percentages of all teams between the 2005-2006 NHL season (since the last labour lockout) and 2011-2012 (the last full NHL season played) to a number of simulated seasons. The observed standard deviation (St.Dev) of win-percentage (win% — the number of games a team wins in the year that they play) over this time is 0.09.

Next, we simulated an NHL season 10,000 times, using the Monte Carlo method and on each iteration every team was given a random strength. When using the rule that the stronger team always wins ("all skill"), the St.Dev of win% is 0.3. When we changed the rule so that each team has a 50% chance of winning ("all-luck") the St.Dev of win% drops to 0.053. This suggests the observed NHL is closer to an "all-luck" league.

We changed the rule to determine who wins a match by varying the amount of random chance ("luck") and skill is required to win a game. If a randomly generated number is less than the pre-determined luck%, then the game has a 50% chance of being won; otherwise the strong team always wins. We varied the amount of luck and skill to win a game and we found the NHL was most similar to a league that is made up of 24% skill and 76% luck. The results of the various skill/luck Monte Carlo iterations can be seen in table 1, as well as the statistical tests to compare similarities to the observed win%.

Table 1. Monte Carlo Results

Luck	Skill	Theoretical Upper Bound	St.Dev Win%	F-Test p-value
0	100	100.0%	0.3000	4.90×10^{-16}
100	0	50.0%	0.0530	0.029
50	50	75.0%	0.1584	0.002
75	25	62.5%	0.0923	0.908
76	24	62.0%	0.8980	0.992
77	23	61.5%	0.0874	0.894

We can use statistical tests to identify which simulated distribution is most simular to our observed distribution. With a p-value of 0.992 it appears that the simulate league with 24% skill and 76% luck is the most similar to our observed data. To use the similar conclusion as [18], "The actual observed distribution of win-loss records in the NHL is indistinguishable from a league in which 76% of games are decided at random and not by the comparative strength of each opponent." What this means for machine learning is that the best classifier would be able to predict 24% of games correctly, and would be able to guess half of the other 76% of games. This suggests there is an upper bound for prediction in the NHL of 24% + (76%/2) = 62%.

4   Data

For the new experiments that we present in this paper, we used the data from all 720 NHL games in the 2012-2013 NHL shortened season, including pre-game texts that we were able to mine from NHL.com. The text report for each game discusses how the teams have been performing in the recent past and their chance of winning the upcoming game. Most reports are composed of two parts, one for each team. This was the case for 708 out of the 720 games. Since we need to extract separate features for each team, we used only these 708 pre-game reports in our current experiments. An example of textual report for one game can be seen in table 2.

We calculated statistical data for each game and team by processing the statistics after each game from the 2012-2013 schedule. As we learned in our previous work [1], the most important features were Goals Against, Goal Differential and Location. Given the difficulty of trying to recreate some of the performance metrics, we only used these three features in the numerical data classifier.

Table 2. Example of Pre-Game text, pre-processed

Text	Label
There are raised expectations in Ottawa as well after the Senators surprised last season by making the playoffs and forcing the top-seeded Rangers to a Game 7 before bowing out in the first round. During the offseason, captain Daniel Alfredsson decided to return for another season. The Senators added Marc Methot to their defense and Guillaume Latendresse up front, while their offensive nucleus should be bolstered by rookie Jacob Silfverberg, who made his NHL debut in the playoffs and will skate on the top line alongside scorers Jason Spezza and Milan Michalek. "I don't know him very well, but I like his attitude – he seems like a really driven kid and I think he wants to do well" Spezza told the Ottawa Citizen.	Win
Over the past two seasons, Ondrej Pavelec has established himself as a No. 1 goaltender in the League, and while Andrew Ladd, Evander Kane, Dustin Byfuglien and others in front of him will go a long way in determining Winnipeg's fortunes this season, it's the 25-year-old Pavelec who stands as the last line of defense. He posted a 29-28-9 record with a 2.91 goals-against average and .906 save percentage in 2011-12 and figures to be a workhorse in this shortened, 48-game campaign.	Loss

For the text classification experiments we used both traditional Natural Language Processing (NLP) features and Sentiment Analysis features. For the NLP features, after experimenting with a number of possibilities, we represented the text using Term Frequency/Inverse Document Frequency (TF-IDF) values, no stemmer and 1,2 and 3 grams. For the Sentiment Analysis we used the AFINN [19] sentiment dictionary to analyze our text. Other sentiment lexicons (MPQA [20] and Bing Lius [21] lexicon) were explored, but it was the AFINN lexicon that led to the best results in early trials. We computed three features: the number of positive words, the number of negative words and the percentage difference between the number of positive and the number of negative words ($(\#positive\_words - \#negative\_words)/\#words$).

As each pre-game report had two portions of text, one for the home team and one for the away team, we had two data vectors to train on for each game. In total, for 708 games, we had 1416 data vectors; each vector was from the perspective of the home and away team, respectively. The team statistical features were represented as the differentials between the two teams, similar to our method in our previous experiments [1].

5   Experiments

For the first experiment, we tried a cascade classifier. In the first layer, we trained separate classifiers on each of the three sets of features: the numerical features, the words in the textual reports, and the polarity features extracted from the textual reports, until the best results were achieved for each set. A number of Weka algorithms were attempted including MultilayerPerceptron (NeuralNetworks), NaiveBayes, Complement Naive Bayes, Multinomial NaiveBayes, LibSVM, SMO (Support Vector Machine), J48 (Decision Tree), JRip (rule-based learner), Logistic Regression, SimpleLog, and Simple NaiveBayes. The default parameters were used, as a large number of algorithms were being surveyed.

As we had 708 games to train on (and 1416 data sets), we split this up into 66% for training and the other 33% for testing. As each game had two data vectors, we ensured that no game was in both the training and the test set. In this way, when we received the output from all three classifiers in the firstlayer, we knew which game the algorithm was outputting its guess for ("Win" or "Loss") and the confidence of the prediction.

The results from all three classifiers were post-processed in a format that Weka can read and was feed back into the Weka algorithms. The features that were used include the confidence of the classifiers' predictions and the label that was predicted. The labels for each game were either "Win" or "Loss". In the second layer of the cascade-classifier, the outputs from all three classifiers were feed back into the Weka algorithms (six features, two from each algorithm) and the new prediction results decided the final output class. We also used two other meta-classifiers: one chose the output based on the majority voting of the three predictions from the first layer and the other chose the class with the highest confidence.

Results of the first layer can be seen in table 3 and results of the second layer can be seen in table 4. Further details of the two layers of the meta-classifiers are presented in the following sections.

5.1   Numeric Classifiers

The numeric classifier used only team statistical data as features for both teams. As we learned from our previous experiments, the most helpful features to use are cumulative Goals Against and Differential, and Location (Home/Away). For each data vector, we represented the values of the teams as a differential between the two values, for each of the three features.

After surveying a number of machine learning algorithms the best results for this dataset came from using the Neural Network algorithm MultilayerPerceptron. The accuracy achieved on the testing data was 58.57%.

5.2   Word-Based Classifier

After experimenting with a number of Bag-of-Word options to represent the text, we settled on using the text-classifier with TF-IDF, no stemmer and 1,2 and 3 grams. Other options that were analyzed included: Bag-of-Words only, various stemmers and with and without bigrams and trigrams. The best result came from this combination.

In pre-processing the text, all stopwords were removed, as well as all punctuation marks. Stopwords were removed based on the Python NLTK 2.0 English stopword corpus. All text was converted to lowercase.

In a similar fashion to the Numeric Classifier, a number of machine learning algorithms were surveyed. The best accuracy came from using JRip, the rule-based learner, on the pre-game texts for both teams. The accuracy achieved on the same test data was 57.32%, just slightly lower than the numeric classifier.

5.3   Sentiment Analysis Classifier

The third and final classifier in the first level of the cascade-classifier is the Sentiment Analysis Classifier. This classifier uses the number of positive and negative words in the pre-game text, as well as the percentage of positive words differential in the text. These three features were feed into the algorithms in a similar fashion and the highest accuracy achieved was from Naive Bayes at 54.39%, lower than the other two classifiers.

Table 3. First Level Classifier Results

Classifier	Algorithm	Accuracy
Numeric	MultilayerPerceptron	58.58%
Text	JRip	57.32%
Sentiment Analysis	NaiveBayes	54.39%

5.4   Meta-classifier

In the second layer of the cascade-classifier, we fed the outputs from each of the three first-level classifiers. As we separated the testing and training data, we were able to label each game with the confidence of the predicted output from the three classifiers, as well as their actual output label. We then experimented with three different strategies. The first was to feed the data into machine learning classifiers, the second was to pick the output with the highest confidence, and the third was to use a majority vote of the three classifiers.

With the first approach, we surveyed a number of machine learning classifiers in the same fashion as the first layer. The highest accuracy came from the Support Vector Machine algorithm SMO and it was 58.58%.

For the next two approaches, we used a Python script to iterate through the data to generate a final decision and compare it to the actual label. In the first method of picking the choice of the highest confidence, the label of the classifier that had the highest confidence in its decision was selected. It achieved an accuracy of 57.53%. In the second approach, the three generated outputs were compared and the final decision was based on a majority vote from the three classifiers. This method returned an accuracy of 60.25%.

Table 4. Second Level Classifier Results

Method	Accuracy
Cascade Classifier using SVM (SMO)	58.78%
Highest Confidence	57.53%
Majority Voting	60.25%

For comparison, we placed all three features sets for each game into a single feature set and fed it into the same machine learning classifiers to see what accuracy is achieved and to compare it to the cascade-classifier. The results can be seen in table 5.

Table 5. All-in-One Classifier Results

Algorithm	Accuracy
NaiveBayes	54.47%
NaiveBayesSimple	58.27%
libSVM	51.56%
SMO	53.86%
JRip	54.62%
J48	50.20%

6   Results and Discussion

In order to put the results into perspective, we need a baseline to compare against. As each game has two data vectors, for win and for loss, a random choice baseline would have an accuracy of 50%. In hockey, there appears to be a home-field advantage where the home team wins 56% of matches; for our dataset, this heuristic would provide a baseline classifier with an accuracy of 56%. With an upper bound of 62% and a baseline of 56%, there is not a lot of room to see improvement with hockey predictions in the NHL. Other hockey leagues have higher upper bounds of prediction, but we could not find pre-game reports for other leagues to run a similar experiment on.

When analyzing the first level results in the cascade-classifier, the accuracy values are not that impressive. Sentiment analysis does worse than always selecting the home team. Using just the pre-game reports does better than the baseline of just selecting the home team, but does not do as well as the numeric data classifier. The classifier based on numerical features performs the best, and it is comparable with the numerical data classifiers that we tested in our provisional work [1], which used many advances statistics in addition to the ones that we selected for the current experiments.

When we look at the results of the second level of the cascade classifier, we see more interesting results. Using the machine learning algorithms on the output from the algorithms in the first layer, we see a little improvement. When we look at the methods of selecting the prediction with the highest confidence and the majority voting, the results improve even more with majority voting, achieving the best accuracy 60.25%.

It was surprising to see that the all-in-one data set did not do very well across all the algorithms that we had earlier surveyed. None of the accuracies were high; except for Naive Bayes Simple[1], none of these algorithms were able to achieve an accuracy higher than selecting the home team. This means that it was a good idea to train separate classifiers on the different features sets. The intuition behind this was that the numerical data provides a different perspective and source of knowledge for each game than the textual reports.

Overall, we feel confident that this method of a cascade classifier to forecast success in the NHL is successful and can predict with a fairly high accuracy, given the small gap of improvement available between 56% (home field advantage) and 62% (the upper bound).

For more comparison, we contrasted our results to PuckPrediction[2] which uses a proprietary statistical model to forecast success in games in the NHL season, each day, and compares their results to "the crowd" (gamblers odds). So far in the 2013-2014 season, PuckPrediction has made predictions on 498 games and the model has guessed 289 correct and 209 incorrectly (58.03%). The crowd has performed slightly better at 296 correct and 202 incorrect (59.44%). While predicting games in different seasons, our cascade-classifier method has achieved an accuracy that is higher than both of their methods. Additionally, their accuracies continue to suggest that it is improbable to predict at an accuracy higher than the upper bound of 62%, as the two external expert models have not broken this bound.

One interesting issue we discovered is which words are adding the most to the prediction. We looked at the top 20 InfoGain values of the word features, with the results seen in table 6. As we did not remove team or city names from the text, it is interesting to see that 7 of the top InfoGain values were referring to players, coaches and cities. This list has picked up on the team of Chicago Blackhawks, who had a very dominant season and ended up winning the NHL post-season tournament, the Stanley Cup Championship. The Pittsburgh Penguins were also considered a top team and had a high InfoGain value. Coach Barry Trotz of the Nashville Predators is a curious pick; it shows up 4 times and although the Nashville Predators were neither a very good or a very bad team in the 2012-2013 season; they did not have any activity that would make them stand out.

This suggests that it would be difficult to train on text across multiple years, as we would start to see evidence of concept drift, where the data the algorithms are learning on changes with time. A team might be really good in one year, but due to losing players in free agency and trade, may be a terrible team the next year. This suggests we should not be training and testing across more than a season or two.

---

[1] A Weka implementation of Naive Bayes where features are modelled with a normal distribution.

[2] http://www.puckprediction.com, accessed 15 December 2013.

Table 6. Info Gain values for the word-based features

Info Gain	ngram	Name/Place?
0.01256	whos hot	No
0.01150	whos	No
0.01124	hot	no
0.00840	three	no
0.00703	chicago	yes
0.00624	kind	no
0.00610	assists	no
0.00588	percentage	no
0.00551	trotz	yes
0.00540	games	no
0.00505	richards said	yes
0.00499	barry trotz	yes
0.00499	barry	yes
0.00499	coach barry	yes
0.00497	given	no
0.00491	four	no
0.00481	pittsburgh penguins	yes
0.00465	body	no
0.00463	save percentage	no

Similarly, we looked at the learnt decision tree from J48 on the pre-game texts and we can see a similar trend. With the top of the tree formed by ngrams of player and city names, this could have dramatic effects if you train on one year where the team is a championship contender and test on the next year when the team may not qualify for the post-season tournament.

7 Conclusion

In these experiments, we built meta-classifiers to forecast the outcome of games in the National Hockey League. In the first step, we trained three classifiers using three sets of features to represent the games. The first classifier was a numeric classifier and used cumulative Goals Against and Differential as well as the location (Home/Away) of both teams. The second classifier used pre-game texts that discuss how well the teams have been performing recently in the season up to that game. We used TF-IDF values on ngrams and did not stem our texts. The third classifier used sentiment analysis methods and counted the number of positive, negative and percentage of positive word differential in the texts.

The outputs were fed into the second layer of the cascade-classifier with the confidence and the predicted output from all three initial classifiers as input. We used machine learning algorithms on this set of six features. In addition, we used two other meta-classifiers, highest confidence and majority voting, to determine the output from the second layer. The best results came from the majority voting within the second layer.

This method returned an accuracy of 60.25% which is higher than any of the results from the first layer, much higher than the all-in-one classifier which uses all the features in a single data set, and it improves on our initial results from the numeric dataset from our previous work.

It is difficult to predict in the NHL as there is not a lot of room for improvement between the baseline and the upper bound. Selecting the home team to always win yields an accuracy of 56%, while the upper bound seems to be around 62%. This leaves us with only 6% to improve our classifier. While our experiments with numerical data from the game statistics were helping in the prediction task, we were happy to see that the pre-game report are also useful, especially when combining the two sources of information.

References

1. Weissbock, J., Viktor, H., Inkpen, D.: Use of performance metrics to forecast success in the national hockey league. In: European Conference on Machine Learning: Sports Analytics and Machine Learning Workshop (2013)
2. Morgan, S., Williams, M.D., Barnes, C.: Applying decision tree induction for identification of important attributes in one-versus-one player interactions: A hockey exemplar. Journal of Sports Sciences (ahead-of-print), 1–7 (2013)
3. Hipp, A., Mazlack, L.: Mining ice hockey: Continuous data flow analysis. In: IMMM 2011, The First International Conference on Advances in Information Mining and Management, pp. 31–36 (2011)
4. Macdonald, B.: An expected goals model for evaluating NHL teams and players. In: Proceedings of the 2012 MIT Sloan Sports Analytics Conference (2012), http://www.sloansportsconference.com
5. Buttrey, S.E., Washburn, A.R., Price, W.L.: Estimating NHL scoring rates. J. Quantitative Analysis in Sports 7 (2011)
6. Jones, M.B.: Responses to scoring or conceding the first goal in the NHL. Journal of Quantitative Analysis in Sports 7(3), 15 (2011)
7. Swartz, T.B., Tennakoon, A., Nathoo, F., Tsao, M., Sarohia, P.: Ups and downs: Team performance in best-of-seven playoff series. Journal of Quantitative Analysis in Sports 7(4) (2011)
8. Huang, K.Y., Chang, W.L.: A neural network method for prediction of 2006 world cup football game. In: The 2010 International Joint Conference on Neural Networks (IJCNN), pp. 1–8. IEEE (2010)
9. Huang, K.Y., Chen, K.J.: Multilayer perceptron for prediction of 2006 world cup football game. Advances in Artificial Neural Systems 2011, 11 (2011)
10. Kahn, J.: Neural network prediction of NFL football games. World Wide Web electronic publication (2003), http://homepages.cae.wisc.edu/~ece539/project/f03/kahn.pdf
11. Purucker, M.C.: Neural network quarterbacking. IEEE Potentials 15(3), 9–15 (1996)
12. Loeffelholz, B., Bednar, E., Bauer, K.W.: Predicting NBA games using neural networks. Journal of Quantitative Analysis in Sports 5(1), 1–15 (2009)
13. Miljkovic, D., Gajic, L., Kovacevic, A., Konjovic, Z.: The use of data mining for basketball matches outcomes prediction. In: IEEE 2010 8th International Symposium on Intelligent Systems and Informatics (SISY), pp. 309–312 (2010)

14. Miljkovic, D., Gajic, L., Kovacevic, A., Konjovic, Z.: The use of data mining for basketball matches outcomes prediction. In: IEEE 2010 8th International Symposium on Intelligent Systems and Informatics (SISY), pp. 309–312 (2010)
15. Yang, J.B., Lu, C.H.: Predicting NBA championship by learning from history data (2012)
16. Wei, N.: Predicting the outcome of NBA playoffs using the naïve bayes algorithms. University of South Florida, College of Engineering (2011)
17. Witten, I.H., Frank, E.: Data Mining: Practical machine learning tools and techniques. Morgan Kaufmann (2005)
18. Burke, B.: Luck and NFL outcomes 3 (2007), http://www.advancednflstats.com/2007/08/luck-and-nfl-outcomes-3.html (accessed February 16, 2014)
19. Nielsen, F.Å.: A new ANEW: Evaluation of a word list for sentiment analysis in microblogs. arXiv preprint arXiv:1103.2903 (2011)
20. Wiebe, J., Wilson, T., Cardie, C.: Annotating expressions of opinions and emotions in language. Language Resources and Evaluation 39(2-3), 165–210 (2005)
21. Hu, M., Liu, B.: Mining and summarizing customer reviews. In: Proceedings of the Tenth ACM SIGKDD International Conference on Knowledge Discovery and Data Mining, pp. 168–177. ACM (2004)

Using Ensemble of Bayesian Classifying Algorithms for Medical Systematic Reviews

Abdullah Aref and Thomas Tran

School of Electrical
Engineering and Computer Science
University of Ottawa
Ottawa, Ontario, Canada

Abstract. Systematic reviews are considered fundamental tools for Evidence-Based Medicine. Such reviews require frequent and time-consuming updating. This study aims to compare the performance of combining relatively simple Bayesian classifiers using a fixed rule, to the relatively complex linear Support Vector Machine for medical systematic reviews. A collection of four systematic drug reviews is used to compare the performance of the classifiers in this study. Cross-validation experiments were performed to evaluate performance. We found that combining Discriminative Multinomial Naïve Bayes and Complement Naïve Bayes performs equally well or better than SVM while being about 25% faster than SVM in training time. The results support the usefulness of using an ensemble of Bayesian classifiers for machine learning-based automation of systematic reviews of medical topics, especially when datasets have a large number of abstracts. Further work is needed to integrate the powerful features of such Bayesian classifiers together.

Keywords: Systematic reviews for the medical domain, Text mining, Ensembles, Performance comparison.

1 Introduction

Systematic reviews (SRs) are the results of reviewing literature on a specific topic with the goal of extracting a targeted subset of data [7]. SRs are considered an essential component in the practice of Evidence-based Medicine (EBM) that attempts to provide better care with better outcomes by basing clinical decisions on solid scientific evidences [8]. Because new information constantly becomes available, medicine is continually changing, and SRs must undergo periodic updates [5]. Applying automated text classification techniques in the stages of probing and screening the possibly pertinent literature can reduce the time it takes to identify and filter pertinent articles for a categorical query [5]. Compared to classical text classification tasks, medical SRs are considered highly imbalanced [12].

In the literature, Support Vector Machine (SVM) was compared to different Bayesian-based classifiers for general textual datasets, and to some stand-alone Bayesian-based classifiers in medical SRs' domain; however, none of these studies consider the use of an ensemble of Bayesian classifiers. The aim of this study

M. Sokolova and P. van Beek (Eds.): Canadian AI 2014, LNAI 8436, pp. 263–268, 2014.

is to compare the performance of combining Discriminative Multinomial Naïve Bayes (DMNB) and Complement Naïve Bayes (CNB) with the performance of linear SVM in the context of medical SRs. The comparison of the performance of DMNB and CNB, with linear SVM is included to clarify the motivation behind combining them together. Multinomial Naïve Bayes (MNB) is included to indicate the consistency of our findings with the most relevant work in [14]. The main contribution of this paper is the finding that the performance of combining DMNB and CNB using a simple fixed rule surpasses that of SVM, as measured by area under the receiver operating curve (AUC), with less computational cost, as measured by training time.

2   Related Work

In the literature, it is generally perceived that SVM classifiers perform well for text classification. However, advanced, heuristic modifications of the classical Naïve Bayes classifier, such as MNB and CNB, perform as well as SVM, while being more expeditious [12]. The utilization of machine learning methods to create SRs discussed in [2] where SVM was found to achieve the best performance for SRs compared to the classical Naïve Bayes (NB) and a specialized AdaBoost algorithm. The use of a Voting Perceptron (VP) algorithm for systematic review was proposed in [4] with the objective of reducing the time spent by experts reviewing journal articles for inclusion in medical SRs. A reduction in the number of articles requiring manual review was found for 11 of the 15 datasets studied. The Factorized version of the Complement Naïve Bayes (FCNB) was proposed in [12] and found to have better performance compared to the VP-predicated classification system proposed in [4]. Generally, SVM outperformed FCNB for SRs using the work preserved over sampling measure [3]. The performance of MNB and SVM was compared in [14] in the context of medical SRs. MNB found to work faster than SVM for SRs without a consequential loss in performance.

A discriminative parameter learning method for Bayesian network classifiers proposed in [18] and its performance was analyzed but not compared to SVM. To the best of our knowledge, the promising DMNB classifier is not compared to SVM in the context of the medical SRs.

Existing surveys like [1] provides more elaboration on different machine learning methods used for text classification in general.

Many studies have shown that combining a set of different classifiers can improve classification performance [11]. The intuition is that, if different classifiers are more likely to make errors in different ways, careful combination of these classifiers can reduce the overall error to improve the performance of the ensemble system [10]. Several fixed rules for combining base classifiers in ensembles were described in [15]; however, the sum (average) rule was reported to outperform other rules because of its resilience to estimation errors.

3   Experimental Results and Discussions

3.1   Experimental Setup

We conduct our experiments using a WEKA [9] which has all the classifiers used in this study pre-implemented. All experiments are performed on a Pentium Dual-Core with 2.3 GHZ CPU, 2G RAM and 32-bit Windows 7 Operating System.

In our experiments, we utilize four medical publication datasets provided by the author of [14] with a total of 3623 labeled articles, each contains the label, the title, the abstract, the publication name, and the MeSHs categories for the article. Each dataset is represented as an unstructured text file with a separator between different articles and another one between different fields within each article. Each article has a label: Include (I) or Exclude (E) and therefore machine learning techniques can be trained and tested. The data sets are highly imbalanced (the ratio of positive (I) class to the entire dataset). Beta Blockers (BB) dataset contains 302 positive articles and 1779 negative articles, Estrogens (ES) dataset contains 80 positive articles and 288 negative articles, Oral-Hypoglycemics (OH) dataset contains 139 positive articles and 364 negative articles, and Triptan (TR) dataset contains 218 positive articles and 453 negative articles

All punctuation, numbers, white spaces and stop list words are abstracted in addition to words with length ≤ 2 and those that do not appear in at least 1% of class type. For document representation, the bag-of-words method is used because of its simplicity for classification purposes [1]. The term frequency and the inverse document frequency (TF x IDF) known as the TF-IDF weighing factor as implemented in R [16] is used to create a term document matrix (TDM). All preprocessing is performed in the R framework [16], and TDMs are preserved and then utilized as input in WEKA [9]. The performance of all classifiers on each dataset is observed via 10 runs of 5*2-fold cross validation.

3.2   Evaluation Measures

Most commonly in bioinformatics text classification research, the performance of a given system on a given task is quantified by precision and recall, which are then cumulated into a single F-measure [5]. AUC is a good measure of the quality of a classifier, which is equivalent to the probability of a randomly chosen positive sample being ranked higher than a randomly chosen negative sample. It is independent of class prevalence and therefore is a good overall measure to utilize when the false positive / negative cost trade-off is not known in advance or when comparing performance on tasks with imbalanced data [5].

3.3   Parameter Selection for CNB and DMNB Algorithms

We notice that the performance of CNB varies depending on the value used for the smoothing parameter, which is used to avoid word probabilities of zero. The default value proposed in [17] is one. For three of the four used datasets, the AUC increases as the value of the smoothing parameter decreases and then reaches a steady state when the value of the smoothing parameter reaches $1*10^{-5}$. Values less than $1*10^{-9}$ are not tested. In this study, we use the smoothing parameter value $1*10^{-5}$. For DMNB, we found that, a very small increase in AUC can be achieved as the number of iterations increases; it is less than 2% as the number of iterations varies from 1 to 100, and most of this enhancement is achieved by the second iteration. However, CPU training time increases significantly as the number of iterations increases. DMNB with one iteration used in this study.

3.4   Performance Comparison

In this subsection, we present the results of comparing the performance of the used algorithims over the four datasets used in this study as well as the ensemble of DMNB and CNB (EDC). The performance comparison presented in this subsection indicated that SVM outperforms DMNB in terms of AUC only in one of the four datasets used in that study, also, the study indicated that SVM outperforms CNB in two of the four datasets. It is interesting to note that there is no intersection between those sets. This suggests that CNB and DMNB are more likely to make errors in different ways, which recommends combining them together to achieve a better performance. Therefore, in addition to comparing the stand alone classifieres, we compare the performance of the EDC on one side with linear SVM on the other side and present the results in this section. To reduce computational complexity, the average of probability is used to combine the base classifieres in EDC rather than useing trained methods. T-test (rectified) with confidence 95% as implemented in WEKA is used in order to find out statistically significant differences.

Table 1 presents the relative performance of used classifiers for different evaluation measures. In terms of AUC, DMNB performs better than SVM in two out of four datasets. CNB performs better than SVM in two out of four datasets (not the same datasets as DMNB). However, SVM performs better than CNB on the ES dataset and better than DMNB on the OH dataset. Addditionally, EDC performs better than SVM in three out of four datasets and there is no statistically significant difference between their performances on the fourth dataset. In terms of CPU training time, all used classifiers in this study performs better than SVM in terms of training time. The EDC is about 23% to 26% faster thean SVM in terms of training time. In terms of CPU testing time, all stand alone algorithms used in this study performs better than SVM. However, in general, there is no statistically significant difference between the EDC and SVM in terms of CPU testing time except for one set, where the EDC is about 50% faster than SVM.

From the results presented, it can be noted that SVM, DMNB and CNB performs very well, in terms of suitability, with small differences. SVM is, however, at a disadvantage, when we consider the training and testing times. Furthermore, the EDC outperforms SVM, in terms of suitability and processing time. While SVM processing time may be acceptable for the datasets used in this study where the size of the largest set is in the order of 10^3, the training and testing

Table 1. Relative Performance of Used Classifiers for Different Evaluation Measures

	SVM	MNB	DMNB	CNB	EDC
AUC					
BB	0.88	0.82 *	0.95 v	0.97 v	0.99 v
ES	0.94	0.85 *	1.00 v	0.78 *	1.00 v
OH	0.94	0.81 *	0.88 *	0.98 v	0.98 v
TR	0.99	0.90 *	0.99	0.97 *	1.00
F_measure					
BB	0.92	0.92	0.96 v	0.99 v	0.99 v
ES	0.90	0.88 *	0.99 v	0.94 v	0.94 v
OH	0.92	0.84 *	0.88 *	0.99 v	0.99 v
TR	0.98	0.81 *	0.97	0.97	0.97
Training Time					
BB	0.97	0.21 v	0.47 v	0.20 v	0.70 v
ES	0.32	0.06 v	0.14 v	0.06 v	0.20 v
OH	0.27	0.05 v	0.12 v	0.05 v	0.19 v
TR	0.28	0.06 v	0.13 v	0.06 v	0.20 v
Testing Time					
BB	0.15	0.07 v	0.03 v	0.07 v	0.10
ES	0.05	0.02 v	0.01 v	0.02 v	0.03
OH	0.04	0.02 v	0.01 v	0.02 v	0.02 v
TR	0.04	0.02 v	0.01 v	0.02 v	0.03
	v: Better compared to SVM		*: Worse compared to SVM		

times of SVM, were unacceptably long for datasets in the order of 10^4 [13]. It was reported in [6] that processing time of SVM tends to grow quadratically with the number of articles in general datasets

4   Conclusions and Future Work

We conclude, from this study, that using TD-IDF to represent term documents coupled with the useof EDC can provide a satisfactory result, at least for the used datasets, while being less computationally expensive than SVM. This can be very useful as the number of articles for SRs increases. Further investigation of integrating powerful features of DMNB and CNB together in a single simple classifier rather than using ensembles to achieve a better performance, analysing the performance of EDC over a wider verity of datasets and integrating them with systems for machine- learning- based SRs are considered as future work.

References

1. Aggarwal, C., Zhai, C.: A survey of text classification algorithms. In: Aggarwal, C.C., Zhai, C. (eds.) Mining Text Data, pp. 163–222. Springer, US (2012)
2. Aphinyanaphongs, Y., Tsamardinos, I., Statnikov, A.R., Hardin, D.P., Aliferis, C.F.: Research paper: Text categorization models for high-quality article retrieval in internal medicine. JAMIA 12(2), 207–216 (2005)
3. Cohen, A.M.: Letter: Performance of support-vector-machine-based classification on 15 systematic review topics evaluated with the wss@95 measure. JAMIA 18(1), 104 (2011)
4. Cohen, A.M., Hersh, W.R., Peterson, K., Yen, P.Y.: Research paper: Reducing workload in systematic review preparation using automated citation classification. JAMIA 13(2), 206–219 (2006)
5. Cohen, A.M., Informatics, D.O.M., Epidemiology, C.: Optimizing feature representation for automated systematic review work prioritization. In: AMIA Annual Symposium Proceedings, pp. 121–125 (2008)
6. Colas, F., Brazdil, P.: Comparison of svm and some older classification algorithms in text classification tasks. In: Bramer, M. (ed.) Artificial Intelligence in Theory and Practice. IFIP, vol. 217, pp. 169–178. Springer, Boston (2006)
7. Frunza, O., Inkpen, D., Matwin, S.: Building systematic reviews using automatic text classification techniques. In: COLING (Posters), pp. 303–311 (2010)
8. Frunza, O., Inkpen, D., Matwin, S., Klement, W., O'Blenis, P.: Exploiting the systematic review protocol for classification of medical abstracts. Artificial Intelligence in Medicine 51(1), 17–25 (2011)
9. Hall, M., Frank, E., Holmes, G., Pfahringer, B., Reutemann, P., Witten, I.H.: The weka data mining software: An update. SIGKDD Explorations 11(1), 10–18 (2009)
10. Lee, K.C., Cho, H.: Performance of ensemble classifier for location prediction task: Emphasis on markov blanket perspective. International Journal of u-and e-Service, Science and Technology 3(3) (2010)
11. Li, Y.H., Jain, A.K.: Classification of text documents. The Computer Journal 41, 537–546 (1998)
12. Matwin, S., Kouznetsov, A., Inkpen, D., Frunza, O., O'Blenis, P.: A new algorithm for reducing the workload of experts in performing systematic reviews. JAMIA 17(4), 446–453 (2010)
13. Matwin, S., Kouznetsov, A., Inkpen, D., Frunza, O., O'Blenis, P.: Letter: Performance of svm and bayesian classifiers on the systematic review classification task. JAMIA 18(1), 104–105 (2011)
14. Matwin, S., Sazonova, V.: Direct comparison between support vector machine and multinomial naive bayes algorithms for medical abstract classification. JAMIA 19(5), 917 (2012)
15. McCallum, A., Nigam, K.: A comparison of event models for naive bayes text classification. In: AAAI 1998 Workshop on Learning for Text Categorization, pp. 41–48. AAAI Press (1998)
16. R Core Team: R: A Language and Environment for Statistical Computing. R Foundation for Statistical Computing, Vienna, Austria (2012), http://www.R-project.org ISBN 3-900051-07-0
17. Rennie, J.D.M., Shih, L., Teevan, J., Karger, D.R.: Tackling the poor assumptions of naive bayes text classifiers. In: Proceedings of the Twentieth International Conference on Machine Learning, pp. 616–623 (2003)
18. Su, J., Zhang, H., Ling, C.X., Matwin, S.: Discriminative parameter learning for bayesian networks. In: ICML, pp. 1016–1023 (2008)

Complete Axiomatization and Complexity of Coalition Logic of Temporal Knowledge for Multi-agent Systems*

Qingliang Chen[1], Kaile Su[2,3], Yong Hu[4], and Guiwu Hu[5]

[1] Department of Computer Science, Jinan University, Guangzhou 510632, China
[2] College of Mathematics, Physics and Information Engineering,
Zhejiang Normal University, Jinhua, China
[3] Institute for Integrated and Intelligent Systems,
Griffith University, Brisbane, Australia
[4] Institute of Business Intelligence and Knowledge Discovery, Department of
E-commerce, Guangdong University of Foreign Studies, Guangzhou 510006, China
[5] School of Mathematics and Statistics,
Guangdong University of Finance and Economics, Guangzhou 510320, China

Abstract. Coalition Logic (CL) is one of the most influential logical formalisms for strategic abilities of multi-agent systems. However CL can not formalize the evolvement of rational mental attitudes of the agents such as knowledge. In this paper, we introduce Coalition Logic of Temporal Knowledge (CLTK), by incorporating a temporal logic of knowledge (Halpern and Vardi's logic of CKL_n) into CL to equip CL with the power to formalize how agents' knowledge (individual or group knowledge) evolves over the time by the coalitional forces and the temporal properties of strategic abilities as well. Furthermore, we provide a complete axiomatization of CLTK, along with the complexity of the satisfiability problem, which is shown to be EXPTIME-complete.

1 Introduction

Multi-agent system has been proved to be a useful conceptual paradigm for distributed intelligent systems, and has received a substantial amount of intensive studies, most of which are dedicated to modeling, representing and reasoning about different agents within a certain environment [1]. There is a strand of research in this community which focuses on the *strategic structure* of the multi-agent system, and explicates the coalitional ability of what a group of agents can accomplish within a system, either together or alone. And this kind of research is

* This work is supported by National Natural Science Foundation of China grant No.61003056,71271061 and 61272415; National Basic Research 973 Program of China grant 2010CB328103; ARC Future Fellowship FT0991785, Fundamental Research Funds for the Central Universities of China grant No.21612413 and 21612414, and Business Intelligence Key Team of Guangdong University of Foreign Studies (TD1202). Corresponding author: Guiwu Hu (guiwuhu@gdcc.edu.cn).

M. Sokolova and P. van Beek (Eds.): Canadian AI 2014, LNAI 8436, pp. 269–274, 2014.
© Springer International Publishing Switzerland 2014

usually closely related to Game theory. As for this kind of researches, Coalition Logic (CL) [2], Alternating-time Temporal Logic (ATL) [3], and STIT [4] logics are arguably the most popular ones in recent years, and many different variants of these logics have been proposed and studied. However, these logics either pay no attention to the mental states and rational attitudes of agents during the cooperations such as knowledge [5], which are important elements in rational behaviors and may affect decision-makings in further moves, or they just ignore their dynamics over the time, or no complete axiomatization has been introduced and justified.

This paper aims to address this deficit. We introduce Coalition Logic of Temporal Knowledge (CLTK), by incorporating the most popular temporal logic of knowledge-Halpern and Vardi's logic of CKL_n [6]-into CL to equip CL with the power to formalize how agents' knowledge (individual or group knowledge) evolves over the time by the coalitional forces and also the temporal properties of their strategic abilities. In this paper, we present the complete axiomatization for CLTK, and characterize the computational complexity of the satisfiability problem. Up to this point, there is no completeness results for CL with temporal and knowledge modalities yet. And this paper is addressing this deficit.

The rest of the paper is organized as follows. Next section will present Coalition Logic of Temporal Knowledge (CLTK), then we introduce a complete axiomatization for CLTK, along with the proof by construction of canonical structures, followed by the study of the complexity for the satisfiability problem. Finally, we discuss the related works and conclude the paper in the last section.

2   Coalition Logic of Temporal Knowledge

Now, we introduce the Coalition Logic of Temporal Knowledge (CLTK), which is obtained by extending the CL [2] language with CKL_n [6]. The temporal logic of knowledge CKL_n, proposed by Halpern and Vardi, is the most popular formalism for temporal knowledge and has received extensive studies. CKL_n is the language of propositional temporal logic augmented by a knowledge operator K_i for each agent i, and common knowledge operators C_G, where G is a group of agents. Given a set Θ of atomic propositions, and a finite set Ag of n agents. Formally, the language of Coalition Logic of Temporal Knowledge (CLTK) is defined as:

$$\phi ::= p \mid \neg\phi \mid \phi \wedge \phi \mid [G]\phi \mid \bigcirc\phi \mid \phi\mathbf{U}\phi \mid K_i\phi \mid C_G\phi$$

where $p \in \Theta$, $i \in Ag$, $G \subseteq Ag$.

CKL_n is interpreted over *interpreted systems* [5], which consists of a pair $(\mathcal{R}, \pi)$, where $\mathcal{R}$ is a set of runs over a set of global states and π a valuation function, which gives the set of primitive propositions true at each point in $\mathcal{R}$ [5]. Thus a global state can be identified by a pair (r, m) (or $r(m)$) consisting of a run r and m as a time point, the first component of $r(m)$ by $r_e(m)$, and for each $i(1 \leq i \leq n)$, the $(i+1)$-th component of the tuple $r(m)$ by $r_i(m)$. In this way, $r_i(m)$ is the local state of agent i in run r at time m. For every agent i, we say that (r, m) and (r', m') are indistinguishable to agent i iff $r_i(m) = r'_i(m')$, and write as $(r, m) \sim_i (r', m')$.

Now we introduce *coalition interpreted system model* to interpret CLTK. A *coalition interpreted system model* $\mathcal{I}$ is defined as a tuple

$$\mathcal{I} = \langle S, \mathcal{R}, \pi, \xi \rangle$$

where

S is a non-empty set of *global states*, consisting of the environment's local state and agent i's local state for every $i \in Ag$;

$\mathcal{R}$ is a set of runs over S and each run $r \in \mathcal{R}$ is a function from the time domain (the natural numbers) to S, so that any $s \in S$ can be labelled as $s = r(u)$ for some $r \in \mathcal{R}$ and some natural number u;

π is a *valuation function*, which assigns each state $s = r(u) \in S$ a set $\pi(r(u)) \subseteq \Theta$;

$\xi : \mathcal{P}(Ag) \times S \longrightarrow \mathcal{P}(\mathcal{P}(S))$ is a *playable effectivity function* [2], where $\mathcal{P}(.)$ is the power set of a given set;

Given a *coalition interpreted system model* $\mathcal{I}$, the semantics of CLTK is defined in a point $(r, u) \in S$ as follows. For simplification, we just use $\models$ to represent $\models_{CLTK}$.

$(\mathcal{I}, r, u) \models \phi$ iff $\phi \in \pi(r(u))$

$(\mathcal{I}, r, u) \models \neg\phi$ iff $(\mathcal{I}, r, u) \not\models \phi$

$(\mathcal{I}, r, u) \models (\phi_1 \wedge \phi_2)$ iff $(\mathcal{I}, r, u) \models \phi_1$ and $(\mathcal{I}, r, u) \models \phi_2$

$(\mathcal{I}, r, u) \models \bigcirc\phi$ iff $(\mathcal{I}, r, u+1) \models \phi$

$(\mathcal{I}, r, u) \models \phi_1 \mathbf{U} \phi_2$ iff $(\mathcal{I}, r, u') \models \phi_2$ for some $u' \geq u$ and $(\mathcal{I}, r, u'') \models \phi_1$ for all $u \leq u'' < u'$

$(\mathcal{I}, r, u) \models K_i\phi$ iff $(\mathcal{I}, r', v) \models \phi$ for all (r', v) such that $(r', v) \sim_i (r, u)$

$(\mathcal{I}, r, u) \models C_G\phi$ iff $(\mathcal{I}, r', v) \models \phi$ for all (r', v) such that $(r', v) \sim_G^C (r, u)$, where $\sim_G^C$ is the transitive closure of $\bigcup_{i \in G} \sim_i$

$(\mathcal{I}, r, u) \models [G]\phi$ iff $\exists T \in \xi(G, r(u))$ such that $\forall r'(u') \in T$, we have $(\mathcal{I}, r', u') \models \phi$

3   Complete Axiomatization for CLTK

3.1   Axiomatic System for CLTK

Axioms:

Prop All tautologies of propositional calculus

K1 $(K_i\varphi \wedge K_i(\varphi \Rightarrow \psi)) \Rightarrow K_i\psi, i = 1, 2, ..., n$

K2 $K_i\varphi \Rightarrow \varphi$

K3 $K_i\varphi \Rightarrow K_i K_i\varphi, i = 1, 2, ..., n$

K4 $\neg K_i\varphi \Rightarrow K_i\neg K_i\varphi, i = 1, 2, ..., n.$

C1 $E_G\varphi \Leftrightarrow \bigwedge_{i \in G} K_i\varphi, , i = 1, 2, ..., n.$

C2 $C_G\varphi \Rightarrow E_G(\varphi \wedge C_G\varphi)$

T1 $\bigcirc(\varphi \Rightarrow \psi) \Rightarrow (\bigcirc\varphi \Rightarrow \bigcirc\psi)$

T2 $\bigcirc\neg\varphi \Rightarrow \neg\bigcirc\varphi$

T3 $\varphi\mathbf{U}\psi \Leftrightarrow \psi \vee (\varphi \wedge \bigcirc(\varphi\mathbf{U}\psi))$

G1 $\neg[G]\bot$

G2 $[G]\top$

G3 $\neg[\emptyset]\neg\varphi \Rightarrow [Ag]\varphi$

G4 $[G](\varphi \wedge \psi) \Rightarrow [G]\psi$

G5 $[G_1]\varphi \wedge [G_2]\psi \Rightarrow [G_1 \cup G_2](\varphi \wedge \psi)$, if $G_1 \cap G_2 = \emptyset$

Inference Rules:

R1 From φ and $\varphi \Rightarrow \psi$ infer ψ
R2 From φ infer $K_i\varphi$
R3 From $\varphi \Rightarrow E_G(\varphi \wedge \psi)$ infer $\varphi \Rightarrow C_G\psi$
R4 From φ infer $\bigcirc\varphi$
R5 From $\varphi' \Rightarrow \neg\psi \wedge \bigcirc\varphi'$ infer $\varphi' \Rightarrow \neg(\varphi'\mathbf{U}\psi)$
R6 From $\varphi \Leftrightarrow \psi$ infer $[G]\varphi \Leftrightarrow [G]\psi$

where $\bot$ expresses **False**, $\top$ expresses **True**, $G \subseteq Ag$ is a coalition of agents while Ag is the complete set of agents.

Lemma 1 (Soundness). *For any CLTK-formula* ϕ, $\vdash \phi$ *implies* $\models \phi$.

3.2   Completeness Proof

In the remainder of this section we show that the above axiomatic system for CLTK is also complete. We will construct a canonical coalition interpreted system model $\mathcal{I}^c$.

Definition 1 (Canonical Model). *The canonical coalition interpreted system model* $\mathcal{I}^c$ *is defined as a tuple*

$$\mathcal{I}^c = \langle S^c, \mathcal{R}^c, \pi^c, \xi^c \rangle$$

where

- S^c *is the set of all maximally CLTK-consistent sets of formulas*
- $\mathcal{R}^c$ *is the set of all possible runs over* S^c, *which should satisfy the following for* $\forall r \in \mathcal{R}^c$:
  - $\bigcirc\varphi \in r(u)$ *iff* $\varphi \in r(u+1)$
  - $\varphi\mathbf{U}\psi \in r(u)$ *iff* $\exists u' \geq u$ *such that* $\psi \in r(u')$ *and* $\varphi \in r(u'')$ *for all* u'' *with* $u \leq u'' < u'$.
- $\pi^c : p \in \pi^c(r(u))$ *iff* $p \in r(u)$ *where* p *is any propositional formula.*
- ξ^c: *if* $[G]\psi \in r(u)$, *then*

$$X \in \xi^c(G, r(u)) \text{ iff } \begin{cases} \{r'(u') \in S^c \mid \psi \in r'(u')\} \subseteq X \text{ if } G \neq Ag; \\ (S^c - X) \notin \xi^c(\emptyset, r(u)), \qquad \text{if } G = Ag. \end{cases}$$

- *For* $\forall i \in Ag$, *agent* i*'s local state in* $r(u)$ *is defined as* $s_i = \{\varphi \mid K_i\varphi \in r(u)\}$. *Therefore,*

$$(r, u) \sim_i (r', u') \text{ iff } \{\varphi \mid K_i\varphi \in r(u)\} = \{\varphi \mid K_i\varphi \in r'(u')\}$$

Firstly, it is quite straightforward to see that

Lemma 2. $\sim_i$ *is an equivalence relation.*

Therefore, the epistemic semantics defined in CLTK is the standard S5 [5].

Now it is not hard to see that ξ^c defined in the canonical model is a *playable effectivity function* [2], which can be proved in the same way of [2].

Lemma 3. ξ^c *defined in the canonical model is a* playable effectivity function.

Lemma 4 (Truth Lemma). *Any CLTK-consistent formula is satisfied in a state in the canonical coalition interpreted system model.*

Proof. By induction on the structure for an arbitrary CLTK-consistent formula φ to show that $(\mathcal{I}^c, r, u) \models \varphi$ iff $\varphi \in r(u)$.   □

Theorem 1. *For any CLTK formula* φ, $\vdash \varphi \Leftrightarrow \models \varphi$.

4   Complexity Analysis

Theorem 2. *The satisfiability problem for CLTK is EXPTIME-complete.*

Proof. Since CLTK subsumes CKL_n logic, which is known to be EXPTIME-complete [6], we immediately have that the satisfiability problem for CLTK is EXPTIME-hard. On the other hand, to prove the satisfiability problem is in EXPTIME, we just devise a comprehensive algorithm to deal with coalition, temporal, knowledge and common knowledge operators. The coalition one has been shown to be in PSPACE [2], temporal one in PSPACE too, the knowledge one in NP, while the common knowledge one in EXPTIME [6]. So generally the algorithm runs in EXPTIME, which subsumes PSPACE and NP.   □

5   Related Works

In recent years, much attention has been paid to the coalitional logic for multi-agent systems, whose focuses are not on the cognitive states of agents, but on the *strategic structure* of the system: what agents can achieve, either individually or in groups. Although many different variants of CL and ATL have been proposed and studied, most of them are dedicated to computational complexity and expressive power. Complete axiomatization of these logics is much fewer because it is technically harder. Only a few works such as Goranko and van Drimmelen's completeness proof for ATL [7], Broersen and colleagues' completeness proofs for different variants of STIT logic [4,8,9], and more recently, Ågotnes and Alechina's completeness proof for epistemic coalition logic [10] are remarkable exceptions. Although [10] has taken the epistemic logic into account, it ignores the temporal dimension and thus fails to model the dynamic evolvement of individual and group knowledge over the time. It is discussed to incorporate time and knowledge in ATL in [11], without the proof of the complete axiomatization however. And the knowledge and time in [4] is quite limited. Up to this point, there are no completeness results for CL with temporal and knowledge modalities yet. And this paper has successfully addressed this deficit.

6   Conclusion

In this paper, we introduce a three-dimensional modal logic of Coalition Logic of Temporal Knowledge (CLTK), by incorporating a temporal logic of knowledge-Halpern and Vardi's logic of CKL_n)-into CL to enable CL to formalize how agents' knowledge (individual or group knowledge) evolves over the time by the coalitional forces and the temporal properties of their strategic abilities as well. Furthermore, we provide a complete axiomatization of the CLTK, along with the complexity of the satisfiability problem, which is shown to be EXPTIME-Complete.

As for the future work, we can take other variants of the epistemic logics into account such as pro attitudes, and also the new semantical model to interpret knowledge of ability "de re" [12].

References

1. Wooldridge, M.: An Introduction to Multiagent Systems, 2nd edn. John Wiley & Sons Press (2009)
2. Pauly, M.: A modal logic for coalitional power in games. Journal of Logic and Computation 12(1), 149–166 (2002)
3. Alur, R., Henzinger, T.A., Kupferman, O.: Alternating-time temporal logic. Journal of the ACM 49(5), 672–713 (2002)
4. Broersen, J.: A complete STIT logic for knowledge and action, and some of its applications. In: Baldoni, M., Son, T.C., van Riemsdijk, M.B., Winikoff, M. (eds.) DALT 2008. LNCS (LNAI), vol. 5397, pp. 47–59. Springer, Heidelberg (2009)
5. Halpern, J.Y., Fagin, R., Moses, Y., Vardi, M.Y.: Reasoning About Knowledge. MIT Press (1995)
6. Halpern, J.Y., Vardi, M.Y.: The complexity of reasoning about knowledge and time: Extended abstract. In: Proceedings of the 18th Annual ACM Symposium on Theory of Computing, Berkeley, California, USA, May 28-30, pp. 304–315 (1986)
7. Goranko, V., van Drimmelen, G.: Complete axiomatization and decidability of alternating-time temporal logic. Theoretical Computer Science 353(1-3), 93–117 (2006)
8. Broersen, J., Herzig, A., Troquard, N.: A normal simulation of coalition logic and an epistemic extension. In: Proceedings of the 11th Conference on Theoretical Aspects of Rationality and Knowledge (TARK 2007), Brussels, Belgium, pp. 92–101 (2007)
9. Broersen, J., Herzig, A., Troquard, N.: What groups do, can do, and know they can do: an analysis in normal modal logics. Journal of Applied Non-Classical Logics 19(3), 261–290 (2009)
10. Ågotnes, T., Alechina, N.: Epistemic coalition logic: Completeness and complexity. In: Proceedings of the Eleventh International Conference on Autonomous Agents and Multiagent Systems (AAMAS 2012), pp. 1099–1106. ACM Press (2012)
11. van der Hoek, W., Wooldridge, M.: Cooperation, knowledge, and time: Alternating-time temporal epistemic logic and its applications. Studia Logica 75(1), 125–157 (2003)
12. Jamroga, W., Ågotnes, T.: Constructive knowledge: What agents can achieve under imperfect information. Journal of Applied Non-Classical Logics 17(4), 423–475 (2007)

Experts and Machines against Bullies:
A Hybrid Approach to Detect Cyberbullies

Maral Dadvar, Dolf Trieschnigg, and Franciska de Jong

Human Media Interaction Group, University of Twente
POBox 217, 7500 AE, Enschede, The Netherlands
{m.dadvar,d.trieschnigg,f.m.g.dejong}@utwente.nl

Abstract. Cyberbullying is becoming a major concern in online environments with troubling consequences. However, most of the technical studies have focused on the detection of cyberbullying through identifying harassing comments rather than preventing the incidents by detecting the bullies. In this work we study the automatic detection of bully users on YouTube. We compare three types of automatic detection: an expert system, supervised machine learning models, and a hybrid type combining the two. All these systems assign a score indicating the level of "bulliness" of online bullies. We demonstrate that the expert system outperforms the machine learning models. The hybrid classifier shows an even better performance.

1 Introduction

With the growth of the use of Internet as a social medium, a new form of bullying has emerged, called cyberbullying. Cyberbullying is defined as an aggressive, intentional act carried out by a group or individual, using electronic forms of contact repeatedly and over time against a victim who cannot easily defend him or herself [1]. One of the most common forms is the posting of hateful comments about someone in social networks. Many social studies have been conducted to provide support and training for adults and teenagers [2, 3]. The majority of the existing technical studies on cyberbullying have concentrated on the detection of bullying or harassing comments [4-6], while there is hardly work on the more challenging task of detecting cyberbullies and studies for this area of research are largely missing. There are few exceptions however, that point out an interesting direction for the incorporation of user information in detecting offensive contents, but more advanced user information or personal characteristics such as writing style or possible network activities has not been included in these studies [7, 8]. Cyberbullying prevention based on user profiles was addressed for the first time in our latest study in which an expert system was developed that assigns scores to social network users to indicate their level of 'bulliness' and their potential for future misbehaviour based on the history of their activities [9]. In the previous work we did not investigate machine learning models. In this study we focus again on the detection of bully users in online social networks but now we look into the efficiency of *both* expert systems and machine learning models for identifying the potential bully users. We compare the performance of both systems for the task of

M. Sokolova and P. van Beek (Eds.): Canadian AI 2014, LNAI 8436, pp. 275–281, 2014.
© Springer International Publishing Switzerland 2014

assigning a score to social network users that indicates their level of bulliness. We demonstrate that the expert system outperforms the machine learner and can be effectively combined in a hybrid classifier. The approach we propose can be used for building monitoring tools to stop potential bullies from conducting further harm.

2   Data Collection and Feature Selection

In this section we will explain the characteristics of the corpus used in this study. We also describe the feature space and the three feature categories that have been used for the design of the expert system and for the machine learning models.

2.1   Corpus

YouTube is a popular user-generated content video platform. Its audience demographics match those from the general internet population [10]. Unfortunately, bullies misuse the platform to victimize their targets through harassing comments and other misbehaviours. To our knowledge no dataset for cyberbully detection is publicly available and therefore we decided to collect our own. To have a reasonable number and variety of cyberbullies we searched YouTube for topics sensitive to cyberbullying. We determined the users who commented on the top three YouTube videos in each topic. A total of 3,825 users were extracted and we collected a log of the users' activities for a period of 4 months (April – June 2012). We also captured profile information of the users, such as their age and the date they signed up. In total there are 54,050 comments in our dataset. On average there are 15 comments per user (StDev = 10.7, Median = 14). The average age of the users is 24 with 1.5 years of membership duration. Two graduate students were employed to independently annotate the users as bullies or non-bullies (inter-annotator agreement = 93%, Kappa = 0.78) based on their comments and the definition of cyberbullying provided earlier. The users that both annotators had labelled as bullying (n=419) were marked as bullies. Disagreements were resolved by the decision of a third annotator. In total, 12% of the users were labelled as bullies.

2.2   Feature Space

We compiled a set of 11 features in three categories to be used in our models. The selection of the features was limited by what is technically possible to extract from YouTube. The three categories representing the actions, behaviour, and characteristics of the users are presented in Table 1.

3   Methods

In this section we introduce the three types of models used for calculating and assigning the bulliness score to the social network users: a multi-criteria evaluation system, a set of machine learning models and two hybrid models that combine the two.

3.1   Multi-Criteria Evaluation System

Multi-Criteria Evaluation Systems (MCES) are a commonly used technique for decision-making in business, industry, and finance [11], but also in fields such as information retrieval [12]. By assigning weights and importance levels to features or criteria, MCES can combine different sources of knowledge to make decisions. In our scenario, a panel of 12 experts in the area of cyberbullying was asked to answer questions about the features presented in Table 1. For each feature they indicated 1) the likelihood that a bully user belongs to a certain category relevant for that feature and 2) the importance of that feature. The likelihood was indicated on four-point scale 'Unlikely', 'Less likely', 'Likely' and 'Very likely' corresponding to values 0.125, 0.375, 0.625 and 0.875 respectively [13]. The 'I don't know' option was also available. The importance was indicated on a four-point scale of 1: not informative, 2: partially informative, 3: informative and 4: very informative. Based on the averaged likelihood and importance weights indicated by the experts a "bulliness score" can be calculated for a user by taking the weighted average of the criteria. Multiple experts indicated combined features to be important for determining bullies. Therefore we added two combined criteria to those explained in section 2.2 which were based on age and profanity (C1), and age and misspellings (C2).

Table 1. The Feature Space

Content features model the content of the user comments	
1	*Number of profane words in the comments*	Based on a dictionary of 414 profane words including acronyms and abbreviations [6]
2	*Length of the comments*	Bullying comments are typically short [5]
3	*First person pronouns*	To detect comments targeted at a specific person
4	*Second person pronouns*	To detect comments targeted at a specific person
5	*Usernames containing profanities*	Users with bad intentions might hide their identity
6	*Non-standard spelling of the words*	Includes misspellings (e.g. 'funy'), or informal short forms of the words (e.g. 'brb')
Activity features model the activity of the user is in the online environment	
7	*Number of uploads*	YouTube users have a public channel, in which their activities such as posted comments and uploaded videos can be viewed. Users can also subscribe to others channels.
8	*Number of subscriptions*	
9	*Number of comments*	
User features model personal and demographic profile information	
10	*Age of the user*	Frequency of bullying incidents as well as choice of words and language structures change in different age groups. Divided into 5 age categories: 13-16, 17-19, 20-25, 26-30, and above 30 years old.
11	*Membership duration of the user*	Divided into 3 groups: less than 1 year, 1- 3 years and above 3 years.

3.2   Machine Learning Approaches

We used three well-known machine learning methods, which use pre-labelled training data for automatic learning: a Naive Bayes classifier, a classifier based on decision trees (C4.5) and Support Vector Machines (SVM) with a linear kernel [14]. The implementation available in WEKA 3 was used [15]. The machine learning models uses the features used by the expert system and, additionally, a number of features that are only interpretable by the machine. These are: (M1) The ratio of capital letters in a comment, to capture shouting in comments; (M2) The number of emoticons used in the comments, to capture explicit emotions; (M3) The occurrence of a second person pronoun followed by a profane word in profanity windows of different sizes (2 to 5 words), to capture targeted harassment; (M4) The term frequency–inverse document frequency (*tfidf*) of frequently repeated words, to capture emphasized content. As the baseline, we trained an SVM classifier (SVM$_B$) using only the content features listed in Table 2.

3.3   Hybrid Approach

Two hybrid approaches were tested to combine the advantages of the expert system with the machine learning models. The expert system is not affected by biased training data and may come up with rules that generalize better to unseen data. Machine learning models on the other hand can detect and analyse complex patterns that experts are not capable of observing. We construct two types of hybrid systems:

H1: Using the outcome of the expert system as an extra feature for training the machine learning models. The hybrid system is formed by adding the following features to the machine learning classifier: 1) the results of the MCES, 2) the features' categories that were used in the expert system as new set of features, and 3) the combined features (C1 and C2).

H2: Using the results of the machine learning model as a new criterion for the expert system. As previously done in the MCES, we assigned equal weights to all the criteria used in the system, including the machine learner criterion.

3.4   Evaluation

In this study the output of the models (i.e. bulliness scores) are probability values ranging from 0 to 1, and not a binary class. Therefore we used a threshold independent measure to evaluate and compare the performance of approaches. We evaluated the discrimination capacity by analysing their receiver operation characteristic (ROC) curves. A ROC curve plots "sensitivity" values (true positive fraction) on the y-axis against "1 - specificity" values (false positive fraction) for all thresholds on the x-axis [16]. The area under such a curve (AUC) is a threshold-independent metric and provides a single measure of the performance of the model. AUC scores vary from 0 to 1. AUC values of less than 0.5 indicate discrimination worse than chance; a score of 0.5 implies random predictive discrimination; and score of 1 indicates perfect discrimination. We used 10-fold cross-validation to evaluate the performance of the machine learning classifiers.

4   Results and Discussion

Based on weights that were assigned to each feature by the experts, profanities and bullying sensitive topics in the history of a user's comments is the most informative feature (average weight equals 3.6) and the least informative feature is the number of non-standard spellings (average weight equals 1.7). The age range between 13 and 16 years was indicated to most likely to contain bullies. Moreover, it is more likely that bully users have a high ratio of second person pronouns and profane words in their usernames. In the Activity features set, a high number of posting comments has the highest likelihood. The discrimination capacity of the MCES was 0.72.

Among the machine learning classifiers the decision tree classifier performed the worst, followed by the SVM classifier. Naive Bayes with discrimination capacity of 0.66 outperformed the other two algorithms. The contribution of each feature in classification task was determined by excluding it from the feature sets. The number of profane words, second person pronouns and pronoun-profanity windows, were the strongest contributing features. The discrimination capacity also improved by adding users' profile information and history of activities to the training features. The MCES outperformed all machine learning models in terms of discrimination capacity. A possible explanation for these differences is the sensitivity of the machine learning methods for class skew which was quite high in our dataset (10% bullying, 90% non-bullying). On the other hand a method based on human reasoning might not be as sensitive to the training data as the machine learning models and therefore give a better performance. For the first scenario of our hybrid approach (H1), the discrimination capacity of all machine learning methods improved. The improvement for the decision tree was smallest. Although the SVM gained the highest improvement in the discrimination capacity, the Naive Bayes classifier still outperformed the others. All measured improvements for H1 were significant (t-test, P<0.05). Also the hybrid model in the second scenario (H2) showed improvement of the discrimination capacity.

Table 2. The discrimination capacity of the tested cyberbully detection methods

	Approach	AUC	Hybrid approach	AUC
Baseline	SVM$_B$	0.57		
	Naive Bayes	0.66	+ MCES (H1)	**0.72**
Machine Learning	Decision Tree	0.52	+ MCES (H1)	**0.54**
	SVM	0.59	+ MCES (H1)	**0.68**
Expert System	MCES	0.72	+ Naive Bayes (H2)	**0.76**

5   Conclusion

In this experiment we developed a multi-criteria evaluation system for identifying potential bully users. We compared this expert system with a variety of machine learning models to assign a 'bulliness' score to YouTube and combined them in two hybrid models. We demonstrated that the expert system outperforms the machine learning models and the hybrid approach results in a further but marginal improvement in prediction performance. The proposed approach is in principle language independent can be adapted to other social networks as well. Spatial features such as location of the users as well as temporal features such as the time of their activities might be useful features to look into. They can provide further information about the pattern of bullying incidents repetitions and common locations and times in which bullying incidents happen.

Acknowledgments. The authors gratefully acknowledge the help of the expert panel for sharing their valuable knowledge and experience with responding to the questionnaire. We are thankful to Professor Dr J. Connolly (York University, Canada), Professor Dr F. Mishna (University of Toronto, Canada), Dr S. Drossaert (University of Twente, the Netherlands), J. van der Zwaan (University of Delft, the Netherlands), and seven other experts who prefer anonymity. The foundation of this study was laid during the visit of M. Dadvar to the York University, Canada, hosted by Professor Dr J. Huang, in January 2013.

References

1. Smith, P.K., et al.: Cyberbullying: Its nature and impact in secondary school pupils. Journal of Child Psychology and Psychiatry 49(4), 376–385 (2008)
2. Perren, S., et al.: Coping with Cyberbullying:ASystematic Literature Review. Final Report of the 'COST'IS 0801' (2012)
3. Campbell, M.A.: Cyber bullying: An old problem in a new guise? Australian Journal of Guidance and Counselling 15(1), 68–76 (2005)
4. Dinakar, K., Reichart, R., Lieberman, H.: Modeling the Detection of Textual Cyberbullying. In: Social Mobile Web Workshop at International Conference on Weblog and Social Media (2011)
5. Yin, D., et al.: Detection of harassment on Web 2.0. In: Proceedings of the Content Analysis in the WEB 2.0 (CAW2.0) Workshop at WWW 2009, Madrid, Spain (2009)
6. Dadvar, M., Trieschnigg, D., Ordelman, R., de Jong, F.: Improving cyberbullying detection with user context. In: Serdyukov, P., Braslavski, P., Kuznetsov, S.O., Kamps, J., Rüger, S., Agichtein, E., Segalovich, I., Yilmaz, E. (eds.) ECIR 2013. LNCS, vol. 7814, pp. 693–696. Springer, Heidelberg (2013)
7. Argamon, S., et al.: Mining the Blogosphere: Age, gender and the varieties of self-expression. First Monday 12(9) (2007)
8. Pazienza, M.T., Tudorache, A.G.: Interdisciplinary contributions to flame modeling. In: Pirrone, R., Sorbello, F. (eds.) AI*IA 2011. LNCS (LNAI), vol. 6934, pp. 213–224. Springer, Heidelberg (2011)
9. Dadvar, M., De Jong, F., Trieschnigg, D.: Expert knowledge for automatic detection of bullies in social networks. In: The 25th Benelux Conference on Artificial Intelligence (BNAIC 2013), Delft (2013)
10. Cha, M., et al.: I tube, you tube, everybody tubes: Analyzing the world's largest user generated content video system. In: Proceedings of the 7th ACM SIGCOMM Conference on Internet Measurement. ACM (2007)
11. Figueira, J., Greco, S., Ehrgott, M.: Multiple criteria decision analysis: State of the art surveys, vol. 78. Springer (2005)
12. Farah, M., Vanderpooten, D.: A multiple criteria approach for information retrieval. In: Crestani, F., Ferragina, P., Sanderson, M. (eds.) SPIRE 2006. LNCS, vol. 4209, pp. 242–254. Springer, Heidelberg (2006)
13. Xu, Z., Khoshgoftaar, T.M., Allen, E.B.: Application of fuzzy expert systems in assessing operational risk of software. Information and Software Technology 45(7), 373–388 (2003)

14. Witten, I.H., Frank, E., Hall, M.A.: Data Mining: Practical Machine Learning Tools and Techniques. Elsevier (2011)
15. Hall, M., et al.: The WEKA data mining software: An update. ACM SIGKDD Explorations Newsletter 11(1), 10–18 (2009)
16. Fielding, A.H., Bell, J.F.: A review of methods for the assessment of prediction errors in conservation presence/absence models. Environmental Conservation 24(1), 38–49 (1997)

Towards a Tunable Framework
for Recommendation Systems Based on Pairwise
Preference Mining Algorithms

Sandra de Amo and Cleiane G. Oliveira

Federal University of Uberlandia, Faculty of Computer Science,
Uberlandia, Brazil
deamo@ufu.br,cleiane.oliveira@ifnmg.edu.br

Abstract. In this article, we present *PrefRec*, a general framework
for developing RS using *Preference Mining* and *Preference Aggregation*
techniques. We focus on Pairwise Preference Mining techniques allowing
to predict which, between two objects, is the preferred one. A preliminary empirical study for analyzing the influence of the different factors
involved in each of the five modules of *PrefRec* is presented.

Keywords: recommender systems, preference mining, contextual preferences, preference aggregation.

1 Introduction

In recent years the use of Recommendation Systems (RS) has become ubiquitous
due to their capacity of predicting and recommending products (items) which
fit the user's needs or taste with great probability. The most popular approach
is the *Collaborative Filtering* (CF) ([8]). The main idea behind CF is that to
recommend items to a given user u (referred as the *active user*) the system
takes into account the past evaluations of *other users* who have similar taste as
u. Although being simple and achieving satisfactory results, CF RS faces some
important challenges. Users normally evaluate few items and so the available
data concerning the active user is *sparse*. Moreover, an item to be recommended
should be evaluated by users having similar taste as the active user (the so-called
item cold-start challenge).

The well-known technique *Content-Based* (CB) ([6]) aims at overcoming the
item cold-start challenge faced by the CF RS. The recommendations are based
on the intrinsic characteristics of the items. CB RS frequently use mining techniques originally designed for classification tasks in order to predict active user's
evaluations of new items not evaluated in the past. Although in a minor scale,
the *item cold-start* challenge is also faced by CB RS.

Most advanced RS used in nowadays commercial applications are not purely
CF or purely CB. They usually employ some kind of hybridization, mixing CF
and CB techniques in such a way that the drawbacks of the former are compen-
sated by the advantages of the latter and vice-versa [2,5,7].

M. Sokolova and P. van Beek (Eds.): Canadian AI 2014, LNAI 8436, pp. 282–288, 2014.

In this article we present **PrefRec**, a general framework for developing hybrid Recommendation Systems based on Pairwise Preference Mining and Preference Aggregation techniques. Such techniques aim at optimizing the item cold-start issue usually faced by recommendation systems, achieving better accuracy then some well-known hybrid RS.

2 The *PrefRec* Framework

The **PrefRec** framework is constituted by five modules, as illustrated in Fig. 1.

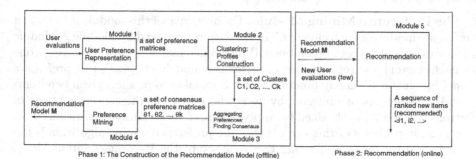

Fig. 1. The *PrefRec* General Schema

1. The Preference Representation Module. Let $\mathcal{U}$ be a set of user identifiers, $\mathcal{I}$ be a set of items and $\mathcal{K}$ be a set of users evaluations of items in $\mathcal{I}$. Each item is a tuple over schema $(item_{id}, A_1, A_2, ..., A_k)$ where $item_{id}$ is the item identifier, and A_i are item attributes. Each user evaluation is a triple (u_i, i_j, r_{ij}) where r_{ij} is the evaluation (rating) given by user u_i to the item with identifier i_j. The input of Module 1 is the set $\mathcal{R}$. The output is a set of matrices, one for each user u. Differently from the usual way as preferences of a user u are represented in state-of-art RS (a set of pairs (item,rating)), we represent the preferences of a user u by a $n \times n$ matrix ($\mathcal{M}pref$), where n is the cardinality of $\mathcal{I}$. The element $\mathcal{M}pref_{i,j}$ is a number $r \in [0,1]$ representing the fact that user u prefers item i to item j with a *degree of preference r*.

There are different functions that execute the task of Module 1. In [3] a family of functions h_n is proposed. The current version of **PrefRec** supports this family of functions given by $h_n(x) = \frac{x^n}{x^n+1}$. The element $\mathcal{M}pref_{i,j}$ of the preference matrix of user u is given by $h(\frac{rating(i)}{rating(j)})$ where $h \in \{h_n : n \geq 1\}$.

2. The Clustering Module. This module receives as input a set of preference matrices (one for each user) and returns a set of clusters $\{C_1, C_2, ..., C_n\}$, where each cluster C_i is constituted by a set of preference matrices which are similar in some sense. Two factors are crucial in Module 2: the distance function and the clustering algorithm. **PrefRec** supports different distance functions (Euclidean,

Manhattan, Cosseno, Dice, Jacquard) and three clustering algorithms based on different clustering paradigms: K-Means, Cure and DBSCAN.

3. The Preference Aggregation Module. The objective of this module is to produce a *preference consensus* for a group of users having similar taste. The module receives as input a set of preference matrices $\mathcal{P}$ and returns a unique preference matrix θ (called the *preference consensus* of the group) obtained by applying an aggregation operator on $\mathcal{P}$. Five aggregation operators are implemented in the current version of **PrefRec**, namely: *arithmetic mean, weighted mean* (the weight of a preference matrix being its silhouette coefficient with respect to the clusters $\{C_1, C_2, ..., C_n\}$), *fuzzy most quantifier, at least half quantifier, as many as possible quantifier* [3].

4. The Preference Mining Module. The objective of this module is to build a preference model for each cluster C_i in the output produced by Module 2. Module 4 receives as input a set of (consensual) preference matrices $\Theta = \{\theta_1, ..., \theta_k\}$ (one for each cluster) and returns a set of preference models $\{\mathcal{P}_1, ..., \mathcal{P}_k\}$. A *preference model* for cluster C_i is any function which is capable to predict, given two items i_1 and i_2, which one is preferred by users whose taste fit the consensual taste of cluster C_i. Each $\mathcal{P}_i$ is obtained by applying a preference mining algorithm on θ_i. The main hypothesis of this work is that the preference mining algorithm is the most important factor impacting the accuracy of the Recommendation Model produced by our methodology **PrefRec**. In fact, as shown in the next section, the large gap in accuracy between CPrefMiner [1] and CPrefMiner* [1], the two preference mining algorithms currently implemented in Module 4, reflects on the accuracy of the Recommendation Model.

Contextual Preference Model. Both algorithms CPrefMiner and CPrefMiner* produces as output a preference model constituted by a (compact) set of *contextual preference rules*. The details on this model as well as on the mining algorithm CPrefMiner can be found in [1].

The Recommendation Model. After the execution of Module 4, the recommendation model is finally built. It is constituted by a set $\mathcal{M} = \{(\theta_1, \mathcal{P}_1), ..., \theta_k, \mathcal{P}_k)\}$, where for each $i = 1, ..., k$, θ_i is the consensual preference matrix associated to cluster C_i and $\mathcal{P}_i$ is the preference model mined from θ_i.

5. The Recommendation Module. The objective of this module is to use the recommendation model $\mathcal{M}$ in order to recommend items to new users. It is executed online, differently from the previous ones which are executed offline. The algorithms used in the following steps are the same used in modules 1, 3 and 4 during the construction of $\mathcal{M}$: (1) Given an active user u and a (small) set $\mathcal{R}_u$ of evaluations provided by u on items in $\mathcal{I}$, the preference matrix $\mathcal{M}pref_u$ associated to $\mathcal{R}_u$ is obtained; (2) The similarity between $\mathcal{M}pref_u$ and each consensual preference matrix θ_i is calculated. Let θ^u be the preference matrix

---

[1] CPrefMiner* is a refinement of CPrefMiner and uses a voting technique for optimizing the predictions.

most similar to $\mathcal{M}pref_u$ and let $\mathcal{P}^u$ its corresponding preference model;(3) $\mathcal{P}^u$ is used to infer the preference between pairs of items in $\mathcal{I}$ which have not been evaluated by the user u in the past. From this set of pairs of items (i, j) indicating that user u prefers item i to item j one can built a *ranking* by applying the algorithm *Order By Preferences* introduced in [4].

3   Experimental Results

The datasets used in the experiments have been obtained from the GroupLens Project[2]. The original data are of the form (UserId, FilmId, rating). Details about the movies, namely Genre, Actors, Director, Year and Language have been obtained from the IMDB website[3]. We chose a set $\overline{U}$ of 296 users and a set $\overline{I}$ of 262 movies. The total amount of user evaluations is 67971. Our basic dataset of user evaluations (with 67971 evaluations) has 12% of sparsity. This dataset is denoted by DB_{100}. We considered five other datasets (Figure 2(a)) obtained from DB_{100} by eliminating from it a certain amount of evaluations.

BD	N. de Evaluations	Sparsity
DB_{100}	67.971	12.35%
BD_{90}	61.143	21.16%
DB_{80}	54.423	29.82%
DB_{70}	47.464	38.80%
DB_{60}	40.831	47.35%
DB_{50}	33.344	57.00%

(a)

DB_{100}	Accuracy		Coverture	
	precisao	*recall*	p_C	r_C
DBScan	**76.10%**	**73.85%**	63.16%	**49.96%**
CURE	75.02%	71.64%	**63.22%**	48.57%

(b)

Fig. 2. (a) The Datasets (b) Finding the *best* clustering algorithm

The framework **PrefRec** has been validated offline under a *5-cross-validation* protocol. Two quality criteria have been considered in our experiments: accuracy and coverture.

3.1   Analyzing the Different Factors

Table I presents the factors impacting at each module of **PrefRec**. The column *Values* shows the implementations currently available in each module concerning the different factors. The column *Basic Default* corresponds to the implementation considered in the tests for finding the best values in previous rows. The value presenting the best results are showed in column *Best* and are used in the experiments for detecting the best values for next factors.

---

[2] www.grouplens.org
[3] imdb.com

DB_{100}	Acuraccy		Coverture	
	precision	recall	precision	recall
A.Mean	**76.10%**	**73.85%**	63.16%	49.96%
W.Mean	75.97%	73.71%	**63.22%**	**50.12%**
F. Most	63.82%	60.69%	56.92%	43.50%
A.L. Half	63.47%	60.41%	57.83%	43.79%
A.M. Poss	63.83%	60.53%	59.05%	44.11%

(a)

DB_{100}	Accuracy		Coverture	
	p_G	r_G	p_C	r_C
CPrefMiner	76.10%	73.85%	63.16%	49.96%
CPrefMiner*	**89.16%**	**88.49%**	**79.30%**	**78.79%**

(b)

Fig. 3. (a) Finding the *best* aggr. operator (b) Finding the *best* mining algorithm

Best Values in Rows 1 and 2. The results of the experiments are not shown since they are not statistically significant. So, we decided to consider the values corresponding to the simpler functions.

Choice of the Clustering Algorithm (Row 3). The results of the experiments are shown in Figure 2(b). The performance of K-Means was much inferior to Cure and DBSCAN, so they are not presented. DBSCAN has been chosen as our "best" alternative for this factor, since it achieved the best results concerning *accuracy* and very similar and satisfactory *coverture precision*.

Choice of the Aggregation Operator (Row 4). The results of the experiments are shown in Figure 5(a). Notice that the arithmetic and weighted mean achieved the best results concerning both accuracy and coverture. The differences between their results are not significant. Based also on its simplicity, we decided to consider the arithmetic mean as the "best" alternative for the aggregation operator.

Choice of the Preference Mining Algorithm (Row 5). The experiments in this module allow to conclude that the most impacting factor in the *PrefRec* framework is the choice of the preference mining algorithm. The results are shown in Figure 5(b). Notice that both quality measures are affected by the better performance of CPrefMiner*. All measures are above 78%.

We call **XPrefRec** the instantiation of the method **PrefRec** with the *best* values obtained at the end of this preliminary empirical study.

Table 1. Factors impacting the quality of the Recommendation Model

Modules	Factor	Values	Basic Default	Best
Module 1	Normalization Function	$h_n (n \geq 1)$	h_1	h_1
Module 2	Distance Function	Euclidean, Dice, Manhattan, Cosseno, Jacquard	Cosseno	Cosseno
Module 2	**Clustering Algorithm**	K-Means, Cure, DBSCAN	K-Means	DBSCAN
Module 3	**Aggregation Operator**	Arith.Mean, Weighted Mean, Fuzzy Most, At Least Half, As Many as Possible	Arith. Mean	Arith. Mean
Module 4	**Pref. Mining Algorithm**	CPrefMiner, CPrefMiner*	CPrefMiner	CPrefMiner*

Table 2. Performance: XPrefRec versus CBCF

Datasets	Accuracy				Coverture			
	precision		recall		precision		recall	
	XPrefRec	CBCF	XPrefRec	CBCF	XPrefRec	CBCF	XPrefRec	CBCF
DB_{100}	**89.16%**	70.64%	**88.49%**	70.64%	**79.30%**	61.19%	**78.79%**	61.16%
DB_{90}	**87.71%**	69.59%	**85.10%**	69.59%	**78.74%**	61.60%	**76.90%**	61.57%
DB_{80}	**83.32%**	64.88%	**83.12%**	64.88%	**73.28%**	61.25%	**73.07%**	61.23%
DB_{70}	**80.38%**	62.47%	**78.39%**	62.47%	**72.84%**	64.72%	**66.21%**	64.72%
DB_{60}	**77.20%**	52.07%	**68.56%**	52.07%	**73.19%**	60.89%	45.83%	**60.87%**
DB_{50}	**75.95%**	37.39%	**66.10%**	37.39%	**74.63%**	51.00%	40.40%	**50.98%**

3.2 Comparing *XPrefRec* and the Baseline CBCF

In order to validate the instantiation *XPrefRec* of our framework, we chose as baseline the CBCF recommendation system introduced in [7]. The reason for this choice are the following: (1) both systems adopt a hybrid approach mixing CF and CB techniques; (2) both systems focus on reducing the *item cold-start* effect; (3) both systems use mining techniques. CBCF uses a *classifier* to predict the missing *ratings* in the input data whereas *XPrefRec* uses a *preference mining* algorithm in order to build a set of preference models which will be used in the final recommendation model.

Accuracy and Coverture. The results about the accuracy and coverture of both methods are shown in Table 2. The experiments have been executed on six different datasets with different sparsity rate.

The accuracy results achieved by *XPrefRec* are much superior than those achieved by CBCF. This is due mainly by the quality of the mining technique adopted by each recommendation method. Concerning the coverture, the capacity of the systems to deal with the item cold-start effect, *XPrefRec* is far better than CBCF for all datasets, except for the two last ones where the coverture recall is better for CBCF.

Execution Time. The execution time of the three steps (building the model, checking the colaborators and making the recommendation of 50 itens among 1300 items) for XPrefRec are 6 min 19 sec, 19 msec and 69 sec 627 msec respectively. For CBCF these steps take 54 min 14 sec, 3 sec 386 msec and 16 min 27 sec respectively.

4 Conclusion

In this paper we proposed *PrefRec*, a general flexible framework for implementing Recommendation Systems based on a hybrid approach using Pairwise Preference Mining and Preference Aggregation techniques. In the short term we intend to elaborate a more rigorous factorial design of *PrefRec*.

Acknowledgements. We would like to thank the Brazilian Research Agencies CAPES, FAPEMIG and CNPq for supporting this work.

References

1. de Amo, S., Bueno, M.L.P., Alves, G., Silva, N.F.: CPrefMiner: An Algorithm for Mining User Contextual Preferences Based on Bayesian Networks. In: ICTAI 2012, pp. 114–121 (2012)
2. Burke, R.: Hybrid Recommender Systems: Survey and Experiments. User Modeling and User-Adapted Interaction 12(4), 331–370 (2002)
3. Chiclana, F., Herrera, F., Herrera-Viedma, E.: Integrating multiplicative preference relations in a multipurpose decision-making model based on fuzzy preference relations. Fuzzy Sets and Systems 122, 277–291 (2001)
4. Cohen, W.W., Schapire, R.E., Singer, Y.: Learning to Order Things. Journal of Artificial Intelligence Research 10(1), 243–270 (1999)
5. Lops, P., de Gemmis, M., Semeraro, G.: A Content-collaborative Recommender That Exploits WordNet-based User Profiles for Neighborhood Formation. User Modeling and User-Adapted Interaction 17, 217–255 (2007)
6. Lops, P., de Gemmis, M., Semeraro, G.: Content-based Recommender Systems: State of the Art and Trends. In: Rec. Systems Handbook, pp. 73–105. Springer, Heidelberg (2011)
7. Melville, P., Mooney, R.J., Nagarajan, R.: Content-boosted Collaborative Filtering for Improved Recommendations, pp. 187–192. AAAI, Menlo Park (2002)
8. Su, X., Khoshgoftaar, T.M.: A survey of collaborative filtering techniques. Journal Advances in Artificial Intelligence 2009 (January 2009)

Belief Change and Non-deterministic Actions

Aaron Hunter

British Columbia Institute of Technology
Burnaby, BC, Canada
aaron_hunter@bcit.ca

Abstract. Belief change refers to the process in which an agent incorporates new information together with some pre-existing set of beliefs. We are interested in the situation where an agent must incorporate new information after the execution of actions with non-deterministic effects. In this case, the observation plays two distinct roles. First, it provides information about the current state of the world. Second, it provides information about the outcomes of any actions that have previously occurred. While the literature on belief change has extensively explored the former, we suggest that existing approaches to belief change have not explicitly considered how an agent uses observed information to determine the effects of non-deterministic actions. In this paper, we propose an approach in which action effects simply progress the agent's underlying plausibility ordering over possible states. In the case of non-deterministic actions, new possible world trajectories are created and then subsequently dismissed as dictated by observations.

1 Introduction

We are interested in the dynamics of belief in domains where actions may have non-deterministic effects. In this context, an agent must perform an action and then observe the effects. As such, the belief change that occurs in this context is actually a kind of *belief revision* that takes into account the possible effects of actions that occurred prior to an observation. In this paper, we develop a simple account of belief change in this context based on the idea that the effect of an action is essentially to "rule out" certain possible states. We illustrate that this approach allows us to model belief change following non-deterministic actions in a Markovian manner, while remaining consistent with our previous work on iterated belief change due to actions [5].

1.1 Motivating Example

Consider a simple scenario involving a single agent and a lamp with a switch for toggling it on or off. When the switch is toggled, the normal result is to change the state of the lamp. The simplest model of this example would involve only deterministic action effects. However, on some occasions, toggling the switch to the on position may actually causes a fuse to break. In this case, the lamp will not turn on. This can be observed by the agent after toggling the switch. It is convenient to assume that the state of the lamp must be explicitly observed after the toggling action, so we may assume that the lamp is inside a box or behind a closed door. We revisit this example throughout the paper.

M. Sokolova and P. van Beek (Eds.): Canadian AI 2014, LNAI 8436, pp. 289–294, 2014.

2   Background

Let $\mathbf{F}$ denote a set of propositional fluent symbols. A *formula* over $\mathbf{F}$ is defined in the usual way, using connectives $\neg$, $\wedge$, $\vee$ and $\rightarrow$. A *belief set* is a set of formulas. A *state* s is a propositional interpretation of $\mathbf{F}$, which can be associated with the subset of elements of $\mathbf{F}$ that are true in s. Let S denote the set of all states. For any set X, we set $\mathcal{P}(X)$ denote the power set of X.

Let $\mathbf{A}$ denote a set of action symbols. The effects of actions are described by a *transition function*, which can be *deterministic* or *non-deterministic*. A deterministic transition function is a function $f : S \times \mathbf{A} \rightarrow S$, so the output is a single state. A non-deterministic transition function is a function $f : S \times \mathbf{A} \rightarrow \mathcal{P}(S)$. Hence, a non-deterministic transition function takes a state and an action as arguments, then it returns a set of states. Informally the output is the set of states that may result from executing the given action in the given state.

Belief change refers to the process in which an agent's beliefs change in response to new information. A *belief revision* operator captures the change that occurs when an agent obtains new information about an unchanged world. One of the most influential approaches to belief revision is the AGM approach. A belief change operator $*$ is said to be an *AGM revision operator* if it satisfies the *AGM postulates*, as specified in [1]. Every operator that satisfies the AGM postulates can be characterized by minimization on an underlying total pre-order on the set of states [9].

Since we are interested in reasoning about action effects, we will also be interested in *belief progression* [10]. Belief progression refers to the process where an agent has some initial belief set that must be changed as a result of an action that is executed.

3   The Approach

We define epistemic states using a variation of the standard definition from [2].

Definition 1. *An epistemic state is a pair* $E = \langle P, \preceq \rangle$ *where* $P \subseteq S$ *and* $\preceq$ *is a total pre-order on* P.

We think of P as the set of states that are possible, and $\preceq$ as a plausibility ordering over the possible states. We define $Bel(E)$ to be the set of $\preceq$-minimal states. If $P = S$, then we simply have a total pre-order over states. However, by allowing P to change as actions are executed, we can provide a more complete model of belief change. In particular, it allows us to distinguish between unlikely and impossible states.

We define an *observation* to be a set of states, and we use the term *belief revision operator* to refer to any total function that maps a total pre-order and an observation to a new total pre-order over states. This is a very wide class of functions that includes all Darwiche-Pearl revision operators.

3.1   Belief Progression and Belief Revision

We define a belief progression operator on pre-orders over states. Let $P \subseteq S$, let f be a deterministic transition function and let A be an action symbol. Then define

$$P_{(f,A)} = \{s \in P \mid s \in f(s_1, A) \text{ for some } s_1 \in P\}.$$

For example, if $P = S$, then $P_{(f,A)}$ is simply the set of possible outcomes of the action A. We will also write $P_{(f,\overline{A})}$ for a sequence $\overline{A}$ of actions, with the obvious meaning.

Definition 2. *Let $\preceq$ be a pre-order over $P \subseteq S$ and let f be a transition function. For any action symbol A define the pre-order $\preceq_A$ over $P_{(f,A)}$ as follows. If $s = f(s_1, A)$ and $t = f(t_1, A)$, then: $s \preceq_A t \iff s_1 \preceq t_1$.*

Clearly $\preceq_A$ is a total pre-order with respect to $P_{(f,A)}$, but it is only a partial pre-order with respect to S. Intuitively, if the action A is executed from a state in P, then the states in $P_{(f,A)}$ are no longer possible. We think of $\preceq_A$ as the ordering obtained by "shifting" $\preceq$ in accordance with the effects of the action A. We can now define the desired progression operator.

Definition 3. *Given a deterministic transition function f and an action symbol A, the progression operator $\diamond$ is the function such that $\langle P, \preceq \rangle \diamond A = \langle P_{(f,A)}, \preceq_A \rangle$.*

Let $*$ be any revision operator that takes two inputs: a total pre-order on states, and an observation expressed as a set of possible states. We extend $*$ to our new conception of an epistemic state as follows:

$$\langle P, \preceq \rangle * \alpha = \langle P, \preceq *(\alpha \cap P) \rangle.$$

Hence, we simply remove all states from α that are not present in P before performing the revision. If P only changes due to action effects, this simply removes all states from our observations that have been ruled out by the actions that have been executed.

Proposition 1. *Let T be a set of states. For any $\preceq$, α we have $\langle T, \preceq \rangle * \alpha = \langle T, \preceq *\alpha \rangle$.*

If T is the set S of all states, then this result implies that our approach is equivalent to regular belief revision in the case where no actions are executed.

For the action sequence $\overline{A}$, we write $\langle \overline{A}, \preceq \rangle$ as a shorthand for $\langle S_{(f,A)}, \preceq \rangle$. Belief progression can then be expressed as concatenation. Given any action sequence $\overline{A}$ and any action B, let $\overline{A} \cdot B$ denote the concatenation of B to the end of $\overline{A}$.

Proposition 2. *For any $\preceq$, $\overline{A}$, B it follows that $\langle \overline{A}, \preceq \rangle \diamond B = \langle \overline{A} \cdot B, \preceq \rangle$.*

The results in this section indicate that belief revision and belief progression can be handled independently in our framework. Sequences of actions just limit the domain of possible states, while revisions operate independently on the states that are permitted.

3.2 The Lamp Problem

Consider the deterministic version of the lamp problem. Our vocabulary involves two fluent symbols: *LampOn* and *FuseBroken*. Let $\preceq$ denote the initial pre-order over states and let α denote an observation. Suppose that the lamp is initially on, the switch is toggled, and then the agent observes that the lamp is off. If we identify a state with its set of true fluent symbols, then we have the following representation of our scenario:

- $\min(\preceq) = \{\{LampOn\}, \{LampOn, FuseBroken\}\}$
- $\alpha = \{\emptyset, \{FuseBroken\}\}$.

Let $*$ be a belief revision operator. If the effects of A are deterministic, we can solve Lamp-type problems by computing $\langle S, \preceq \rangle \diamond A * \alpha$, where A denotes the toggle action.

4   Non-deterministic Action Effects

4.1   Partially Determined Epistemic States

Observations provide information about the state of the world, as well as information about the effects of non-deterministic actions. For non-deterministic actions, $P_{(f,A)}$ is not a single set of states - it is a set of sets of states. Moreover, the ordering $\preceq_A$ is not well defined if A has non-deterministic effects.

To solve these problems, we extend revision and progression to a wider class of epistemic states. We call a set of epistemic states a *partially determined* epistemic state. We associate a belief set with a partially determined epistemic state $\mathcal{D}$ as follows:

$$Bel(\mathcal{D}) = \bigcup \{\min(\preceq) \mid \langle P, \preceq \rangle \in \mathcal{D}\}.$$

Note that the domain in each element of $\mathcal{D}$ may be different, so the believed states may not be possible in every constituent epistemic state. This allows an agent to reason hypothetically about action effects. One is able to believe statements of the form "if the glass broke, then my hand is likely bleeding." The following result is straightforward.

Proposition 3. *If* $\mathcal{D} - \langle \emptyset, \preceq \rangle \neq \emptyset$, *then* $Bel(\mathcal{D}) = Bel(\mathcal{D} - \langle \emptyset, \preceq \rangle)$.

Let f be a non-deterministic transition function, so $f(s, A)$ is a set of states. We extend the definition of $\diamond$ to non-deterministic actions as follows:

$$\langle P, \preceq \rangle \diamond A = \{\langle P, \preceq \rangle \diamond O_A \mid O_A \in f(s, A)\}.$$

So the progression by a non-deterministic action is just the set of all possible progressions corresponding to possible effects of the action. If $\mathcal{D}$ is a partially determined epistemic state, then we define

$$\mathcal{D} \diamond A = \{\langle P, \preceq \rangle \diamond A \mid \langle P, \preceq \rangle \in \mathcal{D}\}$$
$$\mathcal{D} * \alpha = \{\langle P, \preceq \rangle * \alpha \mid \langle P, \preceq \rangle \in \mathcal{D} \text{ and } P \cap \alpha \neq \emptyset\}.$$

Note that we could have simply defined revision in a pointwise manner as we did for progression. But then the resulting epistemic state might contain inconsistent components; by Proposition 3, these inconsistent components do not affect the belief set.

For every A and α it is easy to verify that $|\mathcal{D}| \leq |\mathcal{D} \diamond A|$ and $|\mathcal{D}| \geq |\mathcal{D} * \alpha|$. This seems reasonable. Performing actions adds new outcomes, whereas making observations eliminates branches that are deemed to have not occurred.

4.2   The Lamp Problem, Revisited

We introduce action effects that are non-deterministic, with effects given in Figure 1. Suppose that an agent initially believes the lamp is off and the fuse is unbroken. After toggling the switch, the agent observes that the lamp is still off. We have the following:

- $\min(\preceq) = \{\emptyset\}$
- $\alpha = \{\emptyset, \{FuseBroken\}\}$.

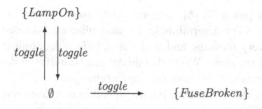

Fig. 1. The Lamp Domain

We are interested in computing $\langle S, \preceq \rangle \diamond A * \alpha$ where A denotes the toggle action. Note that $\langle S, \preceq \rangle \diamond A$ contains two epistemic states. In particular, if O_1 denotes the outcome where the fuse breaks, and O_2 denotes the outcome where the lamp turns on, then

$$\langle S, \preceq \rangle \diamond A = \{\langle \{FuseBroken\}, \preceq \diamond O_1 \rangle, \langle \{LampOn\}, \preceq \diamond O_2 \rangle \}.$$

Note that the intersection $\alpha \cap \{LampOn\} = \emptyset$. As a result, we have:

$$\langle S, \preceq \rangle \diamond A * \alpha = \{\langle \{FuseBroken\}, \preceq \diamond O_1 * \alpha \rangle.$$

Regardless of the ordering, the final belief state will be the singleton $\{FuseBroken\}$.

Prior to the observation that the lamp is off, the partially determined epistemic state includes the state where toggling the switch caused the lamp to turn on. This possible outcome is removed after the contradictory observation. As such, our formal model makes the dual role of observations in the case of non-deterministic actions explicit.

5 Discussion

5.1 Related Work

In previous work, we advocated that expressions of the form $\kappa \diamond A * \alpha$ should be evaluated by regressing α to $\alpha^{-1}(A) = \{s \mid (s \diamond A) \in \alpha\}$ and then revising the initial belief state [5]. The following result indicates that the approach in this paper is consistent with our previous approach in the case of deterministic action effects.

Proposition 4. *Let* $\langle P, \preceq \rangle$ *be an epistemic state, let* $A \in \mathbf{A}$ *be an action with deterministic effects and let* $\alpha \subseteq S$. *Then:* $Bel(\langle P, \preceq \rangle \diamond A * \alpha) = Bel(P) * \alpha^{-1}(A) \diamond A$.

For deterministic actions, the approach taken in this paper is actually very similar to the approach described in [3]. Our notation is different, and we do not make the same assumptions about the effects of actions; but the central idea in both cases is that an action occurrence limits the domain for revision.

5.2 Application: Secure Message Passing

We have previously argued that belief change operators have a role to play in identifying attacks on cryptographic protocols [4], and we have used formal models of action for the verification of protocols [6]. We previously used a model in which the effects of sending a message were deterministic. In particular, we assumed that a malicious intruder intercepts every message and looks at it, before deciding if it should be forwarded to the desired recipient. While this model can be useful, it does not necessarily represent the way agents reason about message exchange.

A more accurate model of message passing involves non-deterministic effects. Consider an agent A that sends an encrypted number N over an insecure network as a challenge to authenticate another agent. The effect of this action is either $has_B(N)$ or has $has_I(N)$, where B is the intended recipient and I is an intruder. In order to interpret the meaning of messages received in the future, A should try to account for this non-determinism by finding evidence that may indicate who received the initial message.

5.3 Conclusion

In this paper, we have explored belief change in the context of non-deterministic action effects. We have suggested that the role played by actions in reasoning about beliefs is essentially to shift the underlying plausibility ordering over states, and also to eliminate certain states from possibility. We have demonstrated that this approach can be particularly useful in situations where action effects may be non-deterministic, as it allows us to make the dual notion of observations explicit: observations give us information about the state of the world, and they also give us information about the effects of preceding actions. Finally, we suggested that this approach may be useful in reasoning about secure message passing systems. In future work, we intend to explore this application.

References

1. Alchourrón, C.E., Gardenfors, P., Makinson, D.: On The Logic of Theory Change: Partial Meet Functions for Contraction and Revision. Journal of Symbolic Logic 50(2), 510–530 (1985)
2. Darwiche, A., Pearl, J.: On the Logic of Iterated Belief Revision. Artificial Intelligence 89(1-2), 1–29 (1997)
3. Delgrande, J.P.: Considerations on Belief Revision in an Action Theory. In: Erdem, E., Lee, J., Lierler, Y., Pearce, D. (eds.) Correct Reasoning. LNCS, vol. 7265, pp. 164–177. Springer, Heidelberg (2012)
4. Hunter, A., Delgrande, J.P.: Belief Change and Cryptographic Protocol Verification. In: Proceedings of AAAI (2007)
5. Hunter, A., Delgrande, J.P.: Iterated Belief Change Due to Actions and Observations. Journal of Artificial Intelligence Research 40, 269–304 (2011)
6. Hunter, A., Delgrande, J.P., McBride, R.: Protocol Verification in a Theory of Action. In: Zaïane, O.R., Zilles, S. (eds.) Canadian AI 2013. LNCS (LNAI), vol. 7884, pp. 52–63. Springer, Heidelberg (2013)
7. Konieczny, S., Pino-Perez, R.: On the Logic of Merging. In: Proceedings of the Conference on Principles of Knowledge Representation and Reasoning (KR 1998), pp. 488–498 (1998)
8. Katsuno, H., Mendelzon, A.O.: On The Difference Between Updating a Knowledge Base and Revising It. In: Proceedings of the Second International Conference on Principles of Knowledge Representation and Reasoning (KR 1991), pp. 387–394 (1991)
9. Katsuno, H., Mendelzon, A.O.: Propositional Knowledge Base Revision and Minimal Change. Artificial Intelligence 52(3), 263–294 (1992)
10. Lang, J.: About Time, Revision, and Update. In: Proceedings of the 11th International Workshop on Non-Monotonic Reasoning, NMR 2006 (2006)

Robust Features for Detecting Evasive Spammers in Twitter

Muhammad Rezaul Karim[1] and Sandra Zilles[2]

[1] Department of Computer Science, University of Calgary, Canada
mrkarim@ucalgary.ca
[2] Department of Computer Science, University of Regina, Canada
zilles@cs.uregina.ca

Abstract. Researchers have designed features of Twitter accounts that help machine learning algorithms to detect spammers. Spammers try to evade detection by manipulating such features. This has led to the design of robust features, i.e., features that are hard to manipulate. In this paper, we propose and evaluate five new robust features.

1 Introduction

Twitter, one of the most popular social networking and microblogging services, is used by spammers to spread malware, to post phishing links as well as for advertising. Researchers have used machine learning to detect spam accounts in Twitter [1, 4, 7, 10–12], mostly relying on features of the spam account. Though most existing features are capable of distinguishing current spammers from legitimate users, they might become less effective as spammers' techniques evolve. For example, evasive spammers can easily manipulate features like *number of followers* [11] or *F-F Ratio* (i.e., the ratio of the number of followings to the number of followers) [4] by purchasing followers from third-party websites or exchanging followers from third-party websites [12]. Empirical results indicate that 30% of the spammers may remain undetected due to their dual behavior [1].

Most of the features studied so far and ranked highly by various feature selection algorithms are not robust [12]. A feature is considered robust if the manipulation of its value is expensive in terms of some cost measures (money, time, effort etc.). In this paper, we propose new robust graph-based features (concerning the Twitter graph), content-based features (concerning the content of the tweets) and neighbour-based features (concerning the followings of an account). The graph-based features are based on the concept of eigenvector centrality [2] and Laplacian centrality [8] of the users in the Twitter graph, while the proposed content-based features are based on the optimal matching [9] and LSA-based semantic similarity of tweets [6]. The neighbour-based feature that we propose is based on the reputation of the accounts a user is following.

We conduct experiments showing that the proposed features, in particular the graph-based ones, can improve the performance of machine learning algorithms when used with some existing robust features.

M. Sokolova and P. van Beek (Eds.): Canadian AI 2014, LNAI 8436, pp. 295–300, 2014.

2   Existing and Newly Proposed Features

In Twitter, users can post short text messages called *tweets*. *Following* an account means subscribing to tweets posted by the account. If user A follows user B, A is a follower of B and as a subscriber, user A can see a tweet posted by user B as soon as it is posted. B is then called a following account of A. Researchers model the Twitter network as a graph, where each user is a node and edges are based on user relationships. In this paper, we follow the directed graph model adopted by Wang [11]. In this model, an edge from A to B means that A follows B.

A feature of a Twitter account is considered robust if it is difficult or expensive to evade [12]. Existing approaches on spam account detection can be classified into two categories: *Standard Spammer Detection Approaches* and *Evasive Spammer Detection Approaches*. Approaches belonging to the first category, such as those reported in [1, 4, 7, 11], ignore the importance of the robustness of the features and assume that spammers do not manipulate their account features.

Each approach in the literature uses a combination of features of different types, for example *profile-based features* [1, 4, 7, 11, 12] (extracted from the user profile information), *content-based features* [1, 4, 7, 11, 12] (relying on the content of the tweets posted by the users), *graph-based features* [4, 12] (extracted from the Twitter graph), and *neighbourhood-based features* [12].

Some of our features use word similarity measures. In natural language processing, various methods are used to compute word-to-word semantic similarity. One approach is to use thesauri; another is to apply statistical methods like, e.g., Latent Semantic Analysis (LSA) [6], which derives a vector representation for each word based on word co-occurrences. The similarity between any two words is then the *cosine* (normalized dot product) of their LSA vectors.

We propose five features for detecting evasive spammers, all of which are hard to manipulate and hence robust.

Eigenvector centrality (graph-based). Eigenvector centrality weighs direct and indirect connections, based on their centralities, and thus takes the full network structure into account [2]. Let $A = [a_{ij}]_{n \times n}$ be the adjacency matrix of a graph G over n nodes. The eigenvector centrality of a node v_i is defined as $\frac{1}{\lambda} \sum_{j=1}^{n} a_{ij} c(v_j)$, where λ is the largest eigenvalue of A, and $c = (c(v_1), \ldots, c(v_n))$ is the corresponding eigenvector [2]. We expect that legitimate users have friends, relatives and colleagues as followers and followings, and that some of their followers have a fair number of connections and high centrality values, which, in turn, contributes to the high eigenvector centrality of a legitimate user. Spammers blindly follow a large number of legitimate users without being followed back. The followers a spammer might buy are expected to have few followers themselves, thus not contributing to the spammer's eigenvector centrality.

Laplacian centrality (graph-based). Laplacian centrality [8] was recently proposed to analyze terrorist networks and is based on the Laplacian energy of a directed graph. The latter reflects internal connectivity [3]. The Laplacian energy for a directed graph is the sum of the squared out-degrees of its nodes. The Laplacian centrality of a node is the decrease of the Laplacian energy in the graph when removing this node. It measures how much "damage" is done to the graph structure due to the removal of the node [8]. Laplacian centrality targets centers of network sub-communities. As spammers probably have poorly connected neighbours and form networks without community structure, their removal from the Twitter graph should have little effect on its Laplacian energy.

Optimal matching similarity (content-based). We compute the average semantic similarity between any two tweets posted by a user by optimal matching [9]. In this approach, the words from one sentence form the set $X = \{x_1, \ldots, x_n\}$, while the words from the other sentence form the set $Y = \{y_1, \ldots, y_n\}$ (preprocessing makes the sentences have equal numbers of words). The optimal matching is a permutation π of $\{1, \ldots, n\}$ that yields the maximum value of the objective function $\sum_{i=1}^{n} w(x_i, y_{\pi(i)})$. Here $w(x, y)$ is the word-to-word similarity between $x \in X$ and $y \in Y$, computed using LIN [5]. Spammers who post tweets with different words but with the same meaning (in order to evade features based on syntactic similarity) will have a high value for this feature.

LSA similarity (content-based). LSA-based word-to-word similarity can be extended to computing the similarity of two sentences. For each sentence, a single vector is computed as a (weighted) sum of the LSA vectors of the individual words [6]. Then the cosine between the resulting vectors is computed.

Average neighbours' reputation (neighbour-based). Inspired by [12], we propose a neighbour-based feature called *average neighbours' reputation*. It refers to the average value of the feature *reputation* [11] over the account's followings. Spammers can try to manipulate this feature value by buying new accounts and following those new accounts. Then they have to change the number of followings and the number of followers of those purchased accounts to make their reputation value look like legitimate users, which requires substantial effort.

3   Experimental Evaluation

We use a random selection of 350 spammers and 400 legitimate users from the labeled data collected by Yang at el. [12]. For most users, the neighbour information required to compute the graph-based feature values is provided in the data. Since a few neighbour entries are missing in the data, the resulting graph-based feature values are only approximations; however, like Yang et al., we assume that not too much information is missing from the data.

We conduct our experiments using five machine learning classifiers: *Sequential Minimal Optimization (SMO), Random Forest, J48, Naive Bayes* and *Decorate* as implemented in WEKA[1]. For each classifier, we use 10-fold cross validation to compute several performance metrics. PAJEK[2] and GEPHI[3] are used to extract graph-based features, while for computing semantic similarity, we use the SEMILAR[4] semantic similarity toolkit. To compute content-based features, we used the 40 most recent tweets posted by each user. To compute tweet similarity, LSA

---

[1] http://www.cs.waikato.ac.nz/ml/weka/

[2] http://pajek.imfm.si/doku.php?id=pajek

[3] https://gephi.org

[4] http://deeptutor2.memphis.edu/Semilar-Web

Table 1. Ranking of all features, sorted by average rank obtained from three ranking methods. The newly proposed features are highlighted in bold.

Feature	Category	Robustness	Reference	InfoGain	ReliefF	Chi-Sq	Avg. Rank
bidirectional links ratio	Graph	Medium	[12]	1	2	1	1.33
bidirectional links	Graph	Low	[4]	2	10	2	1.66
F-F Ratio	Profile	Low	[1, 4, 12]	3	3	3	3.00
reputation	Profile	Low	[11]	4	1	4	3.00
eigenvector centrality	**Graph**	**High**	–	7	4	7	6.00
number of followers	Profile	Low	[1, 11]	5	9	5	6.33
reply ratio	Content	Low	[1, 4, 11]	12	5	10	9.00
Laplacian centrality	**Graph**	**High**	–	9	7	13	9.66
betweenness centrality	Graph	High	[12]	6	19	8	11.00
age	Profile	High	[4, 12]	11	11	11	11.00
local clustering coefficient	Graph	High	[12]	8	22	6	12.00
URL ratio	Content	Low	[1, 4, 11, 12]	10	18	9	12.33
unique URL ratio	Content	Low	[4, 12]	13	13	12	12.66
number of followings	Profile	Low	[1, 11, 12]	14	14	14	14.00
average neighbours' reputation	**Neighbour**	**Medium**	–	18	6	19	14.33
average neighbours' followers	Neighbour	Low	[12]	15	17	15	15.66
average neighbours' tweets	Neighbour	Low	[12]	20	8	20	16.00
optimal matching similarity	**Content**	**High**	–	19	15	18	17.33
tweets per day	Profile	Low	[4]	17	20	16	17.66
tweet similarity	Content	Low	[4, 12]	16	21	17	18.00
hashtag ratio	Content	Low	[1, 11]	22	12	22	18.66
LSA similarity	**Content**	**High**	–	21	16	21	19.33
retweet ratio	Content	Low	[1, 7]	23	23	23	23.00

similarity and optimal matching similarity, pairwise similarities were averaged over all pairs formed from the 40 most recent tweets posted by each user.

Table 1 lists the features we experimented with; for the existing ones, the robustness value is adopted from Yang et al. [12]. We divide our experiments into two groups. Group (A) combines robust and non-robust features. Feature set (A-1) contains only the 18 existing features, while (A-2) has all 23 features. Group (B) uses only the robust features. Set (B-1) consists of 4 existing highly or medium robust features: *age, local clustering coefficient, betweenness centrality, bidirectional links ratio*, while (B-2) has the same 4 existing robust features and the 5 proposed features. (B-3) contains only the newly proposed features.

Table 2 shows the results for (A-1) and (A-2). The addition of our features does not affect the classifier performance much. However, we argue that the existing data still contain relatively few evasive spammers. As spammers start to use more sophisticated methods in order to avoid detection, we will likely need to rely more on robust features. Table 3 reports on group (B). Here, all classifiers improve their performance when our newly proposed robust features are added to the existing robust features (most notably for *SMO* and *Naive Bayes*, where accuracy increases from less than 70% for (B-1) and (B-3) to over 83% for (B-2), and from 76% for (B-1) and roughly 70% for (B-3) to over 80% for (B-2), resp.). Similar improvements are seen in the F-measure and FP rate. These results suggest that, our newly proposed features can help to improve classifier performance when combined with the previously studied features.

It is conceivable that some of our proposed features contribute more to the success of (B-2) than others. We hence ran standard feature ranking algorithms implemented in WEKA, namely *Chi-Square, Information Gain* and *ReliefF*.

Table 2. Accuracy, F-measure and False Positive (FP) Rate using both robust and non-robust features. Highlighted in bold are the best results in the experiment set when the difference is at least 0.005 (i.e., 0.5% for accuracy).

Classifier	Without Our Features (A-1)			With Our Features (A-2)		
	Accuracy(%)	F-measure	FP	Accuracy(%)	F-measure	FP
Random Forest	93.467	0.935	0.065	93.333	0.933	0.066
SMO	**93.867**	**0.939**	**0.059**	92.400	0.924	0.074
J48	90.400	0.904	0.095	90.400	0.904	0.096
Decorate	92.267	0.923	0.075	**93.067**	**0.931**	**0.068**
Naive Bayes	87.866	0.879	0.124	87.600	0.876	0.125

Table 3. Classification results with feature sets containing robust features only

Classifier	Existing Robust Feat. (B-1)			All Robust Feat. (B-2)			New Robust Feat. (B-3)		
	Accuracy(%)	F-measure	FP	Accuracy(%)	F-measure	FP	Accuracy(%)	F-measure	FP
Random Forest	92.400	0.924	0.077	**94.400**	**0.944**	**0.055**	87.600	0.876	0.125
SMO	69.733	0.697	0.293	**83.733**	**0.837**	**0.152**	68.530	0.671	0.284
J48	91.467	0.915	0.084	**92.000**	**0.920**	**0.080**	84.530	0.845	0.156
Decorate	91.467	0.915	0.081	**93.067**	**0.931**	**0.069**	89.200	0.892	0.107
Naive Bayes	76.000	0.755	0.257	**80.533**	**0.805**	**0.189**	69.860	0.694	0.282

The results are shown in Table 1. The proposed graph-based features are ranked highly by all the feature ranking algorithms and are among the top 8 features of all 23 and among the top 3 of the 9 robust features. Our three other proposed features, however, seem not as effective as the proposed graph based features. In order to determine whether or not their addition is useful, we ran experiments on four further features sets: all previously studied features plus our two graph-based features (A-1 graph), all previously studied *robust* features plus our two graph-based features (B-1 graph), the top 8 features in the ranked list of all features (Top 8), and the 4 top-ranked *robust* features (Top 4R). The results are shown in Table 4, in comparison to (A-2) and (B-2). We present only accuracy values here, F-measure and FP rate performances showed similar relationships. Adding only our proposed graph-based features to the set of all existing features yields a performance comparable to that of using all features. When focusing only on robust features though, adding *optimal matching similarity, LSA similarity*, and *average neighbours' reputation* slightly improves the results over (B-1 graph) for four Machine Learning algorithms. The results worsen slightly again when using only the top-ranked features. This suggests that semantic similarity might gain importance when spammers invest more effort into evading detection.

That our semantic similarity features are ranked poorly in Table 1 can be explained by the fact that our used data set contains users who post tweets in different languages and the proposed semantic feature values are computed based on English words only. To test further the effectiveness of these features, we use a reduced data set of 400 users posting English tweets only. In this case, the rank of the feature *optimum matching similarity* improves for all the feature ranking algorithms and for *Information Gain*, it reduces from 19 to 14. Another difficulty with the used data set is that it does not consider advertisers as spammers unless they post malicious links, which also reduces the effectiveness of

Table 4. Comparison of various features sets – accuracy values

Classifier	Including Non-Robust F.			Only Robust F.		
	(A-2)	(A-1 graph)	(Top 8)	(B-2)	(B-1 graph)	(Top 4R)
Random Forest	**93.300**	92.800	89.460	**94.400**	92.930	92.130
SMO	92.400	**92.800**	90.133	**83.733**	82.000	79.860
J48	90.400	90.800	90.800	**92.000**	91.330	90.266
Decorate	93.067	93.333	90.930	93.067	93.460	90.400
Naive Bayes	87.600	88.000	87.460	**80.533**	79.333	75.730

the similarity based features like *tweet similarity* or *optimum matching similarity* in distinguishing spammers from legitimate users.

Note that highly ranked non-robust features like *F-F Ratio, reputation* and *number of followers* may become less effective than robust features over time.

In summary, our proposed robust features can be helpful, when used along with the existing robust features. Feature ranking suggests that the proposed graph-based features are very effective in discovering spammers. It will be interesting to experiment with data sets that contain more evasive spammers, when such data becomes available. We expect all our proposed features to become more effective as spammers use more sophisticated methods for evading detection.

References

1. Benevenuto, F., Magno, G., Rodrigues, T., Almeida, V.: Detecting spammers on Twitter. In: CEAS (2010)
2. Bonacich, P.: Factoring and weighting approaches to status scores and clique identification. J. Math. Soc. 2, 113–120 (1972)
3. Lazic, M.: On the Laplacian energy of a graph. Czech. Math. J. 56, 1207–1213
4. Lee, K., Caverlee, J., Webb, S.: Uncovering social spammers: social honeypots + machine learning. In: ACM SIGIR, pp. 435–442 (2010)
5. Lin, D.: An information-theoretic definition of similarity. In: Proc. 15th International Conference on Machine Learning, pp. 296–304 (1998)
6. Lintean, M.C., Moldovan, C., Rus, V., McNamara, D.S.: The role of local and global weighting in assessing the semantic similarity of texts using latent semantic analysis. In: FLAIRS (2010)
7. McCord, M., Chuah, M.: Spam detection on Twitter using traditional classifiers. In: Calero, J.M.A., Yang, L.T., Mármol, F.G., García Villalba, L.J., Li, A.X., Wang, Y. (eds.) ATC 2011. LNCS, vol. 6906, pp. 175–186. Springer, Heidelberg (2011)
8. Qi, X., Duval, D., Christensen, K., Fuller, E., Spahiu, A., Wu, Q., Wu, Y., Tang, W., Zhang, C.: Terrorist networks, network energy and node removal: A new measure of centrality based on Laplacian energy. Social Networking 2, 19–31 (2013)
9. Rus, V., Lintean, M., Moldovan, C., Baggett, W., Niraula, N., Morgan, B.: SIMILAR Corpus: A resource to foster the qualitative understanding of semantic similarity of texts. In: LREC, pp. 50–59 (2012)
10. Song, J., Lee, S., Kim, J.: Spam filtering in Twitter using sender-receiver relationship. In: Sommer, R., Balzarotti, D., Maier, G. (eds.) RAID 2011. LNCS, vol. 6961, pp. 301–317. Springer, Heidelberg (2011)
11. Wang, A.: Don't follow me: Spam detection in Twitter. In: SECRYPT, pp. 1–10 (2010)
12. Yang, C., Harkreader, R., Gu, G.: Empirical evaluation and new design for fighting evolving Twitter spammers. IEEE TIFS 8, 1280–1293 (2013)

Gene Reduction for Cancer Classification Using Cascaded Neural Network with Gene Masking

Raneel Kumar, Krishnil Chand, and Sunil Pranit Lal

School of Computing, Information, and Mathematical Sciences
University of the South Pacific
Suva, Fiji
{raneel.kumar,krishnil.chand,sunil.lal}@usp.ac.fj

Abstract. This paper presents an approach to cancer classification from gene expression profiling using cascaded neural network classifier. The method used aims to reduce the genes required to successfully classify the small round blue cell tumours of childhood (SRBCT) into four categories. The system designed to do this consists of a feedforward neural network and is trained with genetic algorithm. A concept of 'gene masking' is introduced to the system which significantly reduces the number of genes required for producing very high accuracy classification.

1   Introduction

Early cancer detection is important for the proper treatment of it. However some cancers cannot be easily identified and classified by traditional clinical means. Traditional clinical methods include diagnosis by X—Ray, Magnetic Resonance Imaging (MRI), Computed Tomography (CT), and ultrasonography [1]. Microarray gene profiling is a new way used to improve the accuracy of cancer classification. Microarrays can simultaneously measure the expression level of thousands of genes within a particular mRNA sample [2] but this is a difficult task due to the high dimensionality of gene expression data.

Dimensionality reduction can also be seen as the process of deriving a set of degrees of freedom which can be used to reproduce most of the variability of a data set [3].Researchers have been involved in reducing the high dimensionality of the genes and at the same time preserving the features within the genes that would give a significant increase in accuracy of the classification process.

In this paper we focus on classification of the small round blue cell tumor (SRBCT) into four classes namely neuroblastoma (NB), rhabdomyosarcoma (RMS), non-Hodgkin lymphoma (NHL) and the Ewing family of tumors (EWS). Khan et. al [1] used principal component analysis (PCA) in training artificial neural networks (ANN) to progressively reduce the dimensionality of the SRBCT dataset from 2308 genes to 96 genes. Meanwhile Tibshirani et. al[4] applied nearest shrunken centroid classifier to the same dataset and showed reduction in the number of genes used to 43 with 100% accuracy. The nearest shrunken centroid classifier is essentially an extension of the nearest centroid classifier whereby features which are noisy and have little variation from the overall mean are eliminated using shrinkage factor.

M. Sokolova and P. van Beek (Eds.): Canadian AI 2014, LNAI 8436, pp. 301–306, 2014.

In this paper we extend the classifier proposed by Khan et. al [1] by incorporating a cascaded neural network classifier (Fig. 1) trained using genetic algorithm which leverages on our proposed concept of gene masking to further eliminate the number of genes required for accurate classification.

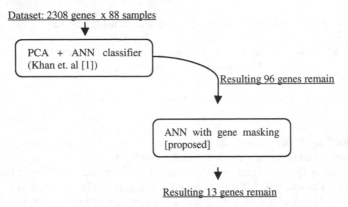

Fig. 1. Conceptual overview of the cascaded classifier

2   Proposed Cascaded Classifier

The cascaded classifier (Fig. 2) has been implemented as a feedforward neural network [5] with 96 inputs expressing the relative red intensity of the corresponding genes obtained from the previous block (Fig. 1). The output layer for the neural network consists of four neurons corresponding to the four cancer types to be classified in this study. The outputs of the four neurons are compared, and final prediction is the cancer type corresponding to the neuron with the largest output. The number of neurons in the hidden layer [6] as well as the choice identity activation function [7] has been determined by empirical methods [8].

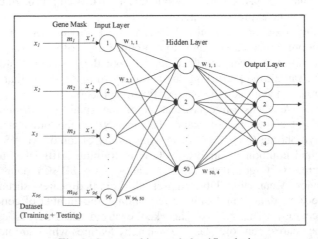

Fig. 2. Gene masking and classifier design

2.1   Feature Selection by Gene Masking

The fundamental concept behind gene masking is that we should be able to eliminate genes which are not important for classification without adversely affecting the classification accuracy. However evaluating classification accuracy for all possible gene elimination combinations is computationally intensive for large number of genes. Therefore we use binary coded genetic algorithm to evolve optimal gene mask. The gene mask is a binary string with length equal to the number of genes being considered. Each gene data x_i is multiplied with the corresponding mask m_i to modify the network input x'_i as follows:

$$x'_i = \begin{cases} x_i, & \text{if } m_i = 1 \\ 0, & \text{otherwise} \end{cases}, \tag{1}$$

3   Training the Classifier with Genetic Algorithm

The learning in neural network takes place by the adjustments of randomly initialized weights such that the classification error is minimized. We trained the neural network using genetic algorithm (GA) to search for optimal weight configuration which minimizes the classification error. GA [9] is a stochastic algorithm based on evolutionary ideas of natural selection and yields itself naturally to realizing the concept of gene masking.

3.1   Chromosome Encoding

Chromosome encoding is the representation of the actual problem into a data structure which can be interpreted by GA. We used binary encoding to represent the gene mask and the weights for the neural network (Fig. 3). The length of the chromosome is 5096 bits where the first 96 bits are for gene masking, the next 4800 bits are weight values of the links from the neurons of the first layer to the neurons of the second layer while the next 200 bits are the weights value for the hidden layer neurons to the output layer neurons.

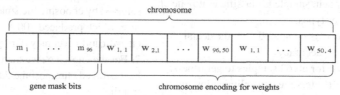

Fig. 3. Representation of a chromosome for the GA with w being weight encoding and m being genes mask

3.2   Fitness Function

The quality of solution represented by a chromosome is captured by its fitness value. Depending on the fitness function those chromosomes which pass the required criterion have more chance to enter the next step in the solution search process. The fitness function formulated for this classification problem takes into account the accuracy of classification, that is, the success of classification of cancer into the right categories, and simultaneously maximizes the number of genes eliminated. The generalized fitness function for the system is given by:

$$\text{Fitness} = (\propto * \text{Accuracy}) + (\beta * \text{Genes Eliminated}), \qquad (2)$$

where $\propto$ is the accuracy to gene elimination ratio in the range (0,1). In our study we set $\beta = (1 - \propto)$ to reduce the computational effort required to tune both the parameters.

4   Experimentation

4.1   SRBCT Dataset

The dataset [1] originally consisted of the expression of 6567 genes measured over 88 samples and was obtained through the cDNA microarray technology. After filtering out noise, 4259 genes were eliminated and the remaining 2308 genes were used as input to the classification scheme proposed by Khan et. al [1]. Our study uses the 96 genes output from [1] as the starting point. The 88 samples are divided in 63 training and 25 testing data samples. The training samples comprise of four tumor types: 8 Burkittlymphoma (BL) samples, 23 Ewing sarcoma (EWS) samples, 12 neuroblastoma (NB) samples and 20 rhabdomyosarcoma (RMS) samples. While the testing samples comprise of 6 BL samples, 3 EWS samples, 6 NB samples and 5 RMS samples with 5 non-SRBCT samples discarded in this study.

4.2   Training and Testing Phases

The training and testing procedure used in this study is captured in Table 1.

Table 1. Training and Testing Phases

Training Phase	Testing Phase
1. A sample, s from 63 training sample is passed through the classifier with initial $\propto = 0.01$	1. The best value of $\propto$ for testing is selected by choosing the smallest value of $\propto$ which produced 100% classification accuracy during training.
2. After an epoch, sample is classified into one of the four cancer types.	
3. Classification is compared with the actual result to get the accuracy	2. For the chosen $\propto$, we then obtain the best evolved chromosome during training.
4. This is done for all 63 samples to get fitness. The fitness considers the gene elimination ratio and classification accuracy	
5. The training is run for 20,000 epochs, using the fitness to guide GA in selecting optimal ANN weights for minimizing the classification error.	3. With the $\propto$ constant and the chromosome fixed we pass each sample from the 20 test samples through the classifier
6. Steps 1 to 5 is carried out for all $\propto$ in $A = \{0.01, 0.05, 0.1, 0.2, 0.3, 0.4, 0.5, 0.6, 0.7, 0.8, 0.9, 0.95, 0.99\}$ such that we get 100% accuracy with maximum number of genes eliminated.	4. The predicted cancer type is compared with the actual cancer type to get the classification accuracy for all the 20 test samples.
7. For each $\propto$, 10 runs are carried out by repeating steps 1-6, the best result is chosen for comparison	

5   Results

The weight parameter $\propto$ assigns relative importance to the accuracy and number of genes eliminated. Large values of $\propto$ (close to 1) results in high accuracy at the expense of smaller number of genes eliminated (Fig. 4a). On the other hand $\propto$ values close to 0 eliminate larger number of genes but the accuracy is sacrificed (Fig. 4b). Therefore in selecting the best value of $\propto$ for testing, we choose the smallest value of $\propto$ which produced 100% classification accuracy during training. Accordingly we choose $\propto = 0.1$. With $\propto$ set to 0.1, all the 20 test samples were correctly classified with 100% accuracy using only 13 genes.

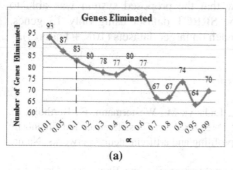 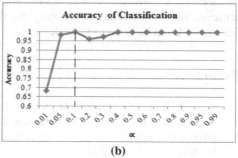

| (a) | (b) |

Fig. 4. Training results for various setting of $\propto$ parameter

5.1   Comparison of Results

The 13 genes which resulted in 100% classification of the test samples were noted and compared with [10]. Table 2 shows 54% of the genes in this experiment were also present in the comparator experiment.

Table 2. The 13 selected genes

Gene	Image ID	Present in [10]
1	296448	✓
2	841641	✗
3	43733	✓
4	629896	✗
5	866702	✓
6	52076	✓
7	563673	✓
8	1469292	✗
9	1409509	✓
10	756556	✗
11	377671	✗
12	325182	✓
13	755599	✗

Table 3. Comparison of results obtained

Method (Classifier)	Genes Remaining
PCA, MLP Neural Network [1]	96
Nearest Shrunken Centroid [4]	43
Gene Masking + ANN (this paper)	13

Comparison of results obtained on the SRBCT dataset with research methods reported in the literature show marked improvement in the number of genes required for accurate classification (Table 3).

6   Conclusion

In this paper we have extended the research reported in [1] by incorporating a cascaded neural network classifier trained using genetic algorithm. By applying gene masking, the learning algorithm was able to significantly reduce the number of genes required for accurate classification. The results show that the proposed system was able to achieve 100% classification accuracy on the SRBCT dataset using only 13 genes. Future work can be done to validate our approach on larger datasets (10K + features).

References

1. Khan, J., Wei, J.S., Ringner, M., Saal, L.H., Ladanyi, M., Westermann, F., Berthold, F., Schwab, M., Antonescu, C.R., Perterson, C., Meltzer, P.S.: Classification and diagnostic prediction of cancers using gene expression profiling and artificial neural networks. Nature Medicine 7, 673–679 (2001)
2. Sarhan, A.M.: Cancer Classification based on Microarray Gene Expression DataUsing DCT and ANN. Journal of Theoretical and Applied Information Technology 6, 208–216 (2009)
3. Ghodsi, A.: Dimensionality Reduction A Short Tutorial. Technical Report 2006-14, Department of Statistics and Actuarial Science, University of Waterloo, Ontario, Canada, pp. 5–6 (2006)
4. Tibshirani, R., Hastie, T., Narasimhan, B., Chu, G.: Diagnosis of multiple cancer types by shrunken centroids of gene expression. Proceedings of the National Academy of Sciences 99, 6567–6572 (2002)
5. Rani, D.K.U.: Analysis of Heart Diseases Dataset Using Neural Network Approach. International Journal of Data Mining & Knowledge Management Process (IJDKP) 1 (2011)
6. Karsoliya, S.: Approximating Number of Hidden layer neurons in Multiple Hidden Layer BPNN Architecture. International Journal of Engineering Trends and Technology 3, 714–717 (2012)
7. Karlik, B., Olgac, A.V.: Performance Analysis of Various Activation Functions in Generalized MLP Architectures of Neural Networks. International Journal of Artificial Intelligence And Expert Systems 1, 111–122 (2011)
8. Singh, S., Chand, A., Lal, S.P.: Improving Spam Detection Using Neural Networks Trained by Memetic Algorithm. In: Proceedings of 5th IEEE International Conference on Computational Intelligence, Mathematical Modeling and Simulation, Seoul, Korea, pp. 55–60 (2013)
9. Goldberg, D.E.: Genetic Algorithms in Search, Optimization, and Machine Learning. Addison-Wesley (1989)
10. Bair, E., Tibshirani, R.: Machine Learning MethodsApplied to DNA Microarray Data Can Improve the Diagnosis of Cancer. Special Interest Group in Knowledge Discovery and Data Mining Explorations 5, 48–55 (2003)

Polynomial Multivariate Approximation with Genetic Algorithms

Angel Kuri-Morales[1] and Alejandro Cartas-Ayala[2]

[1] Instituto Tecnológico Autónomo de México
Río Hondo No. 1
México 01000, D.F.
México
akuri@itam.mx

[2] Instituto Tecnológico Autónomo de México
Río Hondo No. 1
México 01000, D.F.
México
alejandro.cartas@gmail.com

Abstract. We discuss an algorithm which allows us to find the algebraic expression of a dependent variable as a function of an arbitrary number of independent ones where data is arbitrary, i.e. it may have arisen from experimental data. The possibility of such approximation is proved starting from the Universal Approximation Theorem (UAT). As opposed to the neural network (NN) approach to which it is frequently associated, the relationship between the independent variables is explicit, thus resolving the "black box" characteristics of NNs. It implies the use of a nonlinear function (called the activation function) such as the *logistic $1/(1+e^{-x})$*. Thus, any function is expressible as a combination of a set of *logistics*. We show that a close polynomial approximation of *logistic* is possible by using only a constant and monomials of odd degree. Hence, an upper bound (D) on the degree of the polynomial may be found. Furthermore, we may calculate the form of the model resulting from D. We discuss how to determine the best such set by using a genetic algorithm leading to the best L_∞-L_2 approximation. It allows us to find the best approximation polynomial given a selected fixed number of coefficients. It then finds the best combination of coefficients and their values. We present some experimental results.

Keywords: Multivariate polynomials, supervised training, neural networks, genetic algorithms, ascent algorithm.

1   Introduction

The problem of finding a synthetic expression from an experimental set of data for a variable of interest given a matching set of independent variables has repeatedly received attention because of its inherent importance. It may be called multivariate regression or supervised training (ST) depending on the approach taken. As ST it has given rise to a wide sub-area within the realm of artificial neural networks (NN).

M. Sokolova and P. van Beek (Eds.): Canadian AI 2014, LNAI 8436, pp. 307–312, 2014.
© Springer International Publishing Switzerland 2014

Some previous approaches may be found in [1-6]. If seen as a multivariate regression problem other issues may arise (see [7,8,9,10,21]). Here the form of the model is left for the method to determine. The use of NNs has opened a way out of the numerical limitations. The free parameters and architecture of the NN, when adequately determined, yield the desired results [11]. However, the explicit relations between the independent variables remain unknown. We explore an alternative which allows us to keep the closed nature of an algebraic model and its inherent explanatory properties and leaving the task of determining the mathematical model that is best suited for the problem at hand to the method itself.

2   Universal Approximation Theorem for Polynomials

We start by enunciating the Universal Approximation Theorem. The UAT relies on the properties of a well-defined non-linear function [12].

Theorem 1. Let $\varphi(\cdot)$ be a nonconstant, bounded, and monotonically-increasing continuous function. Let I_{mO} denote the m_O-dimensional unit hypercube $[0,1]$. The space of continuous functions on I_{m_O} is denoted by $C(I_{mO})$. Then, given any function $f \in C(I_{mO})$ and $\varepsilon > 0$, there exist an integer M and sets of real constants α_i, b_i and w_{ij}, where $i=1,2,\ldots,m_I$ and $j=1,2,\ldots,m_O$ such that we may define:

$$F(x_1,\ldots,x_{m_O}) = \sum_{i=1}^{m_I}\left[\alpha_i \cdot \varphi\left(\sum_{j=1}^{m_O} w_{ij}x_j + b_i\right)\right] \tag{1}$$

as an approximate realization of the function $\varphi(.)$. The universal approximation theorem is directly applicable to multilayer perceptrons. Perceptrons are computational units which include an extra constant valued input called the *bias*. When including the bias of the output neuron into (1) we get

$$F(x_1,\ldots,x_{m_O}) = \alpha_0 + \sum_{i=1}^{m_I}\left[\alpha_i \cdot \varphi\left(\sum_{j=1}^{m_O} w_{ij}x_j + b_i\right)\right] \tag{2}$$

Now, the equation for the perceptron is given by $y_i=\varphi[\sum_{i=0}^{mO} w_{ij}x_j]$, where, for convenience, we have made $x_0=1$, $w_0=b_0$. We can select $\varphi(x)=logistic(x)$ as the nonlinearity in a neuronal model for the construction of a multi-layer perceptron (MLP) network. We note that (1) represents the output of a MLP described as follows: a) The network has m_O input nodes and a single hidden layer consisting of m_I neurons; the inputs are denoted by $x_1,\ldots,x_{mO}$, b) Hidden neuron i has synaptic weights w_{i1}, w_{i2}, ..., w_{im_O}, c) The network output is a linear combination of the outputs of the hidden neurons, with $\alpha_1,\ldots,\alpha_{ml}$ defining the synaptic weights of the output layer.

The UAT states that *a single hidden layer is sufficient for a multilayer perceptron to compute a uniform ε approximation to a given training set represented by the set of inputs $x_1,\ldots,x_{mO}$ and a desired (target) output $f(x_1,\ldots,x_{mO})$.*

We know [13, 20] that the *logistic* function $1/(1+e^{-x})$ may be approximated by a polynomial $P_n(x)$ of degree n with an error $\varepsilon = |P_n(x)\text{-}logistic(x)|$ where $\varepsilon \to 0$ as $n \to \infty$, i.e. $logistic(x) \approx P_n(x) = \varsigma_0 + \sum_{i=1}^{n} \varsigma_i x^i$. Likewise, we know [14, 15] that a function $y=f(x)$ in $[0,+1)$ may be approximated by a polynomial based on a set of Chebyshev polynomials which achieves the minimization of the least squared error norm and simultaneously approaches the minimum largest absolute error. Finally, an application of the previous concepts allows us to find that the polynomial $P_n(x)$ of theorem (2) may be approximated by a polynomial $Q_l(x) = \beta_0 + \sum_{l=1}^{n} \beta_l x^{2l-1}$ with an error $\varepsilon \to$ 0. The logistic function may be approximated by the polynomial $P_{11}(x) \sim \beta_0 + \sum_{i=1}^{6} \beta_i x^{2i-1}$ with an RMS error $\varepsilon_2 \sim 0.000781$ and a min-max error $\varepsilon_\infty \sim 0.001340$, where $\beta_0 \sim 0.5000$, $\beta_1 \sim 0.2468$, $\beta_2 \sim -0.01769$, $\beta_3 \sim 0.001085$, $\beta_4 \sim -0.000040960$, $\beta_5 \sim 0.0000008215$, $\beta_6 \sim -0.00000000664712$. Therefore, we find that any function as in (3) may be approximately realized with a linear combination of polynomials having a constant plus terms of odd degree.

Theorem 2. Any function of m_O variables may be approximated with a polynomial whose terms are of degree 2^{k-1}.
Proof. We may write,

$$F(x_1,...,x_{mO}) = \alpha_0 + \sum_{i=1}^{m_l} \alpha_i \cdot \varphi\left(\sum_{k=0}^{m_O} w_{ik} x_k\right) = \alpha_0 + \sum_{i=1}^{m_l} \alpha_i \cdot logistic\left(\sum_{k=0}^{m_O} w_{ik} x_k\right)$$

$$= \alpha_0 + \sum_{i=1}^{m_l} \alpha_i \cdot \sum_{j=0}^{\infty} \lambda_j \left(\sum_{k=0}^{m_O} w_{ik} x_k\right)^j = \alpha_0 + \sum_{i=1}^{m_l} \alpha_i \cdot \left[\beta_0 + \sum_{j=1}^{\infty} \beta_j \left(\sum_{k=0}^{m_O} w_{ik} x_k\right)^{2j-1}\right]$$

$$F(x_1,...,x_{mO}) \approx \alpha_0 + \sum_{i=1}^{m_l} \alpha_i \cdot \left[\frac{1}{2} + \sum_{j=1}^{6} \beta_j \left(\sum_{k=0}^{m_O} w_{ik} x_k\right)^{2j-1}\right] \tag{3}$$

3 Approximation Using Genetic Algorithms

The number of terms of (3) is very large, in general. However, the terms corresponding to the hidden neurons may be lumped into terms of degree 0,1,3,...,11 and, subsequently, those at the output neuron will consist of those resulting from the power combinations of degree $(0,1,3,...,11)^1$, $(0,1,3,...,11)^3$,...,$(0,1,3,...,11)^{11}$. In other words, the polynomial at the output neuron will be of the form $P_O = k + \sum_{i=1}^{20} T(i)$, where $T(i) \equiv$ *terms of degree* $L(i)$; $L(i)$ denoting the i-th element in L. However, even if the powers of the monomials are all odd the powers of the variables in the i-th monomial may take any possible combination of the factors of $L(i)$. We define a priori the number (say M) of desired monomials of the approximant and then select which of the p possible ones these will be. There are $C(p, M)$ combinations of monomials and even for modest values of p and M an exhaustive search is out of the question. We used the genetic algorithm (EGA) discussed in [16] [17], as follows. The chromosome is a binary string of size p. Every bit in it represents a monomial ordered as per the sequence of the consecutive powers of the variables. If the bit is '1' it means that the corresponding monomial is retained while if it is a '0' it is discarded. One has to ensure that the number of 1's is equal to M. Assume, for example, that $y = f(v_1, v_2, v_3)$ and that $d_1=1$, $d_2=d_3=2$. In such case the powers assigned to the $2 \times 3 \times 3 = 18$ positions of the genome are 000, 001, 002, 010, 011, 012, 020, 021, 022, 100, 101, 102, 110, 111, 112, 120, 121, 122. The population consists of a set of binary strings of length p in which there are only M 1's. Then the Ascent Algorithm (AA) is applied to obtain the set of M coefficients minimizing $\varepsilon_{MAX}=max(|f_i-y_i|)$. Next, for this set of coefficients ε_{RMS} is calculated. This is the fitness function of the EGA. We retain the individual whose coefficients minimize ε_{RMS} out of those which best minimize ε_{MAX} (from the AA). This is the L_∞-L_2 metric mentioned in the introduction.

3.1   The Ascent Algorithm

The purpose of this algorithm [18] is to express the behavior of a dependent variable (y) as a function of a set of m_O independent variables (v). $y = f(v_1, v_2, ..., v_{mO})$ and $y=f(v)$. The AA has the distinct advantage that the v_i may be arbitrarily defined. This allows us to find the approximation coefficients for different combinations of the powers of the variables. Its goal is to find the values of the coefficients such that the approximated values minimize the difference between the known values of the dependent variable f in the sample and those calculated (say y) for all the objects in the sample. Another distinct advantage of the AA is that, regardless of the value of N, only an MxM set of equations is solved in every step. In our implementation every iteration of the AA requires only $O(M^2)$ flops.

4   Case Studies

To illustrate our method we present two case studies, both taken from the UCI repository [19]. All combinations of degrees were allowed. Case 1 is related to the relative CPU performance of a set of selected ones. There are 209 instances. We found the approximating L_∞-L_2 polynomials of 4, 8, 12 and 20 terms. The EGA ran for only 50 generations and yielded errors smaller than 5% in all cases. Fig. 1 illustrates these results.

The second case corresponds to a wine recognition data base. As before, we found the approximating L_∞-L_2 polynomials of 4, 8, 12 and 20 terms. In Fig. 1b we show the classification results for 20 terms. The line is denoted as "adjusted" in the graph. Since the approximation and original values were so similar, we offset the results by 0.05 so as to make the two lines original/adjusted discernible.

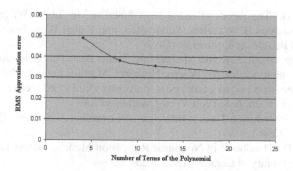

Fig. 1. Approximation error for different number of terms

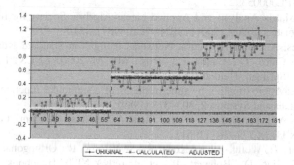

Fig. 2. Classification results for a 20 term polynomial

5 Conclusions

We have shown that any function may be closely approximated using a polynomial of degree 121 whose terms are lumped into no more than 21 monomials. The coefficients of these monomials may be searched using a genetic algorithm whose individuals correspond to polynomials with a predefined number of them. To illustrate we have solved two case problems. We found that both a regression and classification problems may be satisfactorily tackled. One unsolved issue is how to determine the number and maximum degree of the lumped polynomial.

References

[1] Rumelhart, D.E., Hinton, G.E., Williams, R.J.: Learning Internal Representations by Error Propagation. In: Rumelhart, D., McClelland, J., the PDP research group (eds.) Parallel Distributed Processing: Explorations in the Microstructure of Cognition. Foundations, vol. 1. MIT Press (1986)
[2] Haykin, S.: Neural Networks and Learning Machines. In: Multilayer Perceptrons, ch. 4, 3rd edn., Prentice Hall (2009) ISBN-13: 978-0-13-147139-9

[3] Powell, M.J.D.: The theory of radial basis functions. In: Light, W. (ed.) Advances in Numerical Analysis II: Wavelets, Subdivision, and Radial Basis Functions. King Fahd University of Petroleum & Minerals (1992)

[4] Haykin, S.: op. cit, ch. 5. Kernel Methods and Radial Basis Functions (2009)

[5] Cortes, C., Vapnik, V.: Support-Vector Networks. Machine Learning 20 (1995), http://www.springerlink.com/content/k238jx04hm87j80g/

[6] Haykin, S.: op. cit, ch. 6. Support Vector Machines (2009)

[7] MacKay, D.: Information Theory, Inference, and Learning Algorithms, Cambridge (2004) ISBN 0-521-64298-1

[8] Ratkowsky, D.: Handbook of Nonlinear Regression Models. Marcel Dekker, Inc., New York (1990); Library of Congress QA278.2.R369

[9] Beckermann, B.: The condition number of real Vandermonde, Krylov and positive definite Hankel matrices. Numerische Mathematik 85(4), 553–577 (2000), doi:10.1007/PL00005392

[10] Meyer, C.: Matrix Analysis and Applied Linear Algebra, Society for Industrial and Applied Mathematics, SIAM (2001) ISBN 978-0-89871-454-8

[11] Kuri-Morales, A.: Training Neural Networks using Non-Standard Norms- Preliminary Results. In: Cairó, O., Cantú, F.J. (eds.) MICAI 2000. LNCS (LNAI), vol. 1793, pp. 350–364. Springer, Heidelberg (2000)

[12] Cybenko, G.: Approximation by superpositions of a sigmoidal function. Mathematics of Control, Signals and Systems 2, 303–314 (1989)

[13] Bishop, E.: A generalization of the Stone–Weierstrass theorem. Pacific Journal of Mathematics 11(3), 777–783 (1961)

[14] Koornwinder, T., Wong, R., Koekoek, R., Swarttouw, R.: Orthogonal Polynomials. In: Olver, F., Lozier, D., Boisvert, R., et al. (eds.) NIST Handbook of Mathematical Functions. Cambridge University Press (2010) ISBN 978-0521192255

[15] Scheid, F.: Numerical Analysis. In: Least Squares Polynomial Approximation, ch. 21. Schaum's Outline Series (1968) ISBN 07-055197-9

[16] Kuri-Morales, A., Aldana-Bobadilla, E.: The Best Genetic Algorithm Part I. In: Castro, F., Gelbukh, A., González, M. (eds.) MICAI 2013, Part II. LNCS, vol. 8266, pp. 1–15. Springer, Heidelberg (2013)

[17] Kuri-Morales, A.F., Aldana-Bobadilla, E., López-Peña, I.: The Best Genetic Algorithm Part II. In: Castro, F., Gelbukh, A., González, M. (eds.) MICAI 2013, Part II. LNCS, vol. 8266, pp. 16–29. Springer, Heidelberg (2013)

[18] Cheney, E.W.: Introduction to Approximation Theory, pp. 34–45. McGraw-Hill Book Company (1966)

[19] Bache, K., Lichman, M.: UCI Machine Learning Repository. University of California, School of Information and Computer Science, Irvine (2013), http://archive.ics.uci.edu/ml

[20] http://www.math.nus.edu.sg/~matngtb/Calculus/MA3110/Chapter1 0WeierstrassApproximation.pdf

[21] Masry, E.: Multivariate Local Polynomial Regression For Time Series: Uniform Strong Consistency And Rates. J. Time Series Analysis 17, 571–599 (1996)

Improving Word Embeddings
via Combining with Complementary Languages

Changliang Li, Bo Xu, Gaowei Wu, Tao Zhuang,
Xiuying Wang, and Wendong Ge

Institute of Automation, Chinese Academy of Sciences,
Beijing, P.R. China
{changliang.li,xubo,gaowei.wu,tao.zhuang,
xiuying.wang,wendong.ge}@ia.ac.cn

Abstract. Word embeddings have recently been demonstrated outstanding results across various NLP tasks. However, most existing word embeddings learning methods employ mono-lingual corpus without exploiting the linguistic relationship among languages. In this paper, we introduce a novel CCL (Combination with Complementary Languages) method to improve word embeddings. Under this method, one words embeddings are replaced by its center word embeddings, which is obtained by combining with the corresponding words embeddings in other different languages. We apply our method to several baseline models and evaluate the quality of word embeddings on word similarity task across two benchmark datasets. Despite its simplicity, the results show that our method is surprisingly effective in capturing semantic information, and outperforms baselines by a large margin, at most 20 Spearmans rank correlation ($\rho \times 100$).

Keywords: word embeddings, center word embeddings, semantic information.

1 Introduction

A word representation is a mathematical object associated with each word, often a vector. Each dimensions value corresponds to a feature and might even have a semantic interpretation [1]. Word vectors can be used to significantly improve and simplify many NLP applications. So it is significant and rewarding to improve word representations.

An effective approach to word representation is to learn word embeddings. Each dimension of the embeddings represents a latent feature of the word, hopefully capturing useful syntactic and semantic properties [1]. Many approaches have been proposed to learn good performance word embeddings, such as recursive neural language model [2], recurrent neural language model [3], global context-aware neural language model [4] and RNN with morphology [5].

However, despite a lot of work on improving word representations, most existing work shares a common problem that only utilizes mono-lingual corpus

M. Sokolova and P. van Beek (Eds.): Canadian AI 2014, LNAI 8436, pp. 313–318, 2014.

without taking into account of the latent relationship among languages. In fact, all common languages share concepts which are grounded in the real world, and the vector representations of similar words in different languages were related by a linear transformation in vector space [6].

In this paper, inspired by the idea of word embeddings, and based on the linguistic knowledge and regularities, we propose a novel Combination with Complementary Languages(CCL) method to improve word embeddings. Under this method, we replace one words embeddings with its center word embeddings. The center word embeddings are obtained through combination with complementary languages. The obtained word embeddings are represented as the center of the all words embeddings in vector space. So we call it center word embeddings.

This method is in line with the idea of word embeddings. One words embeddings are supposed to capture the words semantic information in language vector space. Meanwhile, its corresponding words embeddings also capture the same or similar semantic information in another language vector space. Based on taking account into the word embeddings and its all corresponding words embeddings, the center word embeddings ability to capture semantic information becomes more reliable. We evaluate the word embeddings obtained through our method on word similarity task across two benchmark datasets. The results show that our method outperforms competitive baselines by a large margin.

The remainder of this paper is organized as follows. In section 2, we discuss related work; Section 3 describes the details of our proposed method; Section 4 illustrates our experiments and results; Section 5 analyzes our method and we conclude in section 6.

2   CCL Method

It was found that the learned word representations capture meaningful syntactic and semantic regularities in a very simple way. And there are many types of similarities among words that can be expressed as linear translations [7]. It was also found that the vector representations of the similar words in different languages were related by a linear transformation (namely, a rotation and scaling) [8].

Inspired by the idea of word embeddings and based on linguistic knowledge and regularities, we propose our CCL method to improve word embeddings ability to capture semantic information. The property of our method is to replace one words embeddings with its center word embeddings. The details of our method are described in this section.

The original idea of word embeddings is that each dimension of the embeddings represents a latent feature of the word, hopefully capturing useful syntactic and semantic properties [1].

Similar to this idea, we propose that one words embeddings in each language capture some useful semantic information of the word. And based on the linguistic knowledge and regularities discussed above, we combine the corresponding words embeddings in different languages, and use center word embeddings to replace the word embeddings.

Firstly, we define that the target language as L_{targ}, N kinds of different complementary languages as L_{trans}^i, $i \in [1, N]$. The word in L_{targ} vocabulary is represented as w_{targ}, the corresponding word in L_{trans}^i is represented as w_{trans}^i. The center word embeddings obtained by CCL method is represented as follows.

$$V'\big(center(w_{targ})\big) = Func\Big\{\alpha V(w_{targ}) + \sum_{i=1}^{N} \beta_i V(w_{trans}^i)\Big\} \qquad (1)$$

Where $Func$ is normalized function; α and β_i are scaling parameters used to prevent either pair of sides from dominating the other. The value of α and β_i indicates the role of corresponding language plays in capturing the words semantic information. We use a dataset annotated by human to learn the parameters α and β_i based on minimum squared error.

From formula (1), we can see that the word embeddings obtained through most existing methods are special cases of our method when $\beta_i = 0$.

Figure 1 illustrate the process of obtaining center word embeddings. We firstly translate one word into different languages, and we look up the embeddings of the word and its corresponding words. Then we combine them and use formula (1) to obtain center word embeddings.

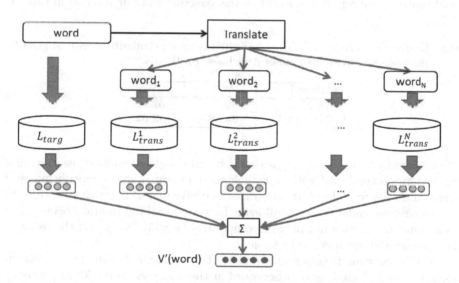

Fig. 1. Process of obtaining center word embeddings

3   Experiment

In order to illustrate the effectiveness of our method, we evaluate the word embeddings obtained through our method on word similarity task on popular WordSim-353 dataset (referred as WS) [9]. In order to avoid over fitting our method to a single dataset, our experiments are conducted on another similar popular benchmark datasets SCWS* [4].

In our experiment, we select English as target language, Chinese as complementary language. We employed Wikipedia[1] documents as English corpus, BaiduBaike[2] documents as Chinese corpus, due to their wide range of topics and word usages, and clean organization of document by topic.

We employed three language models (referred as HSMN, csm-RNN, and skip-gram respectively) as baselines to train word embeddings. The baseline word embeddings are obtained via training baseline language models on English corpus. For the three language models training, we used 50-dimensional embeddings.

3.1   Results

In the word similarity task, the results are evaluated using standard metrics, Spearmans rank correlation ρ. We apply our method to three baselines.

The first baseline is global context-aware neural language model (referred as HSMN). The model defines two scoring components that contribute to the final score of a (word sequence, document) pair. The scoring components are computed by two neural networks, one capturing local context and the other global context. We report the result of this baseline and our method in table 1.

Table 1. Spearmans rank correlation ($\rho \times 100$) between similarity scores assigned by various methods and by human annotators based on HSMN

Method	WS	SCWS*
HSMN	62.58	32.11
HSMN+CCL	**63.88**	**52.02**

The second one is csm-RNN, which is based on global context-aware neural language mode combined with morphology. It operates at the morpheme level instead of at the word level. It integrates recursive neural network structures to neural language model training, allowing for contextual information being taken into account in learning morphemic compositionality [5]. We report the result of this baseline and our method in table 2.

The third baseline is skip-gram model. The model tries to maximize classification of a word based on another word in the same sentence. More precisely, it uses each current word as an input to a log-linear classifier with continuous projection layer, and predicts words within a certain range before and after the current word [10]. One property of this model is low training complexity. We report the result of this baseline and our method in table 3.

From all the results shown above, we can see that our method outperforms each baseline result by a large margin for both datasets, at most 20 Spearmans

---

[1] http://www.wikipedia.org/
[2] http://baike.baidu.com/

Table 2. Spearmans rank correlation ($\rho \times 100$) between similarity scores assigned by various methods and by human annotators based on csm-RNN

Method	WS	SCWS*
csm-RNN	64.58	43.65
csm-RNN+CCL	**68.90**	**59.60**

Table 3. Spearmans rank correlation ($\rho \times 100$) between similarity scores assigned by various methods and by human annotators based on Skip-gram

Method	WS	SCWS*
Skip-gram	63.93	60.81
Skip-gram+CCL	**64.54**	**61.04**

rank correlation ($\rho \times 100$) on SCWS* dataset. So we can make a safe conclusion that our method can improve word embeddings in capturing word semantic information.

4 Analysis

The results highlight the fact that our method is reliable. This might be explained by two reasons: 1) all common languages share concepts which are grounded in the real world; 2) the linear relationship exists among different languages in the same vector space. One word and its corresponding word in another different language have similarity of geometric arrangements. We combine the embeddings of the word and its corresponding words in other languages to build up center word embeddings. Taking into account of the word and its corresponding words semantic information, the center word embeddings are more robust in capturing semantic information.

However, the size of each language corpus impacts the final result of our method. If there is large corpus, comes significant role. Meanwhile, one word often corresponds to more than one word in another language. The number of corresponding words in each language also would impact the final result. This is also related to the quality of translation. Furthermore, it is easy to understand that the more complementary languages employed, the larger the training complexity.

Our method can improve the word embeddings by a large margin. So it can improve many NLP tasks such as machine translation. In this paper we take English and Chinese as example. As our method makes little assumption of language, so it could be generalized to more languages. Our method might be a new clue to improve word representations and to exploit relation among languages.

5 Conclusion

This paper has presented a novel method to improve word embeddings by combining corresponding words embeddings in different languages. It makes full use of linguistic knowledge and regularities in vector space. The experiment results have shown that the center word embeddings obtained via our method are remarkably effective in capturing word semantic information.

We have recognized that the quality of translation impacts the final result of our method. An important direction for our further work is to improve the quality of translation. Meanwhile, training complexity becomes larger while combining more complementary languages. So another promising future work for us is to find an optimal balance between training complexity and the effect of word embeddings.

Acknowledgments. This work is supported by National Program on Key Basic Research Project (973 Program) under Grant No.2013CB329302, National Natural Science Foundation of China (NSFC) under Grants No.61175050 and No.61203281.

References

1. Turian, J., Ratinov, L., Bengio, Y.: Word representations: A simple and general method for semisupervised learning. In: ACL (2010)
2. Socher, R., Pennington, J., Huang, E.H., Ng, A.Y., Manning, C.D.: Semi-supervised recursive auto encoders for predicting sentiment distributions. In: EMNLP (2011)
3. Mikolov, T., Karafi'at, M., Burget, L., Cernock'y, J., Khudanpur, S.: Recurrent neural network based language model. In: INTERSPEECH (2010)
4. E. H. Huang, R. Socher, C. D. Manning, and A. Y. Ng. Improving word representations via global context and multiple word prototypes. In: Annual Meeting of the Association for Computational Linguistics (ACL) (2012)
5. Luong, M., Socher, R., Manning, C.: Better word representations with recursive neural networks for morphology. In: CONLL (2013)
6. Mikolov, T., Le, Q.V., Sutskever, I.: Exploiting Similarities among Languages for Machine Translation. arXiv preprint arXiv:1309.4168 (2013)
7. Mikolov, T., Yih, W.-T., Zweig, G.: Linguistic regularities in continuous space word representations. In: NAACL-HLT (2013)
8. Mikolov, T., Le, Q., Sutskever, I.: Exploiting Similarities among Languages for Machine Translation. Technical report, arXiv (2013)
9. Finkelstein, L., Gabrilovich, E., Matias, Y., Rivlin, H., Solan, Z., Wolfman, G., Uppin, E.: Placing search in context: the oncept revisited. ACM Transactions on Information Systems 20(1), 116–131 (2002)
10. Mikolov, T., Chen, K., Corrado, G., Dean, J.: Efficient estimation of word representations in vector space. arXiv preprint arXiv:1301.3781 (2013)

A Clustering Density-Based Sample Reduction Method

Mahdi Mohammadi[1], Bijan Raahemi[1], and Ahmad Akbari[2]

[1] Knowledge Discovery and Data Mining, University of Ottawa, Lab., Ottawa, Canada
[2] Iran University of Science and Technology, Computer Engineering Department, Tehran, Iran
{mmohamm6,braahemi}@uottawa.ca, akbari@iust.ac.ir

Abstract. In this paper, we propose a new cluster-based sample reduction method which is unsupervised, geometric, and density-based. The original data is initially divided into clusters, and each cluster is divided into "portions" defined as the areas between two concentric circles. Then, using the proposed geometric-based formulas, the membership value of each sample belonging to a specific portion is calculated. Samples are then selected from the original data according to the corresponding calculated membership value. We conduct various experiments on the NSL-KDD and KDDCup99 datasets.

Keywords: Sample reduction, Clustering, Classification, Density-based, Membership function.

1 Introduction

In many real-world applications- such as banking and financial transactions, retail services, network intrusion detection systems, and mobile communications- numerous data are generated. These data can be analyzed to discover useful and actionable patterns. In such cases, the traffic generated by the users exhibit similar behavior in a short limited period of time. This observation is mentioned in the literatures. For instance, the authors in [1] have mentioned the problem of repeated samples in the KDD Cup 99 dataset. Sample reduction is a common approach for addressing these concerns. Several sample reduction methods and their applications are introduced in [2, 3]. The assumption that the instances of the data are not uniformly distributed and some instances are more representative than the others [4] motivates us to use a selection approach. Sample reduction methods encounter more challenges in applications such as intrusion detection in network traffic. For example random sampling a typical network trace containing a large number of probes or DoS attacks packets will dwarf the other smaller attacks or interesting patterns. Our proposed sample reduction is an unsupervised method (i.e. no label is assigned to the samples). First, we group the samples into clusters. Then, the membership of each sample belonging to a specific group is calculated based on the geometric-based equations described in this article. Samples are selected from the input dataset according to the calculated membership value. The rest of the paper is: an overview of related work on sample reduction algorithms is presented in Section 2, followed by introduction to two basic cluster-based algorithms in Section 3. We then describe the details of our

M. Sokolova and P. van Beek (Eds.): Canadian AI 2014, LNAI 8436, pp. 319–325, 2014.

proposed method in Section 4. The experimental results are presented in Section 5. Finally, we conclude the paper in Section 6.

2   Related Work on Sample Reduction Algorithms

There are different supervised methods mentioned in the literatures [5, 6, 7, 8].However, our focus in this paper is on unsupervised methods which do not require labeled data. In the work presented in [9], the authors proposed a density-based unsupervised sample reduction method for reducing the time and memory complexity of the SOM's (Self Organized Map) training phase. We call their approach the DB method (for density-based sample reduction. Two well-known unsupervised sample reduction methods are random sampling (RS), and stratified random sampling (SRS). These two methods are the most commonly applied sample reduction techniques, not only because they are easy to implement, but they are also straightforward and fast. In Stratified random sampling, the input samples are first grouped into different clusters, and then, samples are selected from each cluster. Number of selected samples in each cluster is directly related to the number of samples in that cluster. In the RS method, all samples have the same selection probability. This may eliminate some prominent samples from the input dataset. In this respect, the SRS is more advanced than the RS approach. In this paper, we propose a sampling method which uses the density information of the input data in its sampling process. We introduce an unsupervised, clustering and density-based approach for sample reduction of large datasets. Our method does not require labels of the classes, is as fast as the RS and SRS algorithms, and its memory complexity is significantly lower than the DB method. It is therefore suitable for real time applications.

3   Two Basic Clsuter-Based Sample Reduction Algorithms

In this section, we describe the details of two basic sample reduction. The first phase in these algorithms comprises of clustering. In this step, a clustering technique such as k-means algorithm is applied to the input dataset to produce k clusters. Since the focus is not on the performances of various clustering techniques, we consider k-means clustering for its simplicity without losing generality. According to the number of the samples in each cluster, a fixed number of samples is assigned to each cluster as expressed   in Equation (1):

$$NC_i = \frac{N_i}{N} \times NRS \tag{1}$$

where Ni is the number of samples in cluster ith, N is the total number of samples in the dataset, NRS is number of desired samples in the reduced dataset, and NCi is the total number of samples in cluster i after sample reduction. After the clustering is performed, the following two approaches are considered to select samples from each cluster.

3.1   Near Center-Based Sample Selection (NCB)

In this approach, the samples near the center of the cluster have more chances to be selected for the reduced dataset. The rationale for this approach is that the near center samples are important and meaningful because they represent the core of the cluster, and as such, they are useful for training the classifiers identifying the majority of samples in that cluster. At first, a selection probability, SPj, is calculated for each sample as specified in Equations (2), and (3)

$$SP_j = \frac{R_i - Distance(dj\ , C_i)}{\sum_{k=1,...,NC_i} Distance(d_k{}^{th}, C_i)} \tag{2}$$

In this equation, SP_j is the selection probability of sample dj of the cluster, NC_i is the total number of samples in cluster i, the Distance function calculates the distance between a sample dj and the center of the cluster Ci, and Ri is radius of cluster i calculated based on Equation (3):

$$R_i = \max\left(Distace(dj\ ,)\right)$$
$$j = 1, ..., NC_i \tag{3}$$

The disadvantage of the NCB approach is that the sample scattering of the input datasets and the reduced dataset are not identical.

3.2   Boundary-Based Sample Selection (BB)

In this approach, the boundary samples are more likely to be selected. The rationale for this approach is that the boundary samples represent region of the clusters, and they are useful for better understanding of the cluster boundaries.

Similar to the NCB method, a selection probability is assigned to each sample in each cluster, and some samples are selected to be included in the final reduced dataset with non-uniform probability. The probability in the BB approach is calculated as in Equation (4):

$$SP_j = \frac{Distance(dj, C_i)}{\sum_{k=1,...,NC_i} Distance(d_k{}^{th}, C_i)} \tag{4}$$

In addition to the selected boundary samples, center of the cluster is also included in the reduced dataset.

4   The Proposed Density and Membership-Based Sample Reduction Method

In this section, we propose a new sample reduction approach based on density and membership degree (DMB). Similar to the two previous methods, a selection value for each sample is computed after the clustering phase. Each cluster area is divided into separate regions between the concentric circles with constant width αi, called "portion". Then, a membership value is assigned to each portion which stays the same for all the samples within that portion. Equation (5) shows membership value for the jth portion of the ith cluster:

$$\mu_{ij} = \frac{R_i - (j-1)\alpha_i}{\sum_{k=1}^{P}(R_i - (k-1)\alpha_i)}$$

$$\alpha_i = \frac{R_i}{P}$$

(5)

In this equation, P is number of portions in each cluster which is considered a fixed value in our experiments, μ_{ij} is the membership value for portion j of cluster i , R_i is radius of cluster i and α_i shows the size of the ith portion. In the numerator of this equation, j-1 shows the number of portions between the jth portion and the center of the cluster. Fig.1 visually describes these variables.

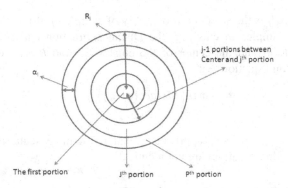

Fig. 1. A visual description of the parameters in the DMB method

For each portion, a membership value is calculated based on Equation (5). This value depends on how far the portion is from the center of the cluster (i.e. the first portion). As shown in Fig.1, there are j-1 portions between the jth portion and the first portion so the distance between this portion and the first portion is (j-1)×α_i. Similarly, $R_i - (j-1)\alpha_i$ shows the distance between the jth portion and the boundary of the cluster. This means that the inner portions have higher membership value in comparison with the outer ones. As such, even some of the samples in the Pth (the last one) portion can be selected for the reduced dataset. The selection value of each sample in a portion is calculated by Equation (6).

$$NCP_{ij} = \frac{n_{ij} \times \mu_{ij}}{\sum_{k=1}^{P} n_{ik} \times \mu_{ik}} \times NC_i$$

(6)

In this equation, n_{ij} is number of samples in portion j of cluster i, and NCP_{ij} is the desired selected number of samples in portion j of clusters i. That means, NCP_{ij} samples are randomly selected from the portion.

4.1   Analysis of the DMB Method

The index of the inner portion is one. The index of the outer portion is P. Based on Equation (6), samples placed in the inner portions have higher membership values, and those placed in outer portions have lower membership values. There is a significant difference between the DMB sample selection and the BB and NCB methods. In the NCB method, the probability of selection for boundary samples is zero and there is no chance for them to be selected. In the BB method, however, the near center samples are eliminated; the boundary samples have chances to be selected, while both of them also eliminate some parts of the samples. This is a major drawback of both algorithms. Based on Equation (5), in the DMB method, the entire boundary and near center samples have a chance to be selected, in contrast with the BB and NCB methods. Although near center samples have more chance to be selected, boundary samples still have not-zero chance to be presented in the reduced dataset.

5 Experimental Results

In this section, we present the experimental results for evaluating the proposed sample reduction method. We employ various performance measurements including Isolation, Davies-Bouldin (DB) indexes and correct recognition rate. The proposed method is evaluated on the NSL-KDD and KDDCup99 datasets. We use Training20%+ as the input training dataset and apply the sample reduction methods on it. For the test phase, we use the test datasets of NSL-KDD, Test+ .

5.1 Experiment-1: DB Index Measure

The DB index is the next validity index we consider to evaluate the proposed reduction methods. The DB index [10] is a function of the ratio of the sum of within-cluster scatter to between-cluster separation as defined in equation (7).

$$DB = \frac{1}{n}\sum_{i=1}^{n}\max_{i \neq j}\left[\frac{S_n(Q_i) + S_n(Q_j)}{S_n(Q_i,Q_j)}\right]$$ (7)

In this equation, n is the number of clusters, S_n is the average distance of samples of a cluster from its center, and $S(Q_i,Q_j)$ is the distance between the clusters centers. The smaller the value of the difference, the better performance of the reduction method. Table (1) shows the results of this experiment on different datasets.

Table 1. The difference between the DB indices of the original reduced datasets on Training 20% dataset

#Original Training Samples =25192							
# Clusters	#Samples	RS	SRS	DB	NCB	BB	DMB
C=5	1000	**1.371363**	1.642232	**1.34621**	3.581681	2.155021	1.184811
	5000	1.322349	**1.300229**	1.395719	1.828994	1.259	**0.112325**
	10000	1.253455	**1.142758**	1.169573	1.714926	1.0623	**0.964372**
	15000	0.806629	**0.627677**	0.69514	0.848748	0.9659	**0.418498**
C=20	1000	1.371363	2.292312	**1.34621**	3.594707	2.1467	**1.318421**
	5000	1.322349	**1.27916**	1.395719	3.028333	2.09594	**1.180376**
	10000	1.253455	**0.987902**	1.169573	1.814933	1.5921	**0.797356**
	15000	0.806629	**0.584308**	0.69514	0.848889	1.03659	**0.496725**

As shown in this table, the DMB method preserves the sample scattering of the original dataset better than the other methods do.

5.2   Experiment-2: Classification Accuracy Measure

In this experiment, we consider a decision tree as the classifier, and evaluate the various sample reduction methods based on the accuracy of the classifier they have trained. In the following tables, the **bold and underlined** values are the best results achieved on each row, while the **bold** values are the second best results.

Table 2. The correct recognition rate of the classifier (decision tree) on KDDCup99 dataset

#Original Training Samples = 400000				Correct Recognition Rate without sample Reduction=97.14			
	#Samples	RS	SRS	DB	NCB	BB	DMB
C=5	10000	91.77	**91.67**	88.97	79.51	75.99	**91.29**
	50000	91.56	**92.44**	90.63	85.14	79.45	**92.19**
	100000	**94.14**	94.12	93.47	90.11	83.47	**94.74**
	200000	94.79	94.67	**94.98**	91.07	87.25	**95.4**
C=20	10000	91.77	**92.07**	88.97	78.29	76.17	**92.26**
	50000	91.56	**93.78**	90.63	86.54	80.4	**93.55**
	100000	94.14	**94.66**	93.47	90.67	83.64	**95.46**
	200000	94.79	95.12	**95.68**	92.41	88.19	**96.14**

Table 3. The correct recognition rate of the classifier (decision tree) on NSL KDD+ dataset

#Original Training Samples =25192				Correct Recognition Rate without sample Reduction=78.0119			
	#Samples	RS	SRS	DB	NCB	BB	DMB
C=5	1000	72.65	**72.68**	70.59	43.19	43.19	**73.21**
	5000	75.31	**75.51**	72.21	59.08	58.69	**75.08**
	10000	76.63	**76.63**	72.95	59.08	60.66	**77.24**
	15000	76.26	**76.99**	74.62	60.41	60.13	**77.93**
C=20	1000	72.65	**72.91**	70.59	56.18	55.34	**74.25**
	5000	75.31	**74.17**	72.21	58.97	58.26	**76.36**
	10000	76.63	**76.96**	72.95	59.02	58.66	**77.44**
	15000	76.26	**77.45**	74.62	59.50	59.75	**77.96**

The RS and DB methods are not based on clustering so the number of clusters does not make any difference for these methods. Accordingly, the results of these methods are the same for C=5 and 20.

As shown in these tables, the DMB is often the best or the second best. This is clear from the tables since in the DMB method, the membership value of samples and the density of samples are important. Moreover, DMB is more capable of preserving the scattering of the training samples in the reduced training dataset.

In the BB method, by eliminating half of the training samples in KDDCup99 dataset (200000 samples are eliminated in this case); the correct recognition rate is reduced about 10%. This value is about 6% for NCB. In DMB, by reducing number of samples for each dataset to half, the correct recognition rate is reduced about 3% in average. When we reduce the number of training samples to 50000, the difference between the DMB and other methods is more significant. The correct recognition is reduced about 18% and 12% for the BB and NCB methods, respectively, while this value is about 5% for the NCB. The average difference between the DMB and SRS is about 0.7%. Although this amount is not that high, since the number of samples in the test phase of KDDCup dataset is very large, the 1% difference means more than 10000 samples. This confirms the higher performance of the NCB in comparison with the SRS.

6   Conclusion

In this paper, we proposed a new sample reduction method, named DMB that is unsupervised, geometric, and density-based. The input dataset is initially divided into clusters, and each cluster is divided into *portions* (the area between two concentric circles). Each portion is assigned a membership value. The samples are then selected according to their assigned membership. We demonstrated that the DMB method generally outperforms the other sampling methods considering various scattering measures such as DB index, while improving the detection rate.

References

[1]   Mahbod, T., Ebrahim, B., Wei, L., Ali, A.G.: A Detailed Analysis of the KDD CUP 99 Data Set. In: Proceeding of Computational Intelligence in Security and Defense Application (CISDA 2009) (2009)

[2]   Cochran, W.G.: Sampling Techniques. John Wiley & Sons (1977)

[3]   Chambers, R.L., Skinner, C.J. (eds.): Analysis of Survey Data. Wiley (2003)

[4]   Liu, H., Motoda, H., Yu, L.: A selective sampling approach to active feature selection. Artificial Intelligence 159(1-2), 49–74 (2004)

[5]   Kim, J.-M.: Calibration approach estimators in stratified sampling. Statistics & Probability Letters 77, 99–103 (2007)

[6]   PeRkalska, E.: Prototype selection for dissimilarity-based classifiers. Pattern Recognition 39, 189–208 (2006)

[7]   Duin, R.P.W., Juszczak, P., Ridder, D., Paclík, D.: PR-Tools, a Matlab toolbox for pattern recognition (2004), http://www.prtools.org

[8]   Cheng, Y.: Mean shift, mode seeking, and clustering. IEEE Trans. Pattern Anal. Mach. Intell. 17(8), 790–799 (1995)

[9]   Xu, Y., Chow, T.W.S.: Efficient Self-Organizing Map Learning Scheme Using Data Reduction Preprocessing. In: Proceedings of the World Congress on Engineering 2010, WCE 2010, London, U.K, June 30-July 2 (2010)

[10]   Davies, D.L., Bouldin, D.W.: A cluster separation measure. IEEE Trans. Pattern Anal. Machine Intell. 1(4), 224–227 (1979)

Web references

[11]   NSL-KKD dataset is available at: http://iscx.ca/NSL-KDD/ (last visit April 2012)

The Use of NLP Techniques in Static Code Analysis to Detect Weaknesses and Vulnerabilities

Serguei A. Mokhov, Joey Paquet, and Mourad Debbabi

Concordia University, Montreal, QC, Canada
{mokhov,paquet,debbabi}@encs.concordia.ca

Abstract. We employ classical NLP techniques (n-grams and various smoothing algorithms) combined with machine learning for non-NLP applications of detection, classification, and reporting of weaknesses related to vulnerabilities or bad coding practices found in artificial constrained languages, such as programming languages and their compiled counterparts. We compare and contrast the NLP approach to the signal processing approach in our results summary along with concrete promising results for specific test cases of open-source software written in C, C++, and JAVA. We use the open-source MARF's NLP framework and its MARFCAT application for the task, where the latter originally was designed for the Static Analysis Tool Exposition (SATE) workshop.

1 Introduction

This work briefly describes the methodology and the corresponding results of application of the machine learning techniques along with simple NLP techniques to static code analysis in search for weaknesses and vulnerabilities in source code. This work is accompanied by a proof-of-concept tool, MARFCAT, a Modular A* Recognition Framework-based Code Analysis Tool [1], first exhibited at the Static Analysis Tool Exposition (SATE) workshop 2010 [2]. We postulate that the presented approach is beneficial in static analysis and routine testing of any kind of code, including source code and binary deployments for its efficiency in terms of speed, relatively high precision. The approach is complementary to others (that do more in-depth semantic analysis) by prioritizing their targets. It can be used in automatic manner in diverse distributed and scalable environments to test the code safety. We use language models to learn and classify such a code.

2 Background and Related Work

A somewhat similar approach independently was presented [3] for vulnerability classification and prediction using machine learning. Statistical NLP techniques are described at length in [4]. We picked the MARF's pipeline and testing methodology for various algorithms in the pipeline adapted to this case found in [1] as it was the easiest implementation available to accomplish the task. The authors also did a binary analysis using NLP+machine learning approach for quick scans for files of known types in a large collection of files is described in [5] as well as the NLP and machine learning for NLP tasks in DEFT2010 [6]. Tlili's PhD thesis covers topics on automatic detection

M. Sokolova and P. van Beek (Eds.): Canadian AI 2014, LNAI 8436, pp. 326–332, 2014.

of safety and security vulnerabilities in open source software [7]. Statistical analysis, ranking, approximation, dealing with uncertainty, and specification inference in static code analysis are found in the work of Engler's team [8]. Kong et al. further advanced static analysis (using parsing, etc.) and specifications to eliminate human specification from the static code analysis in [9]. Some researchers proposed a general data mining system for incident analysis with data mining engines in [10].

3   Methodology

We further describe the methodology of our approach to static code analysis in its core principles, the knowledge base, machine learning categories, and the high-level algorithm.

Core Principles. The core methodology principles include classical n-gram and smoothing techniques (add-δ, Witten-Bell, MLE, etc.) and machine learning combined with dynamic programming. We presently we do not parse or otherwise work at the syntax and semantics levels, however. We treat the source code corpus as "characters", where each n-gram ($n = 1..3$, i.e., two consecutive characters or, more generally, bytes) are used to construct the language model. We show the system the examples of files with known weaknesses and MARFCAT learns them by computing various language models (based on options) from the CVE (Common Vulnerabilities and Exposures)-selected test cases. When the selected techniques are applied the line number information is lost as a part of this process. When we test, we compare trained language models with the unseen code fragments. In part, the methodology can approximately be seen as some n-gram-based "antivirus". We test to find out which algorithm combination gives the highest precision and the best run-time. At the present, however, we are looking at the whole files instead of parsing the finer-grain details of code fix patches and weak code fragments. This aspect lowers the precision, but is relatively fast to scan all the programming language files.

Table 1. CVE Stats for Wireshark 1.2.0

guess	run	algorithms	good	bad	%
1st	1	-nopreprep -char -unigram -add-delta	30	6	83.33
2nd	1	-nopreprep -char -unigram -add-delta	31	5	86.11

guess	run	class	good	bad	%
1st	1	CVE-2009-3829	1	0	100.00
1st	2	CVE-2009-2563	1	0	100.00
1st	3	CVE-2009-2562	1	0	100.00
1st	4	CVE-2009-4378	1	0	100.00
1st	5	CVE-2009-2561	1	0	100.00
1st	6	CVE-2009-4377	1	0	100.00
1st	7	CVE-2009-4376	1	0	100.00
1st	8	CVE-2010-2286	1	0	100.00
1st	9	CVE-2010-0304	1	0	100.00
1st	10	CVE-2010-2285	1	0	100.00
1st	11	CVE-2010-2284	1	0	100.00
1st	12	CVE-2010-2283	1	0	100.00
1st	13	CVE-2009-2559	1	0	100.00
1st	14	CVE-2009-3550	1	0	100.00
1st	15	CVE-2009-3549	1	0	100.00
1st	16	CVE-2010-1455	8	1	88.89
1st	17	CVE-2009-3243	3	1	75.00
1st	18	CVE-2009-3241	2	2	50.00
1st	19	CVE-2009-2560	1	1	50.00
1st	20	CVE-2009-3242	1	1	50.00
2nd	1	CVE-2009-3829	1	0	100.00
2nd	2	CVE-2009-2563	1	0	100.00
2nd	3	CVE-2009-2562	1	0	100.00
2nd	4	CVE-2009-4378	1	0	100.00
2nd	5	CVE-2009-2561	1	0	100.00
2nd	6	CVE-2009-4377	1	0	100.00
2nd	7	CVE-2009-4376	1	0	100.00
2nd	8	CVE-2010-2286	1	0	100.00
2nd	9	CVE-2010-0304	1	0	100.00
2nd	10	CVE-2010-2285	1	0	100.00
2nd	11	CVE-2010-2284	1	0	100.00
2nd	12	CVE-2010-2283	1	0	100.00
2nd	13	CVE-2009-2559	1	0	100.00
2nd	14	CVE-2009-3550	1	0	100.00
2nd	15	CVE-2009-3549	1	0	100.00
2nd	16	CVE-2010-1455	8	1	88.89
2nd	17	CVE-2009-3243	3	1	75.00
2nd	18	CVE-2009-3241	3	1	75.00
2nd	19	CVE-2009-2560	1	1	50.00
2nd	20	CVE-2009-3242	1	1	50.00

CVEs and CWEs—The Knowledge Base. The CVE-selected test cases serve as a primary source of the knowledge base to gather information of how known weak code "looks like" in the language model, which we store clustered per CVE or CWE (Common Weakness Enumeration). Thus, we primarily, we: (a) teach the system from the CVE-based cases; (b) test on the CVE-based cases; (c) test on the non-CVE-based cases.

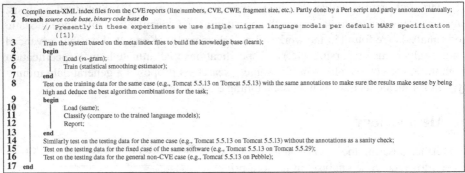

```
 1   Compile meta-XML index files from the CVE reports (line numbers, CVE, CWE, fragment size, etc.). Partly done by a Perl script and partly annotated manually;
 2   foreach source code base, binary code base do
         // Presently in these experiments we use simple unigram language models per default MARF specification
         ([1])
 3       Train the system based on the meta index files to build the knowledge base (learn);
 4       begin
 5           Load (n-gram);
 6           Train (statistical smoothing estimator);
 7       end
 8       Test on the training data for the same case (e.g., Tomcat 5.5.13 on Tomcat 5.5.13) with the same annotations to make sure the results make sense by being
         high and deduce the best algorithm combinations for the task;
 9       begin
10           Load (same);
11           Classify (compare to the trained language models);
12           Report;
13       end
14       Similarly test on the testing data for the same case (e.g., Tomcat 5.5.13 on Tomcat 5.5.13) without the annotations as a sanity check;
15       Test on the testing data for the fixed case of the same software (e.g., Tomcat 5.5.13 on Tomcat 5.5.29);
16       Test on the testing data for the general non-CVE case (e.g., Tomcat 5.5.13 on Pebble);
17   end
```

Fig. 1. High-level static analysis testing algorithm

The introduction by the NIST team of a large synthetic code base with CWEs, serves as a part of knowledge base learning as well.

Categories for Machine Learning. The tow primary groups of classes we train and test our language models on include CVEs [11] and CWEs [12]. The advantages of CVEs are the precision and the associated meta knowledge from [11] that can be all aggregated and used to scan successive versions of the the same software or derived products (e.g., WebKit in multiple browsers). CVEs are also generally uniquely mapped to CWEs. The CWEs as a primary class, however, offer broader categories, of kinds of weaknesses there may be, but are not yet well assigned and associated with CVEs, so we observe the loss of precision. Since we do not parse, we generally cannot deduce weakness types or even simple-looking aspects like line numbers where the weak code may be. Thus, we resort to the secondary categories (that we machine-learn along), that are usually tied into the first two, such as issue types (*sink, path, fix*) and line numbers.

Founding Methodology. The high-level algorithm in NLP and the machine-learning based analysis is in Figure 1. The specific sub-algorithms come from the classical literature and are detailed in [1] and the related works. To be more specific for this work, the loading typically refers to the interpretation of the files being scanned in terms of n-grams and the associated statistical smoothing algorithms, the results of which (a vector, 2D, or 3D matrix) are stored. In our methodology we systematically test and select the best (a tradeoff between speed and accuracy) combination(s) of the

Table 2. CVE Stats for Tomcat 5.5.13

guess	run	algorithms	good	bad	%
1st	1	-nopreprep -char -unigram -add-delta	29	4	87.88
2nd	1	-nopreprep -char -unigram -add-delta	29	4	87.88
guess	run	class	good	bad	%
1st	1	CVE-2006-7197	1	0	100.00
1st	2	CVE-2006-7196	1	0	100.00
1st	3	CVE-2009-2901	1	0	100.00
1st	4	CVE-2006-7195	1	0	100.00
1st	5	CVE-2009-0033	1	0	100.00
1st	6	CVE-2007-1355	1	0	100.00
1st	7	CVE-2007-5342	1	0	100.00
1st	8	CVE-2009-2693	1	0	100.00
1st	9	CVE-2009-0783	1	0	100.00
1st	10	CVE-2008-2370	1	0	100.00
1st	11	CVE-2007-2450	1	0	100.00
1st	12	CVE-2008-2938	1	0	100.00
1st	13	CVE-2007-2449	3	0	100.00
1st	14	CVE-2007-1858	1	0	100.00
1st	15	CVE-2008-4308	1	0	100.00
1st	16	CVE-2008-0128	1	0	100.00
1st	17	CVE-2009-3548	1	0	100.00
1st	18	CVE-2007-5461	1	0	100.00
1st	19	CVE-2007-3382	1	0	100.00
1st	20	CVE-2007-0450	2	0	100.00
1st	21	CVE-2009-0580	1	0	100.00
1st	22	CVE-2007-6286	1	0	100.00
1st	23	CVE-2008-5515	3	1	75.00
1st	24	CVE-2008-1232	1	2	33.33
1st	25	CVE-2009-2902	0	1	0.00
2nd	1	CVE-2006-7197	1	0	100.00
2nd	2	CVE-2006-7196	1	0	100.00
2nd	3	CVE-2009-2901	1	0	100.00
2nd	4	CVE-2006-7195	1	0	100.00
2nd	5	CVE-2009-0033	1	0	100.00
2nd	6	CVE-2007-1355	1	0	100.00
2nd	7	CVE-2007-5342	1	0	100.00
2nd	8	CVE-2009-2693	1	0	100.00
2nd	9	CVE-2009-0783	1	0	100.00
2nd	10	CVE-2008-2370	1	0	100.00
2nd	11	CVE-2007-2450	1	0	100.00
2nd	12	CVE-2008-2938	1	0	100.00
2nd	13	CVE-2007-2449	3	0	100.00
2nd	14	CVE-2007-1858	1	0	100.00
2nd	15	CVE-2008-4308	1	0	100.00
2nd	16	CVE-2008-0128	1	0	100.00
2nd	17	CVE-2009-3548	1	0	100.00
2nd	18	CVE-2007-5461	1	0	100.00
2nd	19	CVE-2007-3382	1	0	100.00
2nd	20	CVE-2007-0450	2	0	100.00
2nd	21	CVE-2009-0580	1	0	100.00
2nd	22	CVE-2007-6286	1	0	100.00
2nd	23	CVE-2008-5515	3	1	75.00
2nd	24	CVE-2008-1232	1	2	33.33
2nd	25	CVE-2009-2902	0	1	0.00

algorithm implementations available to us and then use only those for subsequent testing. The result tables show such ranked combinations. This methodology is augmented
with the cases when the knowledge base for the same code type is learned from multiple
sources (e.g., several independent C test cases).

4 Results

Excerpts of the NLP results are outlined in this section. We summarize the top precisions per test case using NLP-processing of the CVE-based cases and their application
to the general cases. We compare the top results to the similar experiments done for the
signal processing pipeline.

Data and Experiments. We used the NIST data set to practically validate our approach.
Their reference data set contains C/C++, JAVA, and PHP language tracks comprising
CVE-selected cases as well as stand-alone cases. The C/C++ and JAVA test cases of
various client and server OSS software are compilable into the binary code. There are
also synthetic C and JAVA cases generated for various CWE entries. The CVE-selected
cases had a vulnerable version of a software in question with a list of CVEs attached to
it, as well as the most known fixed version within the minor revision number. One of the
goals for the CVE-based cases is to detect the known weaknesses outlined in CVEs using
static code analysis and to verify if they were really fixed in the "fixed version" [2]. The
files from cases with known CVEs and CWEs were used as the training models described
in the methodology. The summary below is a union of the data sets from SATE2010 and
SATE-IV. Top results summary is also there contrasting top NLP vs. signal processing
approaches.

Some select statistical measurements of the precision in recognizing CVEs and CWEs
under different configurations using NLP techniques are shown through the result tables. "Second guess" statistics provided to see if the hypothesis that if our first estimate
of a CVE/CWE is incorrect, the next one in line is probably the correct one. Both are
counted if the first guess is correct. A complete and revised set of statistics along with
the SATE-released data found in [13].

Wireshark 1.2.0 and Wireshark 1.2.18. In Unigram, Add-Delta results on Wireshark
1.2.0's training file for CVEs, the precision is degraded overall compared to the classical
signal processing pipeline. Only 20 out of 22 CVEs are reported, as shown in Table 1.
For Wireshark 1.2.18 the system correctly does *not* report the fixed CVEs.

Chrome 5.0.375.54 and Chrome 5.0.375.70. For NLP processing of Chrome 5.0.375.54
CVEs are reported in Table 3, and CWEs are in Table 4. For Chrome 5.0.375.70, the
system correctly does *not* report the fixed CVEs, so most of the reports come up empty
as they are expected to be for known CVE-selected weaknesses.

CVE-selected test cases:

- C: Wireshark 1.2.0 (vulnerable) and Wireshark 1.2.18 (fixed)
- C: Dovecot0 (vulnerable) and Dovecot17 (fixed)
- C++: Chrome 5.0.375.54 (vulnerable) and Chrome 5.0.375.70 (fixed)
- JAVA: Tomcat 5.5.13 (vulnerable) and Tomcat 5.5.33 (fixed)
- JAVA: Jetty 6.1.16 (vulnerable) and Jetty 6.1.26 (fixed)
- PHP: Wordpress 2.0 (vulnerable) and Wordpress 2.2.3 (fixed)

non-CVE selected and synthetic CWE:

- C: Dovecot 2.0-beta6
- JAVA: Pebble 2.5-M2
- C: Synthetic C covering 118 CWEs and ≈ 60K files
- JAVA: Synthetic JAVA covering ≈ 50 CWEs and ≈ 20K files

Results Summary. Current top precision: Wireshark:

- CVEs (NLP): 83.33%, CWEs (NLP): 58.33%
- CVEs (signal): 92.68%, CWEs (signal): 86.11%,

Tomcat:

- CVEs (NLP): 87.88%, CWEs (NLP): 39.39%
- CVEs (signal): 83.72%, CWEs (signal): 81.82%,

Google Chrome:

- CVEs (NLP): 100.00%, CWEs (NLP): 88.89%
- CVEs (signal): 90.91%, CWEs (signal): 100.00%,

Dovecot (new, 2.x):

- 14 warnings; but it appears all quality or false positive
- (very hard to follow the code, severely undocumented)

Pebble:

- none found during quick testing

Table 3. CVE Stats for Chrome 5.0.375.54 **Table 4.** CWE Stats for Chrome 5.0.375.54

guess	run	algorithms	good	bad	%
1st	1	-nopreprep -char -unigram -add-delta	9	0	100.00
2nd	1	-nopreprep -char -unigram -add-delta	9	0	100.00
guess	run	class	good	bad	%
1st	1	CVE-2010-2304	1	0	100.00
1st	2	CVE-2010-2298	1	0	100.00
1st	3	CVE-2010-2301	1	0	100.00
1st	4	CVE-2010-2295	2	0	100.00
1st	5	CVE-2010-2300	1	0	100.00
1st	6	CVE-2010-2303	1	0	100.00
1st	7	CVE-2010-2297	1	0	100.00
1st	8	CVE-2010-2299	1	0	100.00
2nd	1	CVE-2010-2304	1	0	100.00
2nd	2	CVE-2010-2298	1	0	100.00
2nd	3	CVE-2010-2301	1	0	100.00
2nd	4	CVE-2010-2295	2	0	100.00
2nd	5	CVE-2010-2300	1	0	100.00
2nd	6	CVE-2010-2303	1	0	100.00
2nd	7	CVE-2010-2297	1	0	100.00
2nd	8	CVE-2010-2299	1	0	100.00

guess	run	algorithms	good	bad	%
1st	1	-cweid -nopreprep -char -unigram -add-delta	8	1	88.89
2nd	1	-cweid -nopreprep -char -unigram -add-delta	8	1	88.89
guess	run	class	good	bad	%
1st	1	CWE-399	1	0	100.00
1st	2	NVD-CWE-noinfo	1	0	100.00
1st	3	CWE-79	1	0	100.00
1st	4	NVD-CWE-Other	2	0	100.00
1st	5	CWE-119	1	0	100.00
1st	6	CWE-20	1	0	100.00
1st	7	CWE-94	1	1	50.00
2nd	1	CWE-399	1	0	100.00
2nd	2	NVD-CWE-noinfo	1	0	100.00
2nd	3	CWE-79	1	0	100.00
2nd	4	NVD-CWE-Other	2	0	100.00
2nd	5	CWE-119	1	0	100.00
2nd	6	CWE-20	1	0	100.00
2nd	7	CWE-94	1	1	50.00

Tomcat 5.5.13 and Tomcat 5.5.29. Tomcat 5.5.13 CVE NLP testing shows a higher precision of 87.88%, but the recall is poor, 25/31–6 CVEs are missing out (see Table 2). Subsequent, quick Tomcat 5.5.13 CWE NLP testing was surprisingly poor topping at 39.39% (see Table 5). A quick CVE-based evaluation of Tomcat 5.5.29 is performed to see if the CVEs present in the vulnerable version–and they do not.

CVE-based training and reporting: Line numbers were machine-learned as well as the types of locations and descriptions provided by the SATE organizers and incorporated into the reports via machine learning. This includes the types of locations, such as *fix*, *sink*, or *path* learned from the organizers'-provided data.

CWE-Based Training and Reporting: The CWE-based reports use the CWE as a primary class instead of CVE for training and reporting, and as such currently do not report on CVEs directly (i.e., no direct mapping from CWE to CVE exists unlike in the opposite direction); however, their recognition rates are not very low either in the same locations. The CWE-based training is also used on the testing files, e.g., of Pebble to see if there are any similar weaknesses to that of Tomcat found. CWEs, unlike CVEs for most projects, represent better cross-project classes as they are

Table 5. CWE Stats for Tomcat 5.5.13

guess	run	algorithms	good	bad	%
1st	1	-cweld -nopreprep -char -unigram -add-delta	13	20	39.39
2nd	1	-cweld -nopreprep -char -unigram -add-delta	17	16	51.52
guess	run	class	good	bad	%
1st	1	CWE-16	1	0	100.00
1st	2	CWE-255	1	0	100.00
1st	3	CWE-264	2	0	100.00
1st	4	CWE-119	1	0	100.00
1st	5	CWE-20	1	0	100.00
1st	6	CWE-200	3	1	75.00
1st	7	CWE-22	3	13	18.75
1st	8	CWE-79	1	6	14.29
2nd	1	CWE-16	1	0	100.00
2nd	2	CWE-255	1	0	100.00
2nd	3	CWE-264	2	0	100.00
2nd	4	CWE-119	1	0	100.00
2nd	5	CWE-20	1	0	100.00
2nd	6	CWE-200	4	0	100.00
2nd	7	CWE-22	5	11	31.25
2nd	8	CWE-79	2	5	28.57

largely project-independent. Both CVE-based and CWE-base methods use the same data for training. NLP-based CWE precision in this experiment was quite low ($\approx 39\%$).

5 Conclusion

To our knowledge this was the first time an NLP+machine learning approach was attempted to static code analysis with the first results demonstrated during the SATE2010 workshop [13,2]. The presented experiments use a character model for tokenization, so each printed character is a token. The characters form uni-, bi-, or tri-grams that are counted for their frequencies of occurrence. The language models represent smoothed counts arranged into the respective 1D, 2D, or 3D matrices. The matrices are learned from the known vulnerable code samples and stored. The testing then does the same of looking up the n gram frequencies in the learned matrices instead of updating the matrices for the likely CVE or CWE class. In a way, it's done nearly identically to the classical natural language identification task.

In the experiments as presented, bigrams and trigrams were not used. Unigrams alone have already produced a good precision and were the fastest (but significantly slow than the signal pipeline). The classifiers (or other stage modules) present in the signal pipeline of the tool were not involved in the NLP pipeline of the work, it be interesting to enable such functionality in the future.

References

1. Mokhov, S.A.: Evolution of MARF and its NLP framework. In: C3S2E, pp. 118–122. ACM (2010)
2. Okun, V., Delaitre, A., Black, P.E.: NIST SAMATE: Static Analysis Tool Exposition, SATE (2014), http://samate.nist.gov
3. Bozorgi, M., Saul, L.K., Savage, S., Voelker, G.M.: Beyond heuristics: Learning to classify vulnerabilities and predict exploits. In: Proceedings of the 16th ACM SIGKDD International Conference on Knowledge Discovery and Data Mining, pp. 105–114. ACM, New York (2010)
4. Manning, C.D., Schutze, H.: Foundations of Statistical Natural Language Processing. MIT Press (2002)

5. Mokhov, S.A., Debbabi, M.: File type analysis using signal processing techniques and machine learning vs. `file` unix utility for forensic analysis. In: IMF. LNI, vol. 140, pp. 73–85. GI (2008)
6. Mokhov, S.A.: L'approche MARF à DEFT 2010: A MARF approach to DEFT 2010. In: DEFT, LIMSI / ATALA, pp. 35–49 (2010)
7. Tlili, S.: Automatic detection of safety and security vulnerabilities in open source software. PhD thesis, Concordia Institute for Information Systems Engineering, Concordia University, Montreal, Canada (2009) ISBN: 9780494634165
8. Kremenek, T., Twohey, P., Back, G., Ng, A., Engler, D.: From uncertainty to belief: Inferring the specification within. In: Proceedings of the 7th Symposium on Operating System Design and Implementation (2006)
9. Kong, Y., Zhang, Y., Liu, Q.: Eliminating human specification in static analysis. In: Jha, S., Sommer, R., Kreibich, C. (eds.) RAID 2010. LNCS, vol. 6307, pp. 494–495. Springer, Heidelberg (2010)
10. Eto, M., et al.: NICTER: a large-scale network incident analysis system: case studies for understanding threat landscape. In: BADGERS, pp. 37–45. ACM (2011)
11. NIST: National Vulnerability Database (2014), `http://nvd.nist.gov/`
12. MITRE: Common Weakness Enumeration (CWE) – a community-developed dictionary of software weakness types (2014), `http://cwe.mitre.org`
13. Mokhov, S.A., Paquet, J., Debbabi, M., Sun, Y.: MARFCAT: Transitioning to binary and larger data sets of SATE IV (May 2012), `http://arxiv.org/abs/1207.3718`

Weka-SAT: A Hierarchical Context-Based Inference Engine to Enrich Trajectories with Semantics

Bruno Moreno[1], Amílcar S. Júnior[1],
Valéria Times[1], Patrícia Tedesco[1], and Stan Matwin[2]

[1] Centro de Informática, UFPE, Recife, Brazil
{bnm,asj,vct,pcart}@cin.ufpe.br
[2] Faculty of Computer Science, Dalhousie University, Canada
Institute for Computer Science, Polish Academy of Sciences, Poland
stan@cs.dal.ca

Abstract. A major challenge in trajectory data analysis is the definition of approaches to enrich it semantically. In this paper, we consider machine learning and context information to enrich trajectory data in three steps: (1) the definition of a context model for trajectory domain; (2) the generation of rules based on that context model; (3) the implementation of a classification algorithm that processes these rules and adds semantics to trajectories. This approach is hierarchical and combines clustering and classification tasks to identify important parts of trajectories and to annotate them with semantics. These ideas were integrated into Weka toolkit and experimented using fishing vessel's trajectories.

1 Introduction

Trajectory data are captured in many areas in its raw format as a tuple *(id, x, y, t)*, where *id* is the moving object identifier and *(x, y)* are geographical coordinates collected on time *t*. Due to this lack of semantics, a series of studies are focused on the knowledge discovery based on trajectory data.

Aiming at improving the concept of semantic information to trajectories, Spaccapietra et al. [1] defined a semantic trajectory as being composed by stops and moves. The stops are the most important parts of trajectories while moves are the parts that connect these stops. Current works are capable of identifying important parts of trajectories [2], [3], [4] [5], [6] but they are not capable of determine which activities were performed by the moving object.

The work of Spinsanti et al. [7] proposes a series of rules for the annotation of the whole trajectory with the most probable global activity. However, the limitation of this approach lies in the impossibility of identifying activities at a lower level of detail. To address this issue, we propose a hierarchical approach for providing annotation at episode level. This approach uses a context model to trajectory domain, CMT (Context Model for Trajectories), and is implemented by an algorithm, RB-SAT (Rule Based algorithm for adding Semantic Annotations to Trajectories), that generates rules using context information from CMT.

M. Sokolova and P. van Beek (Eds.): Canadian AI 2014, LNAI 8436, pp. 333–338, 2014.
© Springer International Publishing Switzerland 2014

RB-SAT combines clustering and classification techniques hierarchically and recursively. To increase the utility of our approach, CMT and RB-SAT were integrated to Weka [8], resulting in Weka-SAT (Weka for Semantic Annotation in Trajectories). Weka-SAT was used to execute experiments that added semantics to trajectories. The results were evaluated using *purity* and *coverage* [9].

The rest of this work is organized as follows. Section 2 surveys related work; in Section 3 CMT is presented; Section 4 presents the algorithm to annotate semantics in trajectories; the results of experiments are presented in Section 5 and, finally, Section 6 lists the main conclusions and future works.

2   Related Work

Here, we define *Context of Trajectories* as any information that can be used to characterize the situation of an object in movement. In addition, we adopted also the definition of *Contextual Elements* to refer to data, information and knowledge that are used to define Context [10]. Context of trajectory can be viewed also in terms of the well-known six dimensions of context, these dimensions are named as 5W+1H (*Who, What, Where, When, Why* and *How* dimensions) [10].

Context models are artifacts used to represent domain information considered as context. Vieira et al. [10] defined a generic metamodel for context that assists the creation of context models for any domain. This metamodel asserts that context is composed by *Contextual Elements* (*CEs*) and *Context*. A *CE* is any kind of data, information or knowledge that allows the qualification of an entity for a specific domain; while *Context* is a set of instantiated *CEs* that are used by an agent to execute a relevant task for a specific domain.

The six dimensions of context identify basic units that characterize an entity or a situation. The definition of these dimensions are represented by answering six interrogative pronouns. For the trajectory domain, for example, these answers are given as follows: (1) *Who*, refers to the object's identity that is performing the movement (e.g. car's model, animal species); (2) *What*, refers to a important activity performed by a moving object in a specific part of the trajectory (e.g. tourists visiting a famous place); (3) *Where* refers to the place where the activity was performed; (4) *When* indicates the temporal information about the activity performed; (5) *Why* refers to the reason why a moving object executed an activity or an action; and (6) *How*, refers to the information that indicates the way the context information was acquired (e.g. from sensors, knowledge base).

3   CMT: A Context Model for Trajectories

In this work, we consider that inside the stop the moving object is not necessarily stationary and consequently there may exist another trajectory inside this stop. This subpart of the trajectory is called here as a sub-trajectory. This is an important concept for CMT because this motivates the hierarchical representation adopted. Sub-trajectories may exist inside stops and inside moves as well. Definition 1 describes our sub-trajectory concept.

Definition 1. Sub-trajectory: a sub-trajectory is a list P of spatiotemporal points inside an element E (stop or move). $P = \{p_0 = (x_0, y_0, t_0), p_1 = (x_1, y_1, t_1), ...,$ $p_N = (x_N, y_N, t_N)\}$ where p_0 is the first point of E and p_N is the last point of E. $x_i, y_i \in R$; $t_i \in R^+$ for i = 0, 1, ..., N, and $t_0 < t_1 < t_2 < ... < t_N$.

In this work, a trajectory is segmented in terms of sub-trajectories. In other words, as stops and moves are composed by sub-trajectories, if new segmentations are performed over these sub-trajectories, new context situations can be generated for the application. For example, if we look inside a stop labeled as a shopping center, other sub-trajectories can be detected representing a person performing many activities, like shopping in different stores, working at a store or just stopped watching a movie. Thus, we can split a sub-trajectory into other sub-trajectories recursively and, consequently, organize them hierarchically.

According to Definition 2 and Definition 3, stops and moves captured inside a sub-trajectory are seen as sub-stops and sub-moves, respectively.

Definition 2. Sub-stop: a Sub-stop S is a stop inside a sub-trajectory T where S is a tuple (R_k, t_j, t_{j+N}) such that $R_k = \{(x_j, y_j, t_j), (x_{j+1}, y_{j+1}, t_{j+1}), ...,$ $(x_{j+N}, y_{j+N}, t_{j+N})\}$ and $|t_{j+N} - t_j| < \Delta t$, R_k is the geometry of S and Δt is its minimum time duration.

Definition 3. Sub-move: Sub-move M is a move inside a sub-trajectory T where M is a tuple (R_k, t_j, t_{j+N}) such that $R_k = \{(x_j, y_j, t_j), (x_{j+1}, y_{j+1}, t_{j+1}), ...,$ $(x_{j+N}, y_{j+N}, t_{j+N})\}$ and $|t_{j+N} - t_j| < \Delta t$. R_k is the geometry of M and Δt is its minimum time duration.

Moving Object, Stops, Moves, Sub-stops and *Sub-moves* are considered here as *Contextual Entities*, i.e. they have attributes that allow the identification of contextual information: *Contextual Elements*. Information related to the moving object (e.g. gender of customer) and information processed from raw trajectory (e.g. speed variation) are some examples of our *Contextual Elements*. Figure 1 shows the CMT using the UML notation.

Different instances of *Sub-trajectory*'s *Contextual Elements* represent different meanings in the real-world (i.e. *Contexts*). For *Stop*'s *Sub-trajectory* we have two categories of meanings: *"Place"* or *"Activity"* and for *Move*'s *Sub-trajectory* we have *"Semantic direction"* and *"Transportation mean"*.

According to Figure 1, a *Sub-trajectory, Sub-stops* and *Sub-moves* inherit all Contextual Elements from *Trajectory, Stops* and *Moves*, respectively. The labels next to each concept represent the context dimension correlated to that concept.

Discovering whether the *Contextual Elements* are capable or not of describing different *Contexts* is clearly part of a classification learning task: the *Contextual Elements* are the features used to classify a set of data in different *Contexts*. Thus, we propose the RB-SAT, an algorithm that is capable of adding semantic annotation to trajectories by classifing stops and moves in different classes.

4   The RB-SAT Algorithm

The algorithm RB-SAT (Algorithm 1) basically works as follows. RB-SAT processes each element (stop or move) received as parameter aiming at classifying

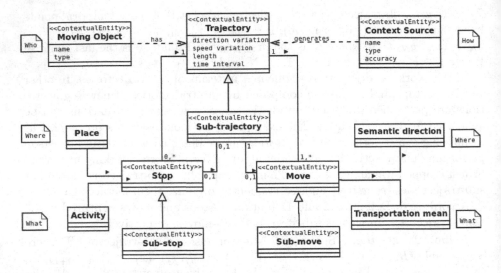

Fig. 1. Trajectory Context Model represented in UML notation

them with labels contained in the outcomes of the rules received as parameter as well (method *checkRules* in line 7). If no outcomes were assigned, clustering algorithms are invoked to generate sub-stops and sub-moves (line 11) and the algorithm is invoked recursively passing the sub-stops and the sub-moves as parameters (line 12). To cluster data, extensions of the CB-SMoT [2] and DB-SMoT [4] were implemented and used allowing an automatic definition of input parameters.

The CMT and the RB-SAT were integrated as an extension of Weka. The goal of this integration is to assist the user in the preprocessing of rules that are required as input by RB-SAT. Our system, named here as Weka-SAT, its source code and a simple tutorial are available for download[1].

5   Results of Experiments

Fishing vessels operating in Brazil are monitored with GPS devices. As a result, large amounts of trajectory data are generated. Fishing activities can be summarized basically by two main steps: (1) release and (2) gather the longline. These two phases compose what is called as "fishing bid".

Vessel's captains commonly write on the board map the coordinates and time instant of each fishing bid. However, due to unkown reasons, this information is sometimes unreadable or absent. Also, it is frequent that some trajectories contain weird movements that are not described on the board map.

To execute the experiments presented here and to generate the needed rules to pass as parameters of RB-SAT, we used the trajectories and the information contained on the board maps from the work of [4]. The rules were generated

---

[1] http://www.brunomoreno.com/softwares/weka-sat

Algorithm 1. RB-SAT

```
1: procedure RB-SAT(T, S, λr, Δt)
2:     result ← ∅ //elements with semantics
3:     aux ← ∅
4:     annotated ← false
5:     for all element s in S do
6:         if time interval of s ≥ Δt then
7:             annotated ← CheckRules(λr, s)
8:             if annotated is true then
9:                 result ← result + s
10:            else
11:                aux ← GenerateSubElements(T, s)
12:                RB-SAT(s.subtrajectory, aux, λr, Δt)
13:            end if
14:        end if
15:    end for
16: end procedure
```

for two different scenarios. In scenario A we considered five different classes: (1) Gathering the longline, (2) Waiting, (3) Releasing the longline, (4) Going fishing and (5) Moving to/from harbor. In the scenario B we considered two classes: (1) fishing and (2) not fishing. The rules passed as parameters to RB-SAT were generated by J48 in Weka-SAT. In addition to these rules, RB-SAT also receives a set of stops and moves as input parameter.

For each scenario, experiments were executed for different combinations of training datasets. As our fishing dataset is composed of four trajectories, our algorithm was trained using all possible combination of two by two and three by three (six and four possibilities, respectively). We tested RB-SAT with two different approaches: (1) training the model using two trajectories and testing with one trajectory and (2) training with three trajectories and testing with one.

Since the trajectories used here are labeled data, we evaluated our results by extending two metrics from [9]: *purity* and *coverage*. Table 1 shows the values of coverage and purity related to our better results. We would like to highlight the relevant results related to scenario B, where the activities performed by a vessel were identified with 86.21% of coverage and with 52.99% of purity on average.

Table 1. Results for purity and coverage

Set type	Scenario	Purity	Coverage
Two trajs.	A	71.08%	30.86%
Three trajs.	A	44.20%	15.51%
Two trajs.	B	88.05%	53.68%
Three trajs.	B	84.37%	52.29%

6   Conclusions

In this paper we presented a model for representing context of trajectories and an algorithm that instantiates this model by using association rules and adds semantics to trajectory data. In addition, we also integrated this algorithm as an extension of Weka. Our motivation is driven by our desire to understand patterns from trajectories domain and improve this data type with semantics.

The main contribution of this work is for the trajectory data mining user, since we are offering an approach that is able to extract trajectories patterns and add this knowledge to raw data. We summarize two main contributions: (1) the definition of a contextual model for trajectories (CMT) and (2) an algorithm that uses these rules from this model to add semantics do trajectories (RB-SAT).

Our approach goes a step further then other approaches ([5], [2], [4]) because it assists the application designer to select features that will be used to classify important parts of trajectories.

Acknowledgment. The authors would like to thank CNPq and the Canadian Bureau of International Education for their financial support.

References

1. Spaccapietra, S., Parent, C., Damiani, M.L., de Macêdo, J.A.F., Porto, F., Vangenot, C.: A conceptual view on trajectories. DKE 65, 126–146 (2008)
2. Palma, A.T., Bogorny, V., Kuijpers, B., Alvares, L.O.: A clustering-based approach for discovering interesting places in trajectories. In: Proceedings of the 2008 ACM Symposium on AC, pp. 863–868. ACM, New York (2008)
3. Zimmermann, M., Kirste, T., Spiliopoulou, M.: Finding stops in error-prone trajectories of moving objects with time-based clustering. In: Tavangarian, D., Kirste, T., Timmermann, D., Lucke, U., Versick, D. (eds.) IMC 2009. CCIS, vol. 53, pp. 275–286. Springer, Heidelberg (2009)
4. Rocha, J.A.M.R.: Db-smot: Um método baseado na direćão para identificaćão de áreas de interesse em trajetórias. Master's thesis, UFPE, Brasil (2010)
5. Alvares, L.O., Bogorny, V., Kuijpers, B., Macêdo, J.A.F., Moelans, B., Vaisman, A.A.: A model for enriching trajectories with semantic geographical information. In: GIS, pp. 22:1–22:8. ACM, New York (2007)
6. Yan, Z., Chakraborty, D., Parent, C., Spaccapietra, S., Aberer, K.: Semitri: a framework for semantic annotation of heterogeneous trajectories. In: 14th International Conference on EDT, pp. 259–270. ACM, New York (2011)
7. Spinsanti, L., Celli, F., Renso, C.: Where you stop is who you are: Understanding peoples activities. In: Proceedings of the 5th Workshop on BMI (2010)
8. Frank, E., Hall, M.A., Holmes, G., Kirkby, R., Pfahringer, B.: Weka - a machine learning workbench for data mining. In: The Data Mining and Knowledge Discovery Handbook, pp. 1305–1314 (2005)
9. Nanni, M., Pedreschi, D.: Time-focused clustering of trajectories of moving objects. J. Intell. Inf. Syst. 27(3), 267–289 (2006)
10. dos Santos, V.V., Brézillon, P., Salgado, A.C., Tedesco, P.: A context-oriented model for domain-independent context management. RIA 22(5), 609–628 (2008)

Inferring Road Maps
from Sparsely-Sampled GPS Traces

Jia Qiu, Ruisheng Wang, and Xin Wang

Department of Geomatics Engineering, University of Calgary, Calgary, Canada
{jiqiu,ruiswang,xcwang}@ucalgary.ca

Abstract. In this paper, we proposed a new segmentation-and-grouping framework for road map inference from sparsely-sampled GPS traces. First, we extended DBSCAN with the orientation constraint to partition the whole point set of traces to clusters representing road segments. Second, we proposed an adaptive k-means algorithm that the k value is determined by an angle threshold to reconstruct nearly straight line segments. Third, the line segments are grouped according to the 'Good Continuity' principle of Gestalt Law to form a 'Stroke' for recovering the road map. Experiment results show that our algorithm is robust to noise and sampling rate. In comparison with previous work, our method has advantages to infer the road maps from sparsely-sampled GPS traces.

Keywords: Map inference, GPS traces, DBSCAN algorithm, k-means clustering, perceptual grouping.

1   Introduction

A road network map is one of the most fundamental geospatial data. Promote and consistent updates of the existing road maps can be more challenging than correcting them [1]. Map inference from GPS traces is a common and efficient method for updating the road network. The purpose is to extract the roads' geometric position, topologic connection, and some attribute information, e.g. lane counts [2]. Most of the existing map inference methods aim at dealing with low-noise, densely-sampled, and uniformly distributed GPS traces [1]. However, GPS probes may reduce the sampling rate due to the energy cost and communication cost [3]. Thus, how to infer high quality road maps from the sparsely-sampled and high-noise GPS traces is an important and challenging topic.

Current road map inferring methods can be roughly classified into three categories: clustering based method, trace merging based method and Kernel Density Estimation (KDE) based method [4]. In addition to these three categories, some other algorithms have also been developed. Karagiorgou and Pfoser [5] proposed an algorithm that use a speed threshold and change in direction as indicators to detect the intersections for road inferring. Biagioni and Eriksson [4] described a hybrid pipeline to infer maps from the traces with noise and disparity. Most of the methods mentioned above mainly address the densely-sampled GPS traces,

M. Sokolova and P. van Beek (Eds.): Canadian AI 2014, LNAI 8436, pp. 339–344, 2014.

in order to overcome the limitation of low GPS sampling rate, we propose a new segmentation-and-grouping framework to generate the road map.

In summary, the contributions of our method are outlined in the following. *First*, a extended DBSCAN (Density-Based Spatial Clustering of Application with Noise) with orientation constraint algorithm is developed for GPS traces segmentation. *Second*, an adaptive k-means clustering method that k is determined by an angle threshold is proposed for nearly straight road segment reconstruction. *Third*, a curve grouping algorithm according to the idea of 'Stroke' [6] is proposed for road segments merger.

The remainder of the paper is organized as follows: Section 2 outlines our map inferring method. Section 3 presents the experiments. Section 4 discusses the conclussion.

2   Methodology

The relation between the sparsely-sampled trace and the base map may have ambiguity errors since segments of the trace does not overlap with the road at the intersection. Therefore, we treat the trace as points with orientation (heading). The orientation can either be read from the database or calculated by consecutive points, since most of the road segments are designed without sharp turning.

In general, we infer the road map by three steps. First, all the points are segmented by their density and orientation to represent road segments. Second, a nearly-straight curve is constructed by the point clusters. Third, curves are grouped according to Gestalt Law [6] to form long roads.

2.1   DBSCAN with Orientation Constraint

One of the most important components of the DBSCAN algorithm is the 'neighborhood of an object' [7]. In classic DBSCAN, the 'ε-neighborhood' is employed to discover the cluster of points. In regard of trajectory data, we define the neighborhood function with two components, which are the distance ε and the difference of orientation α between points, to discover the nearly straight curve from raw GPS traces. The inputs of our algorithm are three parameters–ε, $MinPts$, and α, the output is the point clusters, which denote the road segments.

Definition 1. The $\varepsilon - neighborhood$ $with$ $orientation$ $constraint$ of a point p, denote by N_p, is defined by $N_p = \{q \in P | dis(p,q) \leq \varepsilon \wedge diffAngle(p,q) \leq \alpha\}$, where P is the whole set of points, $dis(p,q)$ is the Euclidian distance between point p and q, and $diffAngle(p,q)$ is the orientation difference of point p and q.

The other definitions and the procedure of our algorithm are same as the classic DBSCAN algorithm. By finding the first 'core point' p_i [7] and its neighbours N_{p_i} w.r.t. ε, $MinPts$, and α, the point cluster, which representing the road segment, is expanded gradually according to the neighbours of points $p_j \in N_{p_i}$.

The choice of the parameters is a trade-off between the detail and error of the road segments: the smaller the ε and the larger the $MinPts$ are, occasionally traveled segments will miss in the final results. In opposite, noise will be introduced. In order to retain more road segments, we set $MinPts$ to 2, and noise will be filtered in subsequent processes. We refer to the work of [9] setting ε to around 15 meters. Since we aim at extracting nearly straight curve from the raw traces, we set α to a small angle (i.e. $1°$). Results based on this setting turns out to be effective and stable in our experiments.

2.2 Nearly Straight Curve Reconstruction

k-means algorithm has wide applications in the field of road centerline generation [4][9]. In their methods, an initial base map is needed before applying k-means. Different from them, we employ k-means to extract the centerlines from the point clusters without base map. Since each cluster generated by DBSCAN represents a nearly straight road segment, we introduce a new parameter–the angle of three consecutive points–as the control parameter to determine the value of k. First, we assign a series of seeds along the diagonal of the Minimal Bounding Rectangle (MBR) of the point cluster. The distance between two seeds equal to a interval l. Second, the seeds are adjusted by k-means algorithm. Third, if any angle of three consecutive seeds is smaller than a angle threshold β, then, we enlarge the interval l to re-assign the seeds and repeat the second step; otherwise, output the sequenced seeds as the centerline of the point cluster.

In the first step, given a point cluster $c_i = \{p_{i,r}|r \in [0, m]\}$, the mean orientation of c_i is $\bar{or}_i = \frac{1}{m}\sum_{r=1}^{m} or_{p_{i,r}}$, the extent of the MBR of c_i is $mbr_i = \{Lat_{top}, Lon_{left}, Lat_{bottom}, Lon_{right}\}$, if $\bar{or}_i \in [0, 90] \cup [180, 270]$, then set the coordinates of the endpoints of the initial centerline as $pc_{i,0} = (Lon_{left}, Lat_{bottom})$, and $pc_{i,n} = (Lon_{right}, Lat_{top})$, where $n = dis(pc_{i,0}, pc_{i,n})/(l \times loop)$, and $loop$ is the repeating time of the second step. Every other points $pc_{i,s}$ are assigned by equal interval between the two endpoints. If $\bar{or}_i \in (90, 180] \cup (270, 360)$, then the initial centerline will be formed by the other two points of the diagonals. In the second step, the procedure of recalculating the centerline is same as the k-means algorithm. The quality of the curve is controlled by β , which range from $0°$ to $180°$. A larger value of this parameter will lead to a straighter curve, while the detail of the curve will be lost. We set the value based on the Gestalt Law, according to [8], the angle threshold for 'Stroke' generation range from $30°$ to $75°$ will produce stable outcomes. Therefore, we set the value of β to $120°$, which means the deflection angle of line segments is $60°$. For the parameter l, we set it to 5 meters, since most of the road is width than this value.

2.3 Line Segments Perceptual Grouping

Road segments generated by previous two steps are trivial, they need to be grouped together to form a long segment to reduce the redundancy as well as overcome the ambiguity error of traces at the intersections. Again, we group the line segments according to the 'Good Continuity' principle of Gestalt Law[6]. The constraints for the grouping of line segments are the closeness, and difference of the mean orientation between two line segments.

The relationship between two nearly straight line segments can be summarized as Fig. 1: (a) l_2 is parallel to l_1; (b) l_2 is the extension of l_1; and (c) l_2 shares a common part with l_1, while is not the extension of l_1. After resampling the segments by adding a point per certain intervals (i.e. 5m), we measure the closeness between two line segments by the parameters $dis(l_i, l_j)$ and $dis_{max}(l_i, l_j)$ describing as following.

Let $l_i = \{p_{i,0}, p_{i,1}, \ldots, p_{i,m}\}$, $l_j = \{p_{j,0}, p_{j,1}, \ldots, p_{j,n}\}$ be two line segments after resampling, and the length of l_i is larger than l_j, then the calculation of $dis(l_j, l_i)$ and $dis_{max}(l_j, l_i)$ from l_j to l_i consists of four steps.

1. $\forall p_{j,s} \in l_j (s \in [0, n])$ find the nearest point $p_{i,r} \in l_i (r \in [0, m])$, and associate the $dis(p_{j,s}, p_{i,r})$ to $p_{i,r}$;
2. $\forall p_{i,r} \in l_i (r \in [0, m])$, select the minimum distance among the distances that are associated to $p_{i,r}$ as the distance of l_j to $p_{i,r}$, representing as $dis(l_j, p_{i,r})$;
3. $\forall p_{i,r} \in l_i (r \in [0, m])$, calculate the mean value of $dis(l_j, p_{i,r})(r \in [0, m])$ where $dis(l_j, p_{i,r}) \neq 0$ as the $dis(l_j, l_i)$.
4. $\forall p_{i,r} \in l_i \land p_{i,r}$ is not the two end points $(r \in [0, m])$, find the maximum distance among the distances that are associated to $p_{i,r} \in l_i (r \in [0, m])$ as $dis_{max}(l_j, l_i)$.

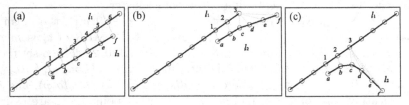

Fig. 1. Illustration of relationship between two line segments. (a) Two segments parallel to each other; (b) a segment is the extension of the other; (c) two segments share a common part.

After calculating the closeness, if $dis(l_2, l_1) > d$, or $dis_{max}(l_2, l_1) > a \times d$, where d is the threshold, and a is a scale factor, then they two segments can not be grouped. In our experiment, we fix a to 2, which is proved to be effective.

The mean orientation or_i of l_i is $or_i = \frac{1}{m} \sum_{r=0}^{m} or_{p_{i,r}}$, where $or_{p_{i,r}}$ is the orientation of $p_{i,r}$. If the orientation difference $\gamma_{or} = |or_i - or_j|$ between two segments is larger than a threshold γ, then they cannot be grouped together.

Each group of the line segments are generated by following. First, get the longest line segment l_i that has not merged to others. Second, find a mergeable line segment l_j w.r.t d and γ, if l_j does not exist, then terminate the algorithm. Third, update the endpoints of l_i, and repeat the second step. The updates of the two endpoints is to replace them by the points that associating to them with the longest distance. Taking Fig. 1(b) as illustration, the end point 3 of l_1 is replaced by the point f of l_2 after merging l_2 to l_1. After obtaining a segment group, the corresponding points are grouped together to construct a new line segment by the k-means clustering method.

In the experiment, short segments degrade the output significantly. Therefore, all the curves will be filtered according to their orientation before the merger. We define the 'Spurious Segment' to filter the curves as Definition 2.

Definition 2. *Spurious Segment*, for a line segment $l_i = \{p_{i,0}, p_{i,1}, \ldots, p_{i,m}\}$, the mean orientation of all points $\overline{or}_P = \frac{1}{m}\sum_{r=0}^{m} or_{p_{i,r}}$, and the mean orientation calculated by consecutive points $\overline{or}_L = \frac{1}{m-1}\sum_{r=0}^{m-1} or_{p_{i,r},p_{i,r+1}}$, where $or_{p_{i,r},p_{i,r+1}}$ is the orientation of the vector $\overrightarrow{p_{i,r},p_{i,r+1}}$, if $|\overline{or}_P - \overline{or}_L|$ is larger than an angle threshold θ, then l_i will be regarded as inferring from noise.

In addition to filter segments, they are merged by two loops. At the first loop, we set a relative small threshold γ_1 (e.g. 5°) to generate relative long segments. At the second loop, the threshold is enlarged to γ_2. In our experiment, γ_2 is set to 45° according to the previous mentioned research [8]. In the line segments filtering process, we set θ to 30°. The other parameter d is set to 20, same as the point cluster generating. After merger, trivial segments whose length are smaller than a certain value (i.e. 20 meters) are removed to clear the road map.

3 Experiments

In order to verify the scalability, effectiveness, and robustness, we use a dataset collected by 12,000 taxies in 4 days to test our method. The location of the taxies is Wuhan, China. Sampling rate of this dataset is sparse, time interval is about 45 second, and total amount of points is more than 3 million.

As we have discussed previously, we set (1) $\varepsilon = 20$, $MinPts = 2$, and $\alpha \approx 1$; (2) $\beta = 120$, $l = 5$; (3) $d = 20$, $\gamma_1 = 5°$, $\gamma_2 = 45°$, and $\theta = 30°$. In the following of this section, we use the above fixed setting to test the algorithm.

The raw GPS traces are displayed as points in Fig. 2(a). Fig. 2(b) is the result of our algorithm. Due to the sparsely sampling rate, some road segments is short. But the algorithm is robust to noise, even the trivial segments match with the

Fig. 2. Result of dataset 2. (a) The raw GPS dataset in Wuhan; (b) road map generated by our method overlapping with the base map.

base map well. As a byproduct, our method can extract the centerline of two opposite directions. However, as a drawback, road segments at the intersections cannot connect with each other. The reason is that in the process of curve reconstruction, since the seed is located in the center of the cluster, the length of curve reconstructed by k-means algorithm is a little shorter than the real one.

4   Conclusion

In this research, we proposed a new segmentation-and-grouping framework for map inference from sparsely-sampled GPS traces. The idea is to partition the points of the whole GPS traces to road segments, and then group the segments to form a long road. The algorithm is robust to noise and sampling rate.

Nevertheless, some problems remaining need to be further researched. The first is the parameters setting, especially the $MinPts$ and α. It is important to adaptively set the value by the feature of the raw dataset. The second is how to build the topological relation of the road segments, which can help to reduce the redundancy of the road map as well as to connect the short segments together and recover the intersections.

References

1. Wang, Y., Liu, X., Wei, H., Forman, G., Chen, C., Zhu, Y.: CrowdAtlas: Self-Updating Maps for Cloud and Personal Use. In: MobiSys 2013. ACM, Taiwan (2013)
2. Schrödl, S., Wagstaff, K., Rogers, S., Langley, P., Wilson, C.: Mining GPS Traces for Map Refinement. Data Mining and Knowledge Discovery 9, 59–87 (2004)
3. Lou, Y., Zhang, C., Zheng, Y., Xie, X., Wang, W., Huang, Y.: Map-Matching for Low-Sampling-Rate GPS Trajectories. In: 17th International Conference on Advances in Geographic Information Systems, pp. 352–361. ACM, New York (2009)
4. Biagioni, J., Eriksson, J.: Map Inference in the Face of Noise and Disparity. In: 20th International Conference on Advances in Geographic Information Systems, pp. 79–88. ACM, New York (2012)
5. Karagiorgou, S., Pfoser, D.: On Vehicle Tracking Data-Based Road Network Generation. In: 20th International Conference on Advances in Geographic Information Systems, pp. 89–98. ACM, New York (2012)
6. Thomson, R.C., Brooks, R.: Exploiting Perceptual Grouping for Map Analysis, Understanding and Generalization: The Case of Road and River Networks. In: Blostein, D., Kwon, Y.-B. (eds.) GREC 2001. LNCS, vol. 2390, pp. 148–157. Springer, Heidelberg (2002)
7. Sander, J., Ester, M., Kriegel, H., Xu, X.: Density-Based Clustering in Spatial Databases: The Algorithm GDBSCAN and Its Applications. Data Mining and Knowledge Discovery 2, 169–194 (1998)
8. Jiang, B., Zhao, S., Yin, J.: Self-Organized Natural Roads for Predicting Traffic Flow: A Sensitivity Study. Journal of Statistical Mechanics: Theory and Experiment, 1–27 (2008)
9. Edelkamp, S., Schrödl, S.: Route Planning and Map Inference with Global Positioning Traces. In: Klein, R., Six, H.-W., Wegner, L. (eds.) Computer Science in Perspective (Ottmann Festschrift). LNCS, vol. 2598, pp. 128–151. Springer, Heidelberg (2003)

Heterogeneous Multi-Population Cultural Algorithm with a Dynamic Dimension Decomposition Strategy

Mohammad R. Raeesi N. and Ziad Kobti

School of Computer Science
University of Windsor
Windsor, ON, Canada N9B 3P4
{raeesim,kobti}@uwindsor.ca

Abstract. Heterogeneous Multi-Population Cultural Algorithm (HMP-CA) is one of the most recent architecture proposed to implement Multi-Population Cultural Algorithms which incorporates a number of heterogeneous local Cultural Algorithms (CAs) communicating with each other through a shared belief space. The heterogeneous local CAs are designed to optimize different subsets of the dimensions of a given problem. In this article, two dynamic dimension decomposition techniques are proposed including the top-down and bottom-up approaches. These dynamic approaches are evaluated using a number of well-known benchmark numerical optimization functions and compared with the most effective and efficient static dimension decomposition methods. The comparison results reveals that the proposed dynamic approaches are fully effective and outperforms the static approaches in terms of efficiency.

1 Introduction

Cultural Algorithm (CA) developed by Reynolds [8] is an Evolutionary Algorithm incorporating knowledge to improve its search mechanism. CA extracts knowledge during the process of evolution and incorporates the extracted knowledge to guide the search direction. Although successful applications of CAs have been reported in various research areas, they are still suffering from immature convergence. One of the most common strategy to overcome this limitation is to incorporate multiple populations. In Multi-Population Cultural Algorithm (MP-CA), the population is divided into a number of sub-populations and each sub-population is directed by a local CA. The local CAs communicate with each other by migrating their extracted knowledge [6].

Different architectures have been proposed to implement MP-CAs. Each architecture has its own strengthes and weaknesses. Homogeneous local CAs is one of the common architectures in which the local CAs have their own local belief spaces and works exactly the same [10,1,5]. Heterogeneous local CAs is another architecture for MP-CAs in which the local CAs are designed to optimize one of the problem dimensions. This architecture is introduced by Lin *et al.* [2] in their proposed Cultural Cooperative Particle Swarm Optimization (CCPSO) to train a Neurofuzzy Inference System.

The most recent architecture is Heterogeneous Multi-Population Cultural Algorithm (HMP-CA) which incorporates only one belief space shared among local CAs instead of one local belief space for each local CA [7]. The shared belief space is responsible

M. Sokolova and P. van Beek (Eds.): Canadian AI 2014, LNAI 8436, pp. 345–350, 2014.

to keep a track of the best parameters found for each dimension. In this architecture the local CAs are designed to optimize different subsets of problem dimensions.

There are various approaches to divide dimensions among local CAs such as assigning one dimension to a local CA [2] and a jumping strategy with various number of local CAs [7]. Recently we investigated various static dimension decomposition approaches including the balanced and imbalanced techniques [4]. The techniques assigning the same number of dimensions to the local CAs are called balanced techniques while the imbalanced ones assign different number of dimensions to the local CAs. As imbalanced techniques, logarithmic and customized logarithmic approaches were investigated with various number of local CAs. The results of this investigation reveals that the imbalanced techniques outperform the balanced approaches by offering 100% success rate and very low convergence ratio.

Although various static strategies have been investigated, it may would be worthwhile to evaluate dynamic strategies. The goal of this article is to design the dynamic dimension decomposition methods.

2 Proposed Method

In this article, two new dynamic dimension decomposition approaches are introduced to improve the efficiency of HMP-CA. The framework of the proposed method is similar to the one incorporated to investigate various static dimension decomposition techniques [4]. Like the published HMP-CA, in the proposed algorithm the local CAs communicate with each other by sending their best partial solutions to the shared belief space and use its updated knowledge to evaluation their partial solutions.

In the published HMP-CA, various static dimension decomposition techniques with various number of local CAs are considered such that the whole population is initially divided among the local CAs evenly. In contrast, our main contribution in this article is to incorporate two dynamic dimension decomposition approaches. Therefore, instead of considering a huge population and dividing it into sub-populations, each local CA gets assigned with a small sub-population of a constant size referred by $PopSize$ parameter. Another parameter is also required as a threshold for the number of generations a local CA cannot find a better solution which is denoted by $NoImpT$ parameter.

In order to design a dynamic dimension decomposition technique, there are two approaches including:

- *Top-Down Strategy*: In this strategy, the idea is to start with a local CA designed to optimize all the problem dimensions and recursively split the dimensions between two newly generated local CAs if a local CA cannot find a better solution for a number of generations. It should be noted here that the decomposed local CA cooperates with the two new local CAs for the next generations such that the decomposed local CA will not be split again. Local CAs with only one assigned dimension will not be decomposed as well.
- *Bottom-Up Strategy*: The idea of this approach is to merge the dimensions of two local CAs when they reach to the no improvement threshold. This approach starts with a number of local CAs, each of which is designed to optimized only one dimension. The number of initially generated local CAs equals to the number of

Table 1. Well-known numerical optimization benchmarks

ID	Function Name	Parameter Range	Minimum Value
f_1	Sphere Model	[-100,100]	0
f_2	Generalized Rosenbrock's Function	[-30,30]	0
f_3	Generalized Schwefel's Problem 2.26	[-500,500]	-12569.5
f_4	Generalized Rastrigin's Function	[-5.12,5.12]	0
f_5	Ackley's Function	[-32,32]	0
f_6	Generalized Griewank's Function	[-600,600]	0
f_7	Generalized Penalized Function 1	[-50,50]	0
f_8	Generalized Penalized Function 2	[-50,50]	0
f_9	Schwefel's Problem 1.2	[-500,500]	0
f_{10}	Schwefel's Problem 2.21	[-500,500]	0
f_{11}	Schwefel's Problem 2.22	[-500,500]	0
f_{12}	Step Function	[-100,100]	0

problem dimensions. These local CAs start to optimize their assigned dimensions until two of them reach to the no improvement threshold. In this stage, a new local CA is generated to optimize all the dimensions of those two local CAs together. Like the top-down approach, each local CA is merged only one time. Therefore, a local CA with all the problem dimensions never get merged.

As the results of the investigation of static dimension decomposition techniques show [4], the imbalanced strategies which assign one dimension to a local CA offer the best efficiency and effectiveness. Therefore, it is expected to see better results from the second approach where the local CAs are initially set to optimize only one dimension compared to the first approach in which it takes a high number of generations to generate the local CAs with only one assigned dimension.

In order to have a fair comparison with static dimension decomposition techniques, the rest of the proposed method is exactly the same as the published HMP-CA [4] such that each local CA incorporates a simple DE with $DE/rand/1$ mutation and binomial crossover operators, and communicates with the shared belief space by sending its best partial solutions. It should be noted here that the belief space contributes to the search process only by providing the complement parameters to evaluate partial solutions.

3   Experiments and Results

Numerical optimization problems are considered to evaluate the proposed dynamic techniques. The goal of these problems is to find the optimal value of a given function. The numerical optimization function is a black box which gets a D-dimensional vector of continuous parameters as input and returns a real number as the objective value for the given parameters. Researchers in this area usually incorporate the set of well-known numerical optimization benchmarks represented in Table 1 [11,9,3]. This table lists 12 benchmark problems with their corresponding parameter ranges and optimal values. These functions are more detailed in the aforementioned papers. In addition to these 12 functions, modified Rosenbrock's function (f_{2M}) [4] is also considered.

Table 2. The average number of FE with respect to each experiments

Style	Top-Down				Bottom-Up			
$PopSize(NoImpT)$	10(5)	10(10)	20(5)	20(10)	10(5)	10(10)	20(5)	20(10)
f_1	123K	120K	235K	215K	**70K**	71K	97K	93K
f_2	391K	**370K**	729K	499K	379K	544K	663K	1,065K
f_{2M}	**986K**	987K	1,586K	1,437K	1,256K	1,192K	1,815K	1,716K
f_3	128K	139K	216K	248K	57K	64K	55K	**46K**
f_4	253K	271K	355K	375K	172K	**149K**	157K	158K
f_5	103K	98K	202K	167K	62K	**61K**	92K	88K
f_6	142K	145K	270K	249K	**97K**	**97K**	116K	110K
f_7	123K	128K	221K	219K	**65K**	67K	96K	91K
f_8	118K	116K	219K	212K	**64K**	66K	93K	89K
f_9	169K	156K	268K	273K	106K	**101K**	123K	103K
f_{10}	197K	196K	357K	353K	**123K**	124K	227K	226K
f_{11}	104K	109K	173K	192K	**55K**	56K	83K	81K
f_{12}	69K	68K	131K	123K	17K	22K	**12K**	13K
Average	224K	223K	382K	351K	**194K**	201K	279K	298K

The proposed dynamic strategies are evaluated with two different values for each of $PopSize$ and $NoImpT$ parameters including 10 and 20 for the former parameter, and 5 and 10 for the latter one. The other parameters are set the same as the ones incorporated to investigate the static dimension decomposition techniques [4]. The number of dimensions is set to 30 which is the common number in the area of numerical optimization and the upper limit for fitness evaluations is considered as 10,000,000. Each experiment is conducted by 100 independent runs to provide statistically significant sample size.

In order to evaluate the effectiveness of the proposed approaches, *Success Rate (SR)* is incorporated which refers to the percentage of the independent runs where an optimal solution is reached. A higher SR means more reliability of the method to find optimal solutions. Therefore, a method with the 100% SR is the most reliable method being able to find an optimal solution within each run. For the efficiency evaluation, the number of *Fitness Evaluations (FE)* is considered in contrast to the number of generations incorporated to evaluate static dimension decomposition approaches [4]. This is due to the fact that dynamic approaches use different number of solutions within each generations.

The experimental results represent that both proposed approaches with all of the four configurations are able to find the optimal solution with respect to each optimization function in every independent run. Therefore, the SR for all the experiments is 100% which implies that both proposed techniques are fully effective methods. The average number of fitness evaluations used to find the optimal solution are illustrated in Table 2 such that the minimum values with respect to each optimization function is emphasized with bold face. The letter K in the table means that the numbers are in thousands and the last two rows represent the average number of fitness evaluation over all the 13 optimization functions.

Table 3. The average CR values to compare dimension decomposition techniques

Static Approaches			Dynamic Approaches							
Logarithmic		Cust	Top-Down				Bottom-Up			
30	60	35	10(5)	10(10)	20(5)	20(10)	10(5)	10(10)	20(5)	20(10)
75.71%	36.30%	29.01%	29.85%	30.06%	84.79%	79.94%	**3.37%**	4.61%	20.09%	21.70%

In order to evaluate the efficiency of the proposed techniques *Convergence Ratio (CR)* measure [4] is incorporate:

$$ConvergenceRatio = \frac{AVG - AVG_{min}}{AVG_{max} - AVG_{min}} \times 100\%$$

where AVG denotes the average number of fitness evaluation required by a specific algorithm to find an optimal solution, and AVG_{min} and AVG_{max} refer to the minimum and the maximum averages obtained by all methods for each optimization function. Therefore, the lower CR value represents the more efficient algorithm.

In order to compare the proposed dynamic approaches with the static techniques, the most efficient ones with 100% success ratio are selected which includes the logarithmic approach with 30 and 60 local CAs and the customized logarithmic approach with 35 local CAs [4]. The CR values are calculated with respect to each optimization function and then they are averaged over all the 13 optimization functions. Therefore, each algorithm gets assigned with a CR value.

Table 3 illustrates the average CR values for both static and proposed dynamic dimension decomposition techniques. The customized logarithmic approach is denoted by Cust in the table. The numbers on the third row specify the configuration of each technique which is the number of local CAs for each static technique and the values for *PopSize* and *NoImpT* parameters for each dynamic method with the format of *PopSize(NoImpT)*. The lowest average CR is represented in bold face.

As expected, the table shows that the best efficiency is offered by the dynamic bottom-up approach with various parameter values. However the best configuration for this strategy is *PopSize* of 10 and *NoImpT* of 5.

4 Conclusions

HMP-CA is the most recent architecture proposed to implement MP-CA, in which local CAs are designed to optimized a subset of the problem dimensions. Therefore, the incorporated dimension decomposition approach has a major effect on the algorithm performance. In this article, two dynamic strategies are proposed including the top-down and bottom-up approaches.

The results of extensive experiments reveals that both dynamic approaches are fully effective to find optimal solutions. In addition to the effectiveness, the results illustrate that the bottom-up approach outperforms the top-down approach by offering the better efficiency. Compared to the most effective and efficient static dimension decomposition methods, the bottom-up approach offers better efficiency while the top-down technique presents competitive efficiency.

As incorporating extra knowledge to choose dimensions for merging or splitting may results in a more efficient method, it has been considered as the future direction for this research area. Furthermore, incorporating more sophisticated optimization functions and higher number of dimensions would be useful to highly evaluate the proposed methods and deeply differentiate them based on their performance.

Acknowledgment. This work is made possible by a grant from the National Science Foundation and NSERC Discovery No. 327482.

References

1. Guo, Y.N., Cheng, J., Cao, Y.Y., Lin, Y.: A novel multi-population cultural algorithm adopting knowledge migration. Soft Computing 15(5), 897–905 (2011)
2. Lin, C.J., Chen, C.H., Lin, C.T.: A hybrid of cooperative particle swarm optimization and cultural algorithm for neural fuzzy networks and its prediction applications. IEEE Transactions on Systems, Man, and Cybernetics, Part C: Applications and Reviews 39(1), 55–68 (2009)
3. Mezura-Montes, E., Velazquez-Reyes, J., Coello Coello, C.A.: A comparative study of differential evolution variants for global optimization. In: Genetic and Evolutionary Computation Conference (GECCO), pp. 485–492 (2006)
4. Raeesi N., M.R., Chittle, J., Kobti, Z.: A new dimension division scheme for heterogenous multi-population cultural algorithm. In: The 27th Florida Artificial Intelligence Research Society Conference (FLAIRS-27), Pensacola Beach, FL, USA, May 21-23 (2014)
5. Raeesi N., M.R., Kobti, Z.: A knowledge-migration-based multi-population cultural algorithm to solve job shop scheduling. In: The 25th Florida Artificial Intelligence Research Society Conference (FLAIRS-25), pp. 68–73 (2012)
6. Raeesi N., M.R., Kobti, Z.: A multiagent system to solve JSSP using a multi-population cultural algorithm. In: Kosseim, L., Inkpen, D. (eds.) Canadian AI 2012. LNCS, vol. 7310, pp. 362–367. Springer, Heidelberg (2012)
7. Raeesi N., M.R., Kobti, Z.: Heterogeneous multi-population cultural algorithm. In: IEEE Congress on Evolutionary Computation (CEC), Cancun, Mexico, pp. 292–299 (2013)
8. Reynolds, R.G.: An introduction to cultural algorithms. In: Sebald, A.V., Fogel, L.J. (eds.) Thirs Annual Conference on Evolutionary Programming, pp. 131–139. World Scientific, River Edge (1994)
9. Schutze, O., Talbi, E.G., Coello Coello, C., Santana-Quintero, L.V., Pulido, G.T.: A memetic PSO algorithm for scalar optimization problems. In: IEEE Swarm Intelligence Symposium (SIS), Honolulu, HI, USA, April 1-5, pp. 128–134 (2007)
10. Xu, W., Zhang, L., Gu, X.: A novel cultural algorithm and its application to the constrained optimization in ammonia synthesis. In: Li, K., Li, X., Ma, S., Irwin, G.W. (eds.) LSMS 2010. CCIS, vol. 98, pp. 52–58. Springer, Heidelberg (2010)
11. Yu, W.J., Zhang, J.: Multi-population differential evolution with adaptive parameter control for global optimization. In: Genetic and Evolutionary Computation Conference (GECCO), Dublin, Ireland, pp. 1093–1098 (2011)

Toward a Computational Model
for Collective Emotion Regulation
Based on Emotion Contagion Phenomenon

Ahmad Soleimani and Ziad Kobti

School of Computer Science, University of Windsor
Windsor, ON, Canada
{soleima,kobti}@uwindsor.ca

Abstract. This paper proposes a novel computational model for emotion regulation process which integrates the traditional appraisal approach with the dynamics of emotion contagion. The proposed model uses a fuzzy appraisal approach to analyze the influence of applying different regulation strategies as directed pro-regulation interventions to the system. Furthermore, the dynamics of changes in the population of emotional and neutral agents were modeled. The proposed model provides an effective framework to monitor and intervene as affect regulator in catastrophic situations such as natural disasters and epidemic diseases.

Keywords: emotion modeling, emotion regulation, emotion contagion.

1 Introduction

Affective Computing (AC) [8] represents the wide-spanned efforts made by AI researchers interested in modeling emotions. An AC system through using a set of rich methods and techniques derived from AI reflects an effective and efficient framework for testing and refining emotion models often proposed within humanistic sciences.

Emotion regulation [3] on the other hand is involved in studying and applying techniques aimed at regulating hyper emotions. According to Gross [3], it includes all of the "conscious and non-conscious strategies we use to increase, maintain, or decrease one or more components of an emotional response".

This article looks at the problem of modeling the process of emotion regulation from the perspective of emotion contagion as a special type of social contagion.With respect to the possible applications for the proposed model, it would appear that an extended version of this model would have the potentials to be used within the field of rescue and disaster management where modeling crowd emotional behavior in catastrophic and time stressed situations such as the resulted panic from earthquakes or epidemic diseases is crucial. In such scenarios, the patterns and rates of emotion contagion in different types of social networks within the stricken community as well as the enforcement of interventions aimed at regulating hyper negatives emotions are of high importance.

M. Sokolova and P. van Beek (Eds.): Canadian AI 2014, LNAI 8436, pp. 351–356, 2014.
© Springer International Publishing Switzerland 2014

2   Related Studies

2.1   Computational Models of Emotion

Different approaches such as appraisal (e.g., [7]), dimensional (e.g., [2]), adaptation (e.g., [6],), etc. can be taken in the process of building a computational model for emotions. Among them, appraisal models are the most widely used approaches to model emotions [5]. At the core of the appraisal theory, a set of dimensions or appraisal variables exist that guides the appraisal processes.

2.2   Social Contagion

In a great deal of scientific research in sociology, the tenet that beliefs, attitudes and behaviors can spread through populations in a similar mechanism to that of infectious diseases is emphasized [11]. Accordingly, the term contagion in psychology is "the spread of a behavior pattern, attitude, or emotion from person to person or group to group through suggestion, propaganda, rumor, or imitation"[4]. Affective behaviors and especially emotions are important examples for social contagion processes. It was shown that emotions spread quickly among the members of social networks with similar dynamics that of infectious diseases . *Hatfield et al.* [4] call this phenomenon as "emotional contagion".

3   Our Approach

In order to model emotion contagion phenomenon, the first step would be to split the population of the individuals within the target society into two sub populations of "emotional" and not-emotional or "neutral" agents. The classification is performed based on the measured emotion response level, ERL of each agent for a given emotion such as *fear*. The ERL is measured in a numeric scale of $[0..1]$ with and imaginary threshold as the cutting edge between the two classes of the population. It needs to be emphasized here that the level of the threshold is completely customizable and that it was not designated specifically for a certain emotion type; serving the generality of the proposed approach.

On the other hand, among the disease spread models, the SIS (Susceptible-Infected-Susceptible) [1] is a well known and widely used model. Accordingly, in the proposed model, a corresponding approach to SIS was introduced and called NEN (Neutral-Emotional-Neutral) to model the emotional transitions (see Fig. 1). Accordingly, the transitioning process in each direction involves one operation that is associated with event appraisal processes and another operation originated in the emotion contagion process (see Fig. 1).

3.1   Problem Formulation

Emotions and Events. As discussed earlier, two out of the four transition operations introduced in the proposed NEN model are associated with a deliberative

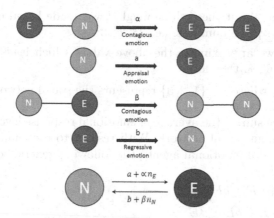

Fig. 1. The NEN model of emotion regulation dynamics. There are four processes by which an emotional state can change. (i) An emotional agent transmits its emotion to a neutral agent with rate α. (ii) A neutral agent becomes emotional as a result of an emotion-triggering event at rate a. (iii) An emotional agent becomes neutral as a result of being exposed to other neutral agents with rate β. (iv) An emotional agent spontaneously becomes neutral as a result of a reversed event at rate b. n_E and n_N represents the population of emotional and neutral agents respectively.

appraisal process of emotion triggering events that take place in the environment of the agent. The full description and formulation of the appraisal processes is provided in [9].

Emotional Contagion. In this section, we focus on formulating the problem of emotion contagion. Accordingly, the change in the emotional response levels of each agent is linked to the "transmissioned quantities" of emotion as a result of interactions between agents. Generally, the changes in the emotional response level of the agents as a result of emotional contagion can be formalized as follows:

$$ERL_{new} = \mu * ERL + (1 - \mu) * E_c$$

where E_c indicates the emotional value that was transmitted through the corresponding contagion process. A persistence factor, μ was introduced to reflect the portion of the current emotional level that will be carried directly to the next time step and consequently will contribute in forming the new emotional response level at the next time step. Hence, $\mu * ERL$ in the above equation reflects the portion of the old ERL that persists in ERL_{new}.

According to *Sutcliff et al.* [10], in social networks, each individual has a intimate support group of five people who are close family members and friends. Furthermore, each individual deals and interacts with about 150 people who know him and he knows them within the community. In addition, we still consider a less influential affective impact for the rest of the population who are out of these two groups. Accordingly, in the proposed model, three different weights corresponding to the three social networks introduced above were considered.

An initial heuristic about these three weights was made based on choosing the first three perfect square numbers, i.e., $1^2, 2^2, 3^2$.

for simplicity, we approximate the above values which leads to $w_1 = 5\%$, $w_2 = 30\%$, $w_3 = 65\%$, therefore,

$$E_c = \sum_i E_{ci} w_i, \text{ where } i \in \{1, 2, 3\} \text{ represents the social network type.}$$

E_{ci} in the above sum is the average of emotional response levels for all agents within the competent social network. With respect to the changes in the population of neutral and emotional agents, the following general equations can be obtained,

$$\frac{dN}{dt} = -aN - \alpha NE + bE + \beta E$$
$$\frac{dE}{dt} = aN + \alpha NE - bE - \beta E$$

where, α reflects the rate of transitioning from neutral agents to emotional agents. Similarly, β is the contagion rate of emotional agents transitioning to neutral ones. Furthermore, a and b are the rates of transitions originated in appraisal processes of occurred events.

In order to find a good approximation for α, it is required that the major factors and parameters that influence this rate to be identified. Here, we introduce a virtual affective channel, CH_{affect} between a pair of agents at which one agent plays the role of affect sender, A_S, and the other agent plays the role of affect receiver, A_R. We argue that α has a close relationship with the strength of the affective channel between the two agents. On the other hand, the strength of this channel has to do with the emotion response level, ERL at both sides of the communication. Hence,

$$\alpha \propto STRENGTH(CH_{affect}) \text{ and } STRENGTH(CH_{affect}) \propto |ERL_S - ERL_R|$$

As an important implication for the proposed approach, it can be inferred that a guided emotion regulation process will be turned into an optimization problem at which the target is to the maximize the population of the neutral agents. By looking at the population equations, such a goal would be achievable by maximizing the rates of b and β through regulation interventions in terms of injected events that are desirable for the neutral agents and preferably undesirable for the emotional agents.

4   Simulation Experiment and Discussion

At this point, we explore one of the simulation experiments that was conducted in order to test the performance of the proposed model. Here, the process of emotion regulation under the influence of emotion contagion as well as interventions in terms of the application of regulation strategies as external events was modeled. Some of experiment conditions were as follows: the population of the agents is chosen to be 1000 distributed equally and randomly between the two groups of emotional and neutral agents, i.e., $P_{E_0} = P_{N_0} = 500$. The cutting edge

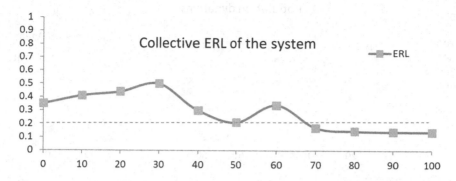

Fig. 2. Collective ERL for all agents of both emotional and neutral margins

(threshold) for the two groups is at $ERL = 0.2$. Emotion under study was *sadness.$E = \{e_1, e_2, e_3, e_4, e_5\}$* is the set of possible events (regulation strategies).

According to Fig. 2, the collective ERL of all agents within both margins started at $ERL = 0.35$. For the first 30 time steps, at the absence of any external regulation interventions, the changes in the ERL is uniquely driven by the spontaneous emotion contagion between the agents. Because of the higher average of the collective ERL for emotional agents (i.e., higher α than β), the collective ERL of the system moved in an anti-regulation direction and at step=30 it touched 0.50. In other words, at this point we ended up with a sadder population. This influence is well reflected in the graph that depicts the population changes of the two groups. Accordingly, in Fig. 3, a huge increase in the population of sad agents during the first 30 time steps can be seen clearly (i.e., $P_E \cong 800, P_N \cong 200$). At step=30, a regulation intervention took place through the injection of e_3 to the system which reflects a very strong in favor of regulation event. The enforcement of this regulation strategy managed to take the collective ERL to a record low amount even below the threshold. Accordingly, the population of sad agents dropped dramatically to around 300 at step=50.

At step=50, the occurrence of e_1 as an anti-regulation event caused the regulation process to flip its direction towards a higher amounts for the collective ERL. This situation did not last long as a result of applying e_2 at step=60. Considering the strong undesirability for this event for the emotional agents, it created a strong wave of spontaneous regulation processes supported by a rapid emotion contagion in favor of regulation that took the ERL to record low of $ERL = 0.145$ at step=80 and further to $ERL = 0.139$ at the end of the simulation. In accordance with such regulation-favored direction, the population of the emotional agents dropped to record low of $P_N = 92$ and later to $P_N = 82$ at the end of the experiment (step=100). Furthermore, it can be seen that the system showed signs of emotional equilibrium starting at step=80 where no significant changes in the collective ERL of the system as well as the population of both groups of agents could be seen.

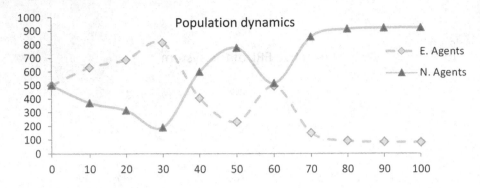

Fig. 3. Dynamics of agent population changes as a result of spontaneous emotion contagion and regulation interventions

5   Conclusion and Future Directions

In this article the process of emotion regulation often aimed at regulating negative hyper emotions from the perspective of emotion contagion was studied and modeled. The proposed model can be expanded in several directions. One direction would be to extend the scope of events as well as the types of social networks between individuals. Furthermore, the dynamism of the system can be increased by considering agent movements which probably make them form clusters of those who have common emotional tendencies.

References

1. Allen, L.: Some discrete-time si, sir, and sis epidemic models. Mathematical Biosciences 124, 83–105 (1994)
2. Gebhard, P.: Alma - a layered model of affect. In: Fourth International Joint Conference on Autonomous Agents and Multiagent Systems (2005)
3. Gross, J., Thompson, R.: Emotion Regulation: Conceptual foundations. In: Handbook of Emotion Regulation. Guilford Press, N.Y (2006)
4. Hatfield, E., Cacioppo, J., Rapson, R.: Emotional contagion. Current Directions in Psychological Science 2, 96–99 (1993)
5. Marsella, S., Gratch, J., Petta, P.: Computational Models of Emotion, for Affective Computing. ict.usc.edu (2010)
6. Marsella, S., Gratch, J.: Ema: A model of emotional dynamics. Cognitive Systems Research 10(1), 70–90 (2009)
7. Ortony, A., Clore, G., Collins, A.: The Cognitive Structure of Emotions. Cambridge University Press (1988)
8. Picard, R.W.: Affective Computing. The MIT Press, MA (1997)
9. Soleimani, A., Kobti, Z.: Event-driven fuzzy paradigm for emotion generation dynamics. In: Proceedings of WCECS 2013, San Francisco, USA, pp. 168–173 (2013)
10. Sutcliff, A., Dunbar, R., Binder, J.: Integrating psychological and evolutionary perspectives. British Journal of Psychology 103, 149–168 (2012)
11. Tsai, J., Bowring, E., Marsella, S., Tambe, M.: Empirical evaluation of computational emotional contagion models. In: Vilhjálmsson, H.H., Kopp, S., Marsella, S., Thórisson, K.R. (eds.) IVA 2011. LNCS, vol. 6895, pp. 384–397. Springer, Heidelberg (2011)

Effects of Frequency-Based Inter-frame Dependencies on Automatic Speech Recognition

Ludovic Trottier, Brahim Chaib-draa, and Philippe Giguère

Department of Computer Science and Software Engineering,
Université Laval, Québec (QC), G1V 0A6, Canada
trottier@damas.ift.ulaval.ca,
{chaib,philippe.giguere}@ift.ulaval.ca
http://www.damas.ift.ulaval.ca

Abstract. The hidden Markov model (HMM) is a state-of-the-art model for automatic speech recognition. However, even though it already showed good results on past experiments, it is known that the state conditional independence that arises from HMM does not hold for speech recognition. One way to partly alleviate this problem is by concatenating each observation with their adjacent neighbors. In this article, we look at a novel way to perform this concatenation by taking into account the frequency of the features. This approach was evaluated on spoken connected digits data and the results show an absolute increase in classification of 4.63% on average for the best model.

Keywords: acoustic model, inter-frame dependencies, speech recognition and automatic digit recognition.

1 Introduction

Automatic speech recognition (ASR) is the transcription of spoken words into text. An ASR model thus takes as input an audio signal and classifies it into a sequence of words. The most common model that is used to perform ASR is the hidden Markov model (HMM) with a Gaussian mixture model as observation density (GMM-HMM). Even though GMM-HMM already showed promising results on various tasks, it is well known that the model has strong assumptions. One of them is the state conditional independence: knowing the latent state at current time makes the current observation independent of all other observations. It has been shown that this hypothesis does not hold when a HMM is used for speech recognition [1].

There are multiple ways in which one can modify the HMM in order to reduce the detrimental effects of state conditional independence on classification performance. Models such as trajectory HMM [3], switching linear dynamical systems [4] and inter-frame HMM [5] were proposed to overcome the assumptions of standard HMM. A recent method (HMM with a deep neural network observation density) used a rectangular window to concatenate each observation with their preceding and succeeding MFCC frames [6]. It has been evaluated

M. Sokolova and P. van Beek (Eds.): Canadian AI 2014, LNAI 8436, pp. 357–362, 2014.

that the window helped the model achieve outstanding results by reducing the effects of the state conditional independence [2]. In this paper, we investigate how the use of a window can help to increase the performance of GMM-HMM by looking at a triangular window that models inter-frame dependencies based on frequency. Signal processing theory suggests that detecting changes is longer for lower frequency signals, and faster for higher frequency signals. Thus, the frequency appears to be a good metric of the speed of variation of information. Therefore, if we create a triangular window that uses information on a longer time-range for low frequencies and a shorter one for high frequencies, we expect to improve the classification of a frame.

The remainder of this paper is structured as follows. Section 2 elaborates on the temporal window and the chosen HMM. In Section 3, the experimental framework and the results of the experiments are described, and we conclude this work in Section 4.

2   Temporal Window

It is normally required that features be extracted from utterances before any classification. Generally, time-based waveform are not as robust as frequency-based features for ASR. For this reason, the most common feature extraction method used for ASR is the Mel frequency cepstral coefficients (MFCC). It uses various transformations, such as the Fourier transform, in order to output a series of frequency-based vectors (frames). As a matter of convention, the coefficients in a frame are ordered from low to high frequency.

In this paper, we propose a novel way of combining feature frames (MFCC) extracted at different times that takes into account the velocity of change at different frequencies. This type of approach that we dubbed triangular window is motivated by signal processing theories which show that the rate at which information changes in signals is proportional to frequency. This is why our window collects frames farther apart for low frequency coefficients and closer for high ones, as depicted in Fig. 1.

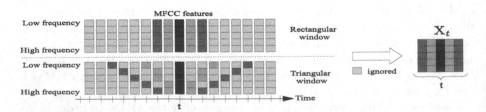

Fig. 1. Example of a rectangular and triangular window of size 5. A given column, on the leftmost figure, represents a frame at a particular time t and the rows are coefficients (MFCCs). Our triangular window approach performs coefficients sub-sampling on each row based on the frequency. The classical rectangular window approach is simply the concatenation of nearby frames together. We denote as X_t the resulting matrix after concatenation.

The difference between rectangular and triangular windows is seen in Fig. 1. The former concatenate coefficients from adjacent frames, whereas the latter concatenate coefficients from other frames based on their frequency. The coefficients at lower frequencies are taken further apart to avoid redundant information. Those at higher frequencies can be made closer because the frames quickly become independent. Importantly, coefficients should not be duplicated in order to avoid redundant information.

In models such as GMM-HMM, where each mixture component is a Gaussian, the resulting matrix X_t can be vectorized to get back a vector $\mathbf{x}_t$. However, $\mathbf{x}_t$ will be high-dimensional if the size of X_t is large. Thus, the evaluation of the observation density would be expensive to compute since it requires the inversion of a high-dimensional covariance matrix Σ. In addition, Σ is more likely to be non-invertible and the learning process could also over-fit it.

For these reasons, the observation density was set to the matrix equivalent of GMM: the matrix normal mixture model (MNMM). In this case, the density function of each component in a MNMM is given by the matrix normal density function:

$$p(X|M,U,V) = \frac{\exp\left(-\frac{1}{2}\operatorname{Tr}\left[V^{-1}(X-M)^{\top}U^{-1}(X-M)\right]\right)}{(2\pi)^{\frac{np}{2}}|V|^{\frac{n}{2}}|U|^{\frac{p}{2}}}, \tag{1}$$

where X and M are $n \times p$ dimensional matrices, U is $n \times n$ and V is $p \times p$. Both models are equivalent if $\Sigma = \operatorname{kron}(V,U)$, the Kronecker product between V and U, and if $\operatorname{vec}(M) = \mu$, the mean of the Gaussian. Most of the time, these models achieve similar results even though MNMM is approximating Σ. However, notice that the $np \times np$ matrix Σ is divided into two smaller matrices U and V which avoids the problems cited earlier.

The Expectation-Maximization (EM) for learning MNMM, as proposed in [7], can be used in our approach to learn MNMM-HMM. To achieve that, we simply replace the posterior probability that observation j belongs to component i $(\tau_{i,j})$ by the posterior probability that frame t belongs to component k of latent state j $(\gamma_t(j,k))$. $\tau_{i,j}$ can be found using component weights whereas $\gamma_t(j,k)$ can be computed using the well-known Forward-Backward recursion. Consequently, the equation for learning the mean of the mixture component k in state j is:

$$\hat{M}_{j,k} = \sum_{t=1}^{T}\gamma_t(j,k)X_t \bigg/ \sum_{t=1}^{T}\gamma_t(j,k) . \tag{2}$$

The equations for the among-row $U_{j,k}$ and among-column $V_{j,k}$ covariance can also be written using the γ smoothing value:

$$\hat{U}_{j,k} = \sum_{t=1}^{T}\gamma_t(j,k)\left(X_t - \hat{M}_{j,k}\right)V_{j,k}^{-1}\left(X_t - \hat{M}_{j,k}\right)^{\top} \bigg/ p\sum_{t=1}^{T}\gamma_t(j,k) , \tag{3}$$

$$\hat{V}_{j,k} = \sum_{t=1}^{T}\gamma_t(j,k)\left(X_t - \hat{M}_{j,k}\right)^{\top}\hat{U}_{j,k}^{-1}\left(X_t - \hat{M}_{j,k}\right) \bigg/ n\sum_{t=1}^{T}\gamma_t(j,k) . \tag{4}$$

The update equations for the mixture weights and the state transitions stay the same. Having now in hands a suitable model that can be used efficiently with the new features X_t, we have elaborated a number of experiments to test the validity of our triangular window.

3   Connected Digits Recognition

The database we used for our experiments is Aurora 2 [8] which contains 11 spoken digits (*zero* to *nine* with *oh*). The digits are connected, thus they can be spoken in any order and in any amount (up to 7) with possible pauses between them. The utterances are noisy and the signal-to-noise ratio (SNR) varies from -5 dB, 0 dB, ..., 20 dB. Different kinds of noise have corrupted the signals such as train, airport, car, restaurant, etc. The training set contains 16,880 utterances, test set A and B have 28,028 and test C has 14,014. On average, an utterance last approximately 2 seconds.

We defined 25 different triangular windows by hands and tested their accuracy with MNMM-HMM. We compared our results with the state-of-the-art GMM-HMM. The latent structure of both HMMs was fixed to 7 states, left-to-right transition only, with additional non-emitting entry and exit. We chose 7 states as the maximum number of states that prevents frames deprivation (more states than frames in HMM). The number of mixture components per state for the emission model was fixed to 4. We used standard 39 dimensional MFCC + Delta + DoubleDelta for GMM-HMM and 13 dimensional MFCC for MNMM-HMM. Both models were trained using EM and the number of iterations it ran was fixed to 10 (we saw a convergence of the log-likelihood after 10 iterations).

Phoneme-based training and testing was performed and 20 different phonemes (including silence) were used to represent the 11 digits. The phonemes were chosen according to The CMU Pronouncing Dictionary[1]. The algorithms were initialized using a standard uniform phoneme segmentation and a silence phoneme was concatenated to the beginning and the end of each utterance. The language model was a uniform bi-gram.

Fig. 2 shows the word accuracy of MNMM-HMM with the best triangular window compared to GMM-HMM. It used coefficients at time $[t - \alpha_i, t - \delta_i, t, t + \delta_i, t + \alpha_i]$ where $\delta = [1, 1, 2, 2, \ldots, 6, 6, 7]$ and $\alpha = [2, 3, 4, 5, \ldots, 12, 13, 14]$ (i iterates from high to low frequency). In addition, the word accuracy computes the number of insertions, deletions and substitutions that are necessary to transform the recognized utterance into the reference. We also added the classification results from the original paper that presented the Aurora 2 database as an additional comparison for our method [8]. We refer to it as HTK since they used the Hidden Markov Model Toolkit to perform the recognition [9]. We can see from the results in Fig. 2 that the triangular window increases the performance of the recognition of about 4.63% on average compared to GMM-HMM.

We notice that the model did not outperform HTK. Indeed, HTK uses a more complex training algorithm that includes, among others, short pause and silence

---

[1] http://www.speech.cs.cmu.edu/cgi-bin/cmudict

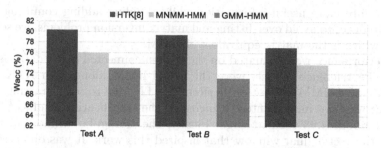

Fig. 2. Word accuracy on Aurora 2 database. For MNMM-HMM, multiple triangular windows were defined and the best one is shown (as explained in the text).

inference, cost function on word transitions and pruning. Instead, a standard EM algorithm was applied on both GMM-HMM and MNMM-HMM. In our preliminary work, we wanted to focus on whether the triangular window could help, or not, the classification. Moreover, the performance increase achieved using a window does not depend on these add-ons. Therefore, the triangular window should still increase the classification results once all the add-ons are implemented.

Moreover, Fig. 3 shows the performance of MNMM-HMM with the best triangular window (depicted earlier) along with rectangular windows of different sizes. We can see that the triangular window helped achieve better accuracy. Also, there is a slight degradation of performance as we increase the size of the rectangular window. This is due to the fact that the coefficients concatenated on both sides are either non informative, redundant or related to other phonemes. This shows that using a triangular window is better than its rectangular counterpart.

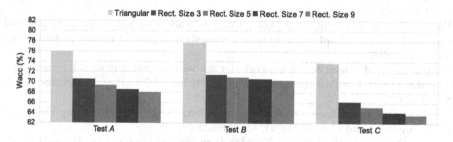

Fig. 3. Comparison for MNMM-HMM of rectangular and triangular windows. The sizes refer to the horizontal size of X_t using a rectangular window. The triangular window is the one that achieved the best performance reported in Fig. 2.

4   Conclusion

In this work, the novel concept of triangular window is investigated. The goal of the triangular window is to take into account the fact that changes in lower-frequency signals cannot be detected rapidly. Therefore, by modifying the concatenation strategy of features depending on their frequency, the model managed to incorporate more useful information while avoiding adding confusing information. In order to avoid over-fitting and matrix inversion problems, we selected MNMM which is the matrix equivalent of GMM.

The performance was evaluated on the spoken connected digits database Aurora 2. The reference models were the GMM-HMM along with HTK. Even though MNMM-HMM did not outperformed HTK, due to some training procedures that were not implemented, it increased the classification results of GMM-HMM by about 4.63% on average. Finally, the triangular window was compared to the rectangular window that inspired this work. It was observed that MNMM-HMM had better results when incorporating frequency dependencies to the concatenation of adjacent frames.

We plan to apply our approach to other existing models such as HTK and deep belief network HMM. For DBN-HMM, the triangular window could reduce the number of neurons on the input level to allow deeper networks to be trained.

References

1. Rabiner, L., Juang, B.H.: Fundamentals of Speech Recognition. Prentice Hall, United States edition (1993)
2. Pan, J., Liu, C., Wang, Z., Hu, Y., Jiang, H.: Investigation of deep neural networks (DNN) for large vocabulary continuous speech recognition: Why DNN surpasses GMMS in acoustic modeling. In: ISCSLP, pp. 301–305. IEEE (2012)
3. Zen, H., Tokuda, K., Kitamura, T.: Reformulating the HMM as a trajectory model by imposing explicit relationships between static and dynamic feature vector sequences. Computer Speech & Language 21(1), 153–173 (2007)
4. Mesot, B., Barber, D.: Switching linear dynamical systems for noise robust speech recognition. IEEE Transactions on Audio, Speech, and Language Processing 15(6), 1850–1858 (2007)
5. Hanna, P., Ming, J., Smith, F.J.: Inter-frame dependence arising from preceding and succeeding frames - Application to speech recognition. Speech Communication 28(4), 301–312 (1999)
6. Hinton, G., et al.: Deep neural networks for acoustic modeling in speech recognition: The shared views of four research groups. IEEE Signal Processing Magazine 29(6), 82–97 (2012)
7. Viroli, C.: Finite mixtures of matrix normal distributions for classifying three-way data. Statistics and Computing 21(4), 511–522 (2011)
8. Pearce, D., Hirsch, H.: Ericsson Eurolab Deutschland Gmbh: The Aurora Experimental Framework for the Performance Evaluation of Speech Recognition Systems under Noisy Conditions. In: ISCA ITRW ASR 2000, pp. 29–32 (2000)
9. Young, S.J., Evermann, G., Gales, M.J.F., Hain, T., Kershaw, D., Moore, G., Odell, J., Ollason, D., Povey, D., Valtchev, V., Woodland, P.C.: The HTK Book, version 3.4. Cambridge University Engineering Department, Cambridge, UK (2006)

Social Media Corporate User Identification Using Text Classification

Zhishen Yang[1], Jacek Wołkowicz[1,2], and Vlado Kešelj[1]

[1] Faculty of Computer Science
Dalhousie University, Halifax, Canada
{zyang,jacek,vlado}@cs.dal.ca
[2] LeadSift Inc.
Halifax, Canada
jacek@leadsift.com

Abstract. This paper proposes a text classification method for identifying corporate social media users. With the explosion of social media content, it is imperative to have user identification tools to classify personal accounts from corporate ones. In this paper, we use text data from Twitter to demonstrate an efficient corporate user identification method. This method uses text classification with simple but robust processing. Our experiment results show that our method is lightweight, efficient and accurate.

Keywords: Social media, Twitter, Market analysis, Text classification.

1 Introduction

Everyday, users produce vast amounts of information through various social media platforms. This makes social media a great source for data mining [5]. Social media platforms give users a place to express their attitudes and opinions. Collection of personal data is significant for companies who want to increase their products' reputation in a short time. This data is also essential for on-line marketing analysis. Kazushi et al. [4] expressed that comparing to traditional marketing and reputation analysis approaches, marketing and reputation analysis technologies using social media data are low cost, real time and high volume.

There are however several challenges with collecting data from social media platforms for marketing purposes. First, one should always avoid collecting data from users who can not be categorized as potential or target customers. Second, one should decide which kind of data is important in order to realize our marketing purposes.

With the rapid growth of social media business promotion, many corporations have social media accounts which promote their products and services. These accounts are not potential targets for social media e-commerce systems, because these accounts do not represent customers. In order to filter out these corporate users for marketing analysis, one needs a method of distinguishing these users

M. Sokolova and P. van Beek (Eds.): Canadian AI 2014, LNAI 8436, pp. 363–368, 2014.
© Springer International Publishing Switzerland 2014

from personal users. In this paper, we propose a text-based method to identify social media corporate users. This method uses text classification and small amount of data from social media users' profiles.

In our method, we use Twitter as our data source. Twitter is a major social media platform with over six hundred million active users. These corporate or personal users produce numerous tweets (micro-blog entries) per day, which makes Twitter a good data mining resource [5]. Meanwhile, there are no obvious methods for distinguishing between corporate and personal Twitter profiles. This is not the case for some other social media platforms, such as Facebook.

The rest of this paper is organized as follows. Section 2 outlines related work. Section 3 describes proposed method for identifying social media corporate users. Section 4 contains presentation and analysis of experiment results. Section 5 concludes this paper.

2   Related Work

There are many author profiling methods that relate to the problem we approach in this paper. Estival et al. [3] estimated author's age, gender, nationality, education level, and native language from Arabic email collection by using text attribution tool. Argamon et al. [1] introduced a text-based classification method that can estimate author's age, gender, native language, and personality from their blogs and essays. The restriction of these methods is that they use blog, email, and essays as information sources. In our circumstances, information is retrieved from social media platform which is more dynamic and inconstant. Ikeda et al.[4] proposed a text-based and community-based method for extracting demographic from Twitter users. They used their method to estimate user's gender, age, area, occupation, hobby, and marital status. Their evaluation based on a large dataset demonstrates their method has good accuracy and even worked for users who only tweet infrequently. In particular, this method is not good for building a light and efficient solution to our problem. The reason for this is that this method needs large scale data not only retrieved from users but their friends. In the case of Twitter, we want to build a solution which only requires what is readily available. Although these previous studies have high accuracy in dealing author profiling problem by extracting demographic data, but acquiring demographic information is not crucial in identifying corporate users.

There are many text classification technologies that can be applied to our method, such as Decision Trees, Support Vector Machines (SVM), Neural Networks, K-nearest Neighbours and Naïve Bayes. Bobicev et al. [2] compared performance of these text classification technologies on many Twitter data sets, they showed that Naïve Bayes performs more efficiently and nearly as effective as SVM.

3 Method

The method for social media corporate user identification consists of a training phase and a prediction phase. In the training phase, we use unsupervised text classification which produces two classifiers: name classifier and biography classifier. In the prediction phase, we use two classifiers which are produced in the training phase to score users. User's score is the criteria for identifying whether the user is corporate user.

3.1 Training Phase

Data Selection. The purpose of this step is to prepare data for unsupervised text classification.

In Twitter, every user has an unique user name and biography (brief profile description). Twitter name is a personal identifier. Normally, personal users would use personal names or nicknames, while corporate users would use corporate names or business names. A user's biography is the short text description of an individual Twitter user. Most corporate users will use words describing their business in their Twitter biography, while most individual users will talk about their personal preferences or social status.

We decided to use only name and biography as input data for our method. Retrieving more complex information becomes infeasible in real time applications. Name and biography is also sufficient for humans to determine whether they are dealing with a corporate account or not.

Feature Extraction. The purpose of this step is to build a feature extractor model to extract features from training data.

We applied bag of terms approach to extract both *unigrams* and *bigrams* from input data sets to generate feature sets. This model simplifies feature extraction by disregarding grammar and word order in a text, apart from their immediate vicinity.

Example:

String: Social media analysis
Unigrams: {Social, media, analysis}
Bigrams: {(Social, media), (media, analysis)}

Feature Selection. The goal of this step is to eliminate low information features. We applied feature selection on both name and biography feature sets.

We used chi-square as metric to filter out low information features to improve performance of text classification. Chi-square finds how common a *unigram* or a *bigram* is in positive training data, compared to how common it occurs in negative data.

NLTK (Natural Language Toolkit) is the tool we used in the following steps. NLTK has libraries to support text classification including chi-square.

1. We calculated frequency of each *unigram* (Overall frequency and its frequency in both positive and negative feature sets) using *FreqDist* method from NLTK. *ConditionalFreqDist* method is employed for frequency in positive and negative feature sets. These frequencies are input for next step.
2. We used *BigramAssocMeasures.chisq* method for *unigram* features and *bigram_finder.nbest* method for *bigram* features to find top N important features. In the experiments section, we show how to find number N and how this step improves the accuracy of text classification.

Unsupervised Text Classification. The final step is text classification. We used Naïve Bayes classifier due to its efficiency and performance. NLTK already provides Naïve Bayes classification function and we directly used it in this step. As mentioned earlier, we created two separate classifiers for name and biography fields.

$$Name\ Classifier = NaiveBayesClassifier.train(Name\ Training\ set)$$
$$Biography\ Classifier = NaiveBayesClassifier.train(Biography\ Training\ set)$$

3.2   Prediction Phase

The purpose of prediction phase is to generate the score of a test user based on Twitter name and biography. The score indicates how likely a given user is a corporate user. We used the same feature extractor from training phase to generate test user's feature sets without any additional feature selection.

Passing name feature set to name classifier calculates the likelihood that this user is a corporate user from Twitter name aspect. Similarly, passing biography feature set to biography classifier calculates the likelihood that this user is a corporate user from Twitter biography aspect. Then we weight those two values to give out the final user score. In the next section, we show how to weight these two values and the criterion, i.e. threshold, of judging whether a user is a corporate user.

4   Experiments

4.1   Data-Sets

To train our method, we downloaded 136 corporate users' profiles and 208 personal users' profiles. For test purposes, we collected a random and representative sample of 116 corporate and 316 personal profiles. We labelled our test data manually.

4.2   Parameters

We defined the following formula to weight name score and biography score to generate final score:

$$score_d = \alpha \times namescore_d + (1 - \alpha) \times bioscore_d \tag{1}$$

Where d is the user profile, $namescore_d$ denotes name score which is the probability produced from name classifier, $bioscore_d$ denotes biography score which is the probability produced from biography classifier and $\alpha \in [0,1]$ denotes balance parameter to weight name score and biography score.

Next, we identified corporate users. We used $t \in [0,1]$ as the threshold. If $score_d$ calculated from user using Formula 1 is greater than t then system identifies this user as a corporate user, otherwise as a personal user:

$$result_d = \begin{cases} corporate & \text{if} \quad score_d >= t \\ personal & \text{otherwise} \end{cases} \tag{2}$$

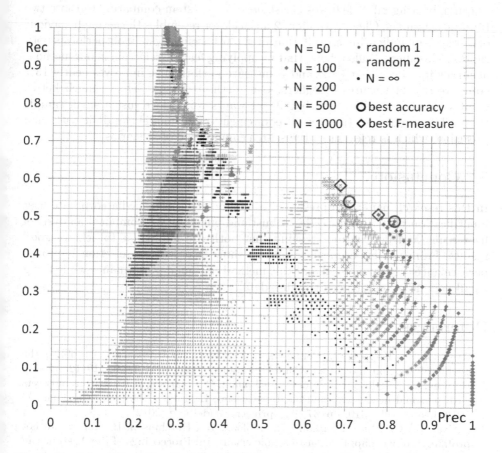

Fig. 1. Precision (P) — Recall (R) graph for our method for various N features classifiers comparing to random classifiers. Each cloud of points represent a system setup with varying α and t parameters, hence different precision and recall values. We indicated points of highest accuracy and F-measure in the graph for 100 and 500 features classifier.

4.3 Experiment Result

We built confusion matrix for all *alpha*, t values with step of 0.01 and various N number of features. Having that, we calculated accuracy, precision, recall and F-measure performance. The highest accuracy of 83% was achieved for $\alpha = 0.27$, $t = 0.58$ and $N = 100$. The highest F-measure (0.64) was obtained for $\alpha = 0.57$, $t = 0.53$ and $N = 500$.

We compared the system to two random classifiers, the first one (`random 1`) was returning a random result from the set $\{corporate, personal\}$ with even probability. The second one (`random 2`) was returning a random score given Formula 1 where $namescore_d$ and $bioscore_d$ were drawn randomly. We also used the full featureset ($N = \infty$) as our baseline. Figure 1 shows precision-recall graph indicating all N features classifiers of the system comparing to those two random classifiers. Clearly, `random 2` much closer resembles the actual behaviour of our system, indicating as well, that even our baseline system outperforms both random classifiers. It also shows that by applying feature selection in our method, accuracy improves nearly 6% and F-measure performance improves nearly 13% compare to full features classifiers. Our method doesn't achieve much better results for high recall values, but if one wants to retrieve just a small sample from a bigger data set (low recall requirements), our model can do it with high precision, which is relevant to the search engine type of scenarios.

5   Conclusion

In this research paper, we presented a method for social media corporate user identification that is lightweight, robust and efficient. We combined analysis of user name and description showing that both are important to achieve good results. From our experiment result, we show that feature selection improved performances of our method and this method has good performance while using small volume of unbalanced training data.

References

1. Argamon, S., Koppel, M., Pennebaker, J., Schler, J.: Automatically profiling the author of an anonymous text. Communications of the ACM 52(2), 119–123 (2009)
2. Bobicev, V., Sokolova, M., Jafer, Y., Schramm, D.: Learning sentiments from tweets with personal health information. In: Kosseim, L., Inkpen, D. (eds.) Canadian AI 2012. LNCS, vol. 7310, pp. 37–48. Springer, Heidelberg (2012)
3. Estival, D., Gaustad, T., Pham, S., Radford, W., Hutchinson, B.: Tat: an author profiling tool with application to arabic emails. In: Proceedings of the Australasian Language Technology Workshop, pp. 21–30 (2007)
4. Ikeda, K., Hattori, G., Ono, C., Asoh, H., Higashino, T.: Twitter user profiling based on text and community mining for market analysis. Knowledge-Based Systems 51, 35–47 (2013)
5. Khan, F., Bashir, S., Qamar, U.: Tom: Twitter opinion mining framework using hybrid classification scheme. Decision Support Systems (2013)

Convex Cardinality Restricted Boltzmann Machine and Its Application to Pattern Recognition

Mina Yousefi, Adam Krzyżak, and Ching Y. Suen*

Department of Computer Science and Software Engineering
Concordia University
Montreal, Quebec, Canada H3G 1M8
mi_yous@encs.concordia.ca,
krzyzak@cs.concordia.ca,
suen@cse.concordia.ca

Abstract. The Restricted Boltzmann machine is a graphical model which has been very successful in machine learning and various applications. Recently lots of attention has been devoted to sparse techniques combining a cardinality potential function and an energy function. In this paper we use a convex cardinality potential function for increasing competition between hidden units to be sparser. Convex potential functions relax conditions on hidden units, encouraging sparsity among them. In this paper we show that combination of convex potential function with cardinality potential produces better results in classification.

Keywords: Cardinality Restricted Boltzmann machine, cardinality potential function, convex function.

1 Introduction

A Deep belief network (DBN) is a generative neural network model with many hidden layers that can learn internal representation. Training deep structure algorithm appears to be a very difficult optimization problem [2].

The Restricted Boltzmann machine (RBM) is a particular form of a Markov Random Field. One of the successful usages of the RBM is in DBN [11]. One of the key advantages of Deep RBM's is that inference can be done very easily [9]. Hinton presented a greedy layer-wise algorithm that uses a stack of RBM's, while all stack has been considered as one probability model [13]. Disconnect between the RBM and the discriminative part is a disadvantage of this method [12]. There are many methods for improving performance the RBM [8].

Performance of the DBN can be improved by applying sparse approach to it [2]. Sparsity can be obtained directly by sparse coding, or by adding penalty functions [12]. Sparse methods applied directly on features do not work very

---

* This research was supported by the Natural Sciences and Engineering Research Council of Canada.

M. Sokolova and P. van Beek (Eds.): Canadian AI 2014, LNAI 8436, pp. 369–374, 2014.
© Springer International Publishing Switzerland 2014

well. It was shown that when data set has high number of active inputs a sparse RBM does not work very efficiently [12]. Therefore another method should be used for the RBM to produce sparse hidden layer.

One of the ways to achieve sparsity in graphical models is counting the number of hidden units that have value one (cardinality) and using it as a parameter of high order potential function [12]. In a Cardinality RBM (CaRBM), a high order potential function that uses cardinality of hidden units is combined with the energy function of RBM. This combination restricts the number of active hidden units to a value that does not exceed a constant number, so it yields the sparse RBM.

In this paper, we want to increase the flexibility of cardinality potential function by using convex potential functions. We introduce some convex potential functions and combine them with the RBM. All these functions have only one parameter (number of active hidden units). Then we compare the results with the standard CaRBM. These functions encourage hidden units to become sparse, unlike the CaRBM that enforces them to become sparse.

2   Background

The RBM is a graphical model that has some unique attributes. It has two types of units: visible and hidden ones. Hidden units and visible units are fully connected together (all visible units to all hidden units,) but there are no other connections

The RBM is an energy based models. We define v as the binary visible vector, h the binary hidden vector, $\{b, c, w\}$ model parameters, w the weight matrix between visible and hidden layers, b and c are biases of visible and hidden vectors, where $w \in R^{N_v * N_h}, c, h \in R^{N_h}, b \in R^{N_v}$. During learning process weights (w) and biases are learned by the gradient descent [5].

2.1   The Cardinality Restricted Boltzmann Machine

Potential functions have been introduced to machine learning recently. One of the important higher order potential functions is the cardinality potential function. This function counts the number of 1's in a vector, so it is called cardinality function. If we combine a cardinality potential function with the RBM potential we obtain the following probability function:

$$P(v, h) = \frac{1}{z} \exp(-E(v, h)) \cdot \psi_k(\sum_{j=1}^{H} h_j) = \frac{1}{z} \exp(b^T v + c^T h + h^T w v) \cdot \psi_k(\sum_{j=1}^{H} h_j)$$

where H is the number of hidden units, z the normalization factor and $\psi_k(\cdot)$ is a higher potential function that depends only on the number of active hidden units and it is defined as follows $\psi_k(x) = 1$ if $x \leq k$ and it is zero otherwise [12].

It is obvious that RBM with ψ_k is constrained to have only k active hidden units. Therefore, it can be considered as sparse RBM. If the CaRBM is used in the DBN one important consideration is fine-tuning by adding a nonlinear function. In a standard binary-binary RBM, the sigmoid nonlinearity is typical nonlinearity that used for calculating marginals and sampling. But in the CaRBM we should define another nonlinear function μ as follows [12]:

$$\mu\left(w^T v + c\right) = E_{P(h|v)}[h] \tag{1}$$

Finding the Jacobian J of (1) with respect to w is necessary for devising a learning scheme. Since $\left(w^T v + c\right)$ is symmetric and it can be multiplied by Jacobian of any function in view of results of Luckily and Domke [3], we obtain the following formula [12]:

$$J\left(w^T v + c\right) l \approx \frac{\mu\left(w^T v + c + \varepsilon.l\right) - \mu\left(w^T v + c - \varepsilon.l\right)}{2\varepsilon}$$

3   The Cardinality Restricted Boltzmann Machine with Convex Potential Function

CaRBM can be encouraged to be sparse by using other high order potential functions in hidden layers. One choice is convex cardinality potential, i.e., a convex function that can cause competition among binary unites [6]. There are seven attributes for choosing a convex function introduced by Stevenson in [7]. We consider three convex potential functions, namely the Bouman and Sauer, the Shulman and Herve and the Green functions [10].

3.1   Combining Bouman and Sauer with Cardinality as Potential Function

Choosing the Bouman and Sauer function (convex function) and combining it with the energy function produces a new energy function.

$$E\left(v, h\right) = -v^T wh - v\left(\sum h\right).$$

If we introduce some changes in the Bouman and Sauer function we obtain a new one as follows:

$$v\left(\sum h\right) = \begin{cases} |\sum h|^p & \text{if } \sum h \leq c \\ -\infty & \text{otherwise} \end{cases} \tag{2}$$

where $p \in [1, 2]$. A general formula for calculating a new joint probability distribution is

$$P(v, h) = \frac{\exp(v^T wh + v\left(\sum h\right))}{z} = \begin{cases} \frac{\exp(v^T wh + v(\sum h) + |\sum h|^p)}{z} & \text{if } \sum h \leq c \\ 0 & \text{otherwise.} \end{cases}$$

3.2 Combining the Shulman and Herve function with Cardinality as Potential Function

After a small modification we can use the Shulman and Herve function instead of the cardinality potential function. Similar to the Bouman and Sauer function we combine this function with energy energy function and the joint probability function is defined as:

$$P(v,h) = \begin{cases} \frac{\exp(v^T wh + |\sum h|^2)}{z} & \text{if} \quad |\sum h| \leq c \\ \frac{\exp(v^T wh + 2a|\sum h| - a^2)}{z} & \text{if} \quad c < |\sum h| < a \\ 0 & \text{otherwise.} \end{cases} \tag{3}$$

3.3 Combining the Green Function with Cardinality as Potential Function

The third potential function uses the Green function. Similar to the last two functions we introduce some changes and then obtain a joint probability distribution as:

$$P(v,h) = \begin{cases} \frac{\exp(-E(v,h))}{z} = \frac{e^{v^T wh}}{z e^{2c^2} \frac{\exp\left(\frac{\sum h}{c}\right) + \exp\left(\frac{-\sum h}{c}\right)}{2}} & \text{if} \quad \sum h \leq c \\ 0 & \text{otherwise.} \end{cases}$$

This joint probability distribution function is smoother than the CaRBM. They have all attributes of the CaRBM with extra conditions that help them to produce more competition among hidden nodes improving sparsity.

Inference for such potential functions can be done by some algorithms [4], [1]. But we use the sum-product algorithm [3] because computing the marginal distribution and sampling are done more clearly. The same approach is used in the CaRBM [12]. For using the convex CaRBM in the DBN as a block, we should define a non-linear function for doing backpropagation in the neural network [12]. Because we use the cardinality potential with the same convex function as in the CaRBM we can use the non-linear function used in the CaRBM and discussed in Section 2.

4 Experimental Results

4.1 Classification Results on MNIST

The MNIST is the handwritten digit data set. It includes 60000 training images and 10000 testing images. Images in this data set are binary black and white 28x28 images. In first experiment, we trained the RBM combined with the convex potential function and then compared the results with the CaRBM. For training the first convex function, i.e., the Bouman and Sauer function, we need to define another parameter p which represents power of the number of active hidden units. We set p=1.5 in all experiments. The classification error does not vary much with different convex functions, and it is similar to the CaRBM error. But there are small changes when the second convex function is used. Results are shown in Table 1. We compare the number of dead nodes and lifetime for showing competition between hidden nodes for more sparsity, see Figure 1.

Table 1. Result on data sets

Function	BoumanRBM p=1.5	GreenRBM	ShulmanRBM	CaRBM
Classification Error on MNIST	0.0496	0.0495	0.0495	0.0495
Classification Error on CIFAR 10	0.636	0.6424	0.6357	0.6423
Classification Error on Caltech 101	0.5097	0.5097	0.5088	0.5173

4.2 Classification Results on CIFAR 10 and Caltech 101

The CIFAR 10 contains 60000 32x32 color images in 10 classes. There are 50000 training images and 10000 test images. We change this data set to 16x16 white and black images and choose 10 batches for each image. Images are clustered to 10 groups as (airplane, automobile, bird, cat, deer, dog, frog, horse, ship, truck).

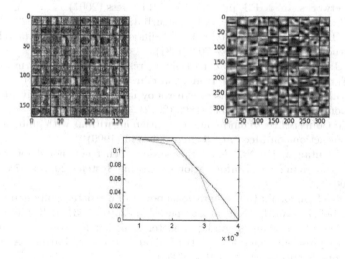

Fig. 1. Classification results with the Shulman and Herve convex potential function on Caltech 101 and CIFAR 10 (*left and right*). Picture on second row shows life time of hidden nodes on MNIST (blue line for CaRBM and green line for Shulman and Herve convex CaRBM).

The Caltech 101 includes 101 category of color images with size 300x200 pixels. Most categories have about 50 images. We choose 10 categories with most images then pick 75% of images for training and 25% of images for testing. We change images to black and white and resize them to 16x16. The best results on both data sets are obtained with the Shulman and Herve CaRBM and worst with the CaRBM, see Table 1. The best result is shown in Figure 1.

5 Conclusions

We introduced some convex potential functions and combined them with the RBM. Likewise the CaRBM, performance of convex CaRBM is similar to the sparse RBM. An important difference between the convex CaRBM and the CaRBM is that CaRBM forces units to be sparse while the convex CaRBM encourages them to become sparse. Best results on all data sets are obtained by the CaRBM with the Shulman and Herve potential function. Results on different data sets show that performance of the convex CaRBM is better than CaRBM. Convex functions could be combine with the RBM and they encourage hidden nodes to become sparse thus showing that encouraging sparsity works better than enforcing it.

References

[1] Barlow, R.E., Heidtmann, K.: Computing k-out-of-n system reliability. IEEE Transactions on Reliability 33, 322–323 (1984)

[2] Bengio, Y., Lamblin, P., Popovici, D., Larochelle, H.: Greedy layer-wise training of deep networks. In: NIPS, pp. 153–160. MIT Press (2007)

[3] Domke, J.: Implicit differentiation by perturbation. In: NIPS, pp. 523–531 (2010)

[4] Gail, M., Lubin, J.H., Rubinstein, L.: Likelihood calculations for matched case-control studies. Biometrika, 703–707 (1981)

[5] Hinton, G.E.: A practical guide to training restricted boltzmann machines. In: Neural Networks: Tricks of the Trade, 2nd edn., pp. 599–619 (2012)

[6] Hinton, G.E.: Training products of experts by minimizing contrastive divergence. Neural Computations 14(8), 1771–1800 (2002)

[7] Lange, K.: Convergence of em image recognition algorithms with gibbs smoothing. IEEE Transactions on Medical Imaging, 439–446 (1990)

[8] Lee, H., Chaitanya, E., Ng, A.Y.: Sparse deep belief net model for visual area v2. In: Advances in Neural Information Processing Systems 20, pp. 873–880. MIT Press (2008)

[9] LeRoux, N., Bengio, Y.: Representational power of restricted boltzmann machines and deep belief networks. Neural Computations 20(6), 1631–1649 (2008)

[10] Li, S.: Closed-form solution and parameter selection for convex minimization-based edge preserving smoothing. IEEE Transactions on Pattern Analysis and Machine Intelligence 20(9), 916–932 (1998)

[11] Salakhutdinov, R., Hinton, G.: Deep boltzmann machines. In: Proceedings of the International Conference on Artificial Intelligence and Statistics, vol. 5, pp. 448–455 (2009)

[12] Swersky, K., Tarlow, D., Sutskever, I., Salakhutdinov, R., Zemel, R., Adams, R.: Cardinality restricted boltzmann machines. In: NIPS, pp. 3302–3310 (2012)

[13] Tang, Y., Salakhutdinov, R., Hinton, G.E.: Robust boltzmann machines for recognition and denoising. In: CVPR, pp. 2264–2271. IEEE (2012)

Task Oriented Privacy (TOP) Technologies

Yasser Jafer

University of Ottawa, Canada
yjafe089@uottawa.ca

Abstract. One major shortcoming with most of the exiting privacy preserving techniques is that, they do not make any assumption about the ultimate usage of the data. Therefore, they follow a 'one-size-fits-all' strategy which usually results in an inefficient solution and consequently leads to over-anonymization and information loss. We propose a Task Oriented Privacy (TOP) model and its corresponding software system which incorporates the ultimate usage of the data into the privacy preserving data mining and data publishing process. Our model allows the data recipient to perform privacy preserving data mining including data pre-processing using metadata. It also provides an intelligent privacy preserving data publishing technique guided by feature selection and personalized privacy preferences.

Keywords: Privacy, Task Oriented Privacy, Data mining, Data publication, Classification.

1   Introduction

The large amount of collected digital information creates tremendous opportunities for knowledge discovery and data mining. However, these datasets may contain sensitive information and therefore, necessary measures should be put in place to ensure that the privacy of individuals is protected.

In a typical scenario, the participators of the privacy preserving data mining/publishing process include a data holder and a data recipient. The data holder holds the dataset (which usually consists of identifiers, quasi-identifiers, sensitive information, and non-sensitive information). The data recipient, on the other hand, aims to perform data analysis on the dataset. Before releasing the dataset, the data holder must ensure that the privacy of individuals in the dataset is protected. Two possible solutions exist: The *first solution* includes modifying the dataset and releasing a modified version of it. The *second solution* is to have the data holder to build the model based on the original data and then to release the model to the data recipient. Both solutions, however, have shortcomings associated with them. The problem with the *first solution* is that, data modification usually results in information loss and degradation in the quality of the data for analysis purposes. The problem with the *second solution* is twofold: Firstly, in most cases, the data holder is not a data mining expert. Secondly, since the model is built using the original data, the model itself may violate the privacy. We propose the TOP model to address the shortcomings associated with these two solutions.

M. Sokolova and P. van Beek (Eds.): Canadian AI 2014, LNAI 8436, pp. 375–380, 2014.

2   The TOP Model

The TOP model supports two complimentary scenarios, i.e., *interactive* and *non-interactive*. In the *interactive* scenarios, we propose a technique to achieve privacy without modifying the data itself, and as such address the challenges associated with the *second solution*. In this approach we aim to build privacy preserving data mining models using the original data. The difference is that, in this approach, the data holder does not publish the data and remains in full control of it; the data holder only grants access to the data to the recipient based on the required analysis task. In other words, we "share the knowledge not the data itself". The data recipient, on the other side, does not see the data and is only able to perform remote data mining. Our approach guarantees that the built model itself does not breach the privacy since the model will be built inside a virtual 'black box' controlled by the data owner. After all, the model(s) will be used by the data holder and therefore, will not be released. In this approach, the data recipient does not use the metadata to build the model and therefore the solution is still based on the semantics of the data. The metadata is used only to provide statistical information about various attributes in the dataset. The *interactive* setting proposed here is different from the interactive settings in, say, differential privacy [1] in which queries are submitted to the data curator in an interactive manner. We do not address query answering; rather, we aim at privacy preserving data mining. Furthermore, our interactive scenario does not apply anonymization operations such as generalization and suppression used in syntactic paradigms or perturbation which is used in differential privacy.

In our TOP model, we also consider the *non-interactive* scenario in which the data publisher publishes, rather, a special version of the anonymized dataset based on the intended analysis task. In general, in privacy preserving data publishing, since the original dataset is modified, eliminating the shortcoming associated with the *first solution* is not possible. It is, however, possible to find methods to protect privacy without negatively impacting the quality of analysis. We follow the observation outlined by different studies that the commonly followed strategy of applying general purpose modification of the dataset via 'one-size-fits-all' approach usually leads to over-anonymization of the dataset [2]. Different privacy models have been proposed in the past. These privacy models address number of attack models such as record linkage, attribute linkage, table linkage, and probabilistic attack. In the *non-interactive* scenario, we aim at enhancing some of the current syntactic privacy models and to show how the ultimate analysis task could be employed to change the dimensionality of the data and construct datasets that, while preserving the privacy according to the selected model, also preserve the utility for a given analysis task such as classification or clustering.

The TOP model considers both the TOP data publishing and data mining and supports the following real world scenarios. In one scenario, the Data Holder (DH) wants to perform data mining but does not have in-house expertise and needs to outsource the data mining task to the Data Recipient (DR). In the second scenario, the DH wants to publish a task-oriented dataset that is tailored for a specific analysis task. The abstract view of the TOP framework is illustrated in Figure 1. The Virtual Secure Environment (VSE) is a virtual 'black box' under full control of DH and is protected against unauthorized accesses. The DR interacts with the *VSE* via interface *I*. In other words, the DR's view of the 'black box' is limited, and dictated by the DH, and the DH is fully responsible and in charge of controlling the nature and the amount of information that is revealed to the DR.

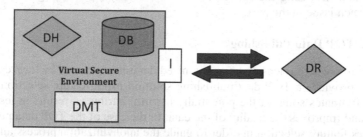

Fig. 1. Abstract view of the TOP model. **DB**: Database, **I**: Interface, **DMT**: Data Mining Tool.

2.1   The TOP Data Mining

In the case of TOP data mining, the process is initiated by the DR sending a request to the DH to obtain general statistical information about the dataset, a.k.a. metadata. The DH sends this metadata including number of attributes, type of attributes, number of tuples, range of numerical attributes, cardinality of nominal/categorical attributes, etc to the DR. This metadata consists of essential information that shape the DR's under-standing of the dataset at hand. In the next step, the DH provides the DR with infor-mation about missing values, noisy data, inconsistent data, outliers, etc. The DR makes necessary decision about how to perform preprocessing and updates the dataset accordingly. For instance, it may decide to remove records with missing values or to replace missing values with other values. In the above steps, appropriate measures are taken into consideration so that revealing the metadata does not breach the privacy. After this step, the DR uses the data mining tool in order to perform remote data min-ing, and to build and evaluate the model.

The proposed TOP data mining technique allows the DR to build a classification model using the original data and not the modified data. To the best of our know-ledge, no work, so far, has considered using metadata to build classification models. Due to the ongoing flow of data, datasets are, in general, dynamic and records are updated, added to, and removed from the database on continuous basis. In addition to the benefits of using data in its pristine condition, working with up-to-date data elimi-nates the risk of building outdated models.

The TOP framework provides a practical solution for an important problem which has been overlooked in the privacy preserving data publishing/data mining literature. Most of the works in these areas consider some benchmarking datasets in the UCI directory such as the *adult* dataset [3] when testing their methods. They, however, ignore one essential requirement of the data mining process with regard to pre-processing. The real-world data is usually highly susceptible to inconsistency, error, missing values, etc. Low quality data eventually leads to low quality mining results. Once again, the assumption that the data custodians are mostly not data mining ex-perts creates another challenge with respect to pre-processing the data. On the one hand, the DH does not have expertise to perform pre-processing; on the other hand, mining data without preprocessing has a negative impact on the results. To address this challenge, the model allows the DR to perform data pre-processing using metadata and without accessing the raw data. To the best of our knowledge, no study has addressed such issue in the past.

2.2   The TOP Data Publishing

There are cases where releasing an anonymized dataset is required by law/regulations. Our model provides a TOP data publishing solution using feature selection. Feature selection eliminates some of the potentially harmful attributes, results in less anonymization, and improves the quality of the data. In the case of the TOP data publishing, the DH uses feature selection in order to guide the anonymization process and to generate an anonymized dataset that is specifically tailored and customized to satisfy the given analysis task at hand. In [4], we show that when feature selection is combined with anonymization, while satisfying k-anonymity, better classification accuracy comparable with (or in some cases even better than) the ones associated with the original data is obtained. Some of the experimental results are shown in Figure 2. In this figure, a comparison between the classification accuracy with/without feature selection for four UCI datasets is illustrated. Details of these experiments are discussed in [4].

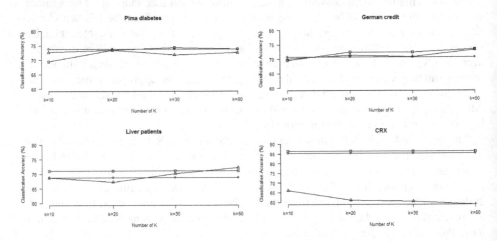

Fig. 2. Classification accuracy of C4.5 for the models built using the original dataset (circle), k-anonymous dataset using Mondrian (triangle), and a combination of Wrapper Feature Selection (WFS) and Mondrian (rectangle).

3   Future Work

In the next step, we will be implementing the remote data mining technique. This includes algorithms and software that implement the data pre-processing step, the remote data mining process, and defining the functions of the user interface in checking and controlling the amount of information that is being exchanged between the DH and the DR. We will be considering the following cases:

1- We will be implementing a privacy preserving data preprocessing technique.
2- After ensuring privacy preserving data preprocessing, the DH outsources the data mining task to the DR. The DR builds the model remotely. Eventually, the model will be used by the DH.
3- We will also consider the case where the DH provides the DR with metadata and the DR uses such metadata to generate synthetic dataset which will be used to build the model.

With respect to the TOP data publishing, we already consider anonymization combined with wrapper feature selection [4] as it was mentioned in Section 2.2. In the future, we aim to consider a feature selection algorithm that is not only guided by the classification accuracy but a trade-off between classification accuracy and privacy. We call it *privacy preserving feature selection*.

The wrapper feature selection approaches result in higher performance compared with filter approaches. This performance is achieved at higher computational cost. By including the privacy cost as another factor, filter approaches may provide very useful results. This is specially the case, when using wrapper approach is not applicable. That is, when the data recipient is aware of classification as the ultimate analysis but does not know which classification algorithm should be used. The TOP data publishing approach should consider the multiple release publishing, sequential release publishing, and continuous data publishing into consideration as well.

We will incorporate personalized privacy into the TOP model. This includes defining a privacy gain/loss cost vs. benefit by the individuals. Finally, we will consider integrating the TOP model into the existing data mining standards such as CRISP-DM that are already vastly adapted and accepted by business.

4   Related Works

Incorporating the ultimate analysis goal into privacy preserving data mining/publishing has lead to two categories of works. The first category [5, 6] considers guiding the anonymization process based on the application purpose. These works consider prioritization of the attributes according to the target class. However, such prioritization process involves user intervention and this turns out to be infeasible in many scenarios. In [6], this intervention includes manually assigning weights to all of the attributes according to the target class. In [5], this intervention consists of identifying the most and the least important attributes with respect to the target class followed by exploiting the independency between attributes in order to determine the priority for the remaining ones. In either case, there is a lack of an effective and fully automated technique for identifying relevant attributes with respect to the target class. In another closely related research dimension, number of works [7] considered the notion of personalized privacy in which individuals (with records in the dataset) are given the option to choose various privacy levels. In other words, those privacy levels are set according to the individuals' privacy preferences. Personalized privacy could be thought of as a natural complement of application purpose-based anonymization since when the individuals know how their data will eventually be used, they will be able to make wiser decisions about their privacy. Combining application purpose-based anonymization with personalized privacy has not been addressed in the past and can be potentially very useful. The second category of works (e.g. [8, 9]) consider performing anonymization for a certain utility function such as classification and cluster analysis amongst others. These works aim at optimizing anonymization according to the predefined data mining task and show that such guided anonymization approaches protect the privacy while improving the data utility for classification purpose. However, in all of these works the models are built based on anonymized datasets and none of them considers using the raw dataset to build the model.

Acknowledgement. I would like to thank my supervisors Dr. Stan Matwin and Dr. Marina Sokolova. I would also like to thank the reviewers for their comments and suggestions. The work is supported by NSERC Strategic Projects Grant.

References

1. Dwork, C.: Differential privacy. In: Bugliesi, M., Preneel, B., Sassone, V., Wegener, I. (eds.) ICALP 2006. LNCS, vol. 4052, pp. 1–12. Springer, Heidelberg (2006)
2. LeFevre, K., DeWitt, D.J., Ramakrishnan, R.: Workload-aware anonymization. In: Proceedings of the 12th ACM SIGKDD International Conference on Knowledge Discovery and Data Mining, Philadelphia, USA, pp. 277–286 (2006)
3. UCI repository, http://archive.ics.uci.edu/ml/
4. Jafer, Y., Matwin, S., Sokolova, M.: Task oriented privacy preserving data pub-lishing using feature selection. To Appear in Proceedings of the 27th Canadian Conference on Artificial Intelligence, Montreal, Canada (2014)
5. Sun, X., Wang, H., Li, J., Zhang, Y.: Injecting purpose and trust into data ano-nymization. Computers & Security 30, 332–345 (2011)
6. Xiong, L., Rangachari, K.: Towards application-oriented data anonymization. In: Proceedings of the 1st SIAM International Workshop on Practical Privacy-Preserving Data Mining, Atlanta, US, pp. 1–10 (2008)
7. Xiao, X., Tao, Y.: Personalized privacy preservation. In: Proceedings of the 2006 ACM SIGMOD International Conference on Management of Data, Chicago, Illinois, US, pp. 229–240 (2006)
8. Fung, B., Wang, K., Yu, P.: Top-down specialization for information and privacy preserva-tion. In: Proceedings of the 21st International Conference on Data Engineering, pp. 205–216 (2005)
9. Wang, K., Yu, P., Chakraborty, S.: Bottom-up generalization-A data mining solu-tion to privacy protection. In: Proceedings of the 4th IEEE International Conference on Data Mining, Brighton, UK, pp. 249–256 (2004)

Graph-Based Domain-Specific Semantic Relatedness from Wikipedia

Armin Sajadi

Faculty of Computer Science, Dalhousie University, Halifax B3H 4R2, Canada
sajadi@cs.dal.ca
https://www.cs.dal.ca

Abstract. Human made ontologies and lexicons are promising resources for many text mining tasks in domain specific applications, but they do not exist for most domains. We study the suitability of Wikipedia as an alternative resource for ontologies regarding the Semantic Relatedness problem.

We focus on the biomedical domain because (1) high quality manually curated ontologies are available and (2) successful graph based methods have been proposed for semantic relatedness in this domain.

Because Wikipedia is not hierarchical and links do not convey defined semantic relationships, the same methods used on lexical resources (such as WordNet) cannot be applied here straightforwardly.

Our contributions are (1) Demonstrating that Wikipedia based methods outperform state of the art ontology based methods on most of the existing ontologies in the biomedical domain (2) Adapting and evaluating the effectiveness of a group of bibliometric methods of various degrees of sophistication on Wikipedia for the first time (3) Proposing a new graph-based method that is outperforming existing methods by considering some specific features of Wikipedia structure.

Keywords: Semantic Relatedness, Wikipedia Mining, Biomedical Domain.

1 Introduction

Measuring Semantic Relatedness is of great interest to philosophers and psychologists for its theoretic importance[7] as well as computer scientists for its various applications such as information retrieval[2]. Studies in both general domain and biomedical domain show that ontology-based methods outperform corpus based methods and are even more practical[4]. On the other hand lexical resources are expensive and labor-intensive, hence not available for most domains.

The first motivation for this project was assessing the suitability of Wikipedia as an alternative resource to domain specific human made ontologies for semantic relatedness problem by evaluating it in the biomedical domain, a domain with well known ontologies and successful graph based methods.

Wikipedia graph is not taxonomic, not even hierarchical. Most of the methods developed to work with knowledge bases are assuming this hierarchical structure

M. Sokolova and P. van Beek (Eds.): Canadian AI 2014, LNAI 8436, pp. 381–386, 2014.

and hence not applicable. On the other hand, this structure is more similar to citation graph extracted from scientific papers. The similarity between the two graphs was our second motivation to evaluate a class of similarity methods originally proposed for similarity in bibliometrics. We tuned and modified some of the methods to make them applicable on Wikipedia.

We also propose a new similarity method which considers a special feature of Wikipedia graph. Our experiments with bibliometric methods show that the increase in the usage of the graph structure results in a decrease in performance. We believe that it is a result of the dense structure of Wikipedia and its small diameter. Many nodes have more than thousands neighbors and in all of these methods, any neighbor plays the same role in calculating relatedness. There is more evidence backing up this claim [12] and this phenomena motivated us to propose a new algorithm which takes into account this fact by ordering the nodes by their role in the graph for calculating relatedness.

2   Methodology

A Wikipedia *page* is associated to each *concept*, so a directed graph $G_b(V_b, E_b)$ can be obtained with nodes (V_b) representing concepts and $(u, v) \in E_b$ if there is a text segment in the page associated with u pointing to the article associated with v. We call this *Basic Wikipedia Graph (G_b)*.

There are two types of edges, regular edges and *redirect* $E_r \subset E_b$ edges. Redirecting denotes synonymy. Let the *epsilon closure of a node* v, denoted by $\varepsilon(v)$, be the set of the nodes (concepts) accessible from v by travelling along redirect links. We use this concept to formally define the *Synonym Ring* of a node v, denoted by $sr(v)$, as the set of nodes equivalent to v. Finally the *Wikipedia Graph* can be formally defined by abstracting from this redirection using Synonym Ring. In this paper, by referring to a node associated with a concept v, we always mean $sr(v)$.

To compute the relatedness between concepts, we start from simple and well known graph-based methods. The class of methods we want to use are all based on this idea: *"Two objects are similar if they are related to similar objects"* [5].

2.1   Basic Bibliometrics and Simrank-Based Methods

Due to the similarity of the structure of Wikipedia to scientific papers, we propose evaluating well known bibliometrics on Wikipedia. Using bibliographic similarity terminology, we can count the portion of common incoming neighbors (*co-citation*), common outgoing neighbors (*coupling*), and combination of both through a weighted average (*amsler*).

The methods mentioned so far focus on only incoming or outgoing links and ignore the whole structure of the graph. If we want to go a step further and consider the relationships among the neighbors, one possibility is using a well known algorithm called SimRank[5]. If the rationale behind the recursion is generalized to outgoing links it is called *rvs-SimRank*[13] and if done for both directions, it is called *P-Rank*[13]. Due to scalability limitations and the huge size of Wikipedia, SimRank is not runnable on demand (it runs globally), hence we do the recursions locally by defining the concept of *joint-neighborhood graph* for two nodes, which would be much a smaller graph.

2.2 Our Proposed Method: HITS Based Similarity

We believe that the unexpected decrease of performance with the increase of incorporating graph structure is a result of the density of the wikipedia graph and non-taxonomic relations: nodes mostly have high degrees (eg., USA has more than 500,000 neighbors) and therefor, not all edges have the same importance. For example, *September 2008* and *Clozapine* are both connected to *Schizophrenia*, while the former is just a date that among many other things, some new statistics about behavioural disorders was published, and the later is a drug to treat *Schizophrenia*.

Our idea is to rank the neighbors of each node based on the role they play in its neighborhood. We use Hyperlink-Induced Topic Search (HITS) [6], a well known concept in information retrieval originally developed to find authoritative sources in hyperlinked document set. We incorporate it in our class of similarity calculation methods, and refer to them by *HIT-Based methods* in this project. To find authoritative pages, it gives every node two scores: *hub score* and *authority score* and using the *mutual reinforcement* between the two, it iteratively updates them. So if it is run over a graph consisting of pages related to a concept (*focused graph* in HITS terminology), the final product of the algorithm is two ranked lists: authoritative pages and those which are good hubs to the authoritative pages.

Having two ordered lists, computing a real number showing the similarity between the two concepts can be done by comparing the two lists using Kendall's tau Distance[3]. Algorithm 1 is our proposed similarity method.

3 Evaluation and Conclusion

The standard datasets are a set of paired concepts with a human-assigned score which is considered to be the ground truth for their relatedness. The more scores reported by the automatic system correlate with the ground truth, the better the system is. The preferred method for this task is Spearman rank correlation (a.k.a Spearman's ρ). We base our evaluations on the most reliable dataset in biomedical domain, Pedersen et al.[11], known as Pedersen Benchmark. The dataset consists of 29 pairs scored by two different groups of physicians and medical coders. We compare our results to two reference reports, (1) McInnes et al.[8] (2) Garla et al.[4] which includes state of the art ontology based methods based on different ontologies such as SNOMED-CT, MeSH and *umls* (the aggregation of these two and 60 restriction free terminologies). Tables 1 and 2 reports the correlation between different automatic systems and human-scores.

From our experiments the following conclusions can be drawn:

Algorithm 1. HITS Based Similarity Computation

1: **function** $HITS\text{-}Sim(a,b)$
Input: : a,b, two concepts
Output: : Similarity between a and b
2:　　$N[a] \leftarrow$ Extract a neighborhood graph for a
3:　　$N[b] \leftarrow$ Extract a neighborhood graph for b
4:　　$L[a] \leftarrow HITS(N[a])$
5:　　$L[b] \leftarrow HITS(N[b])$
6:　　$L'[a] \leftarrow append(L[a], reverse(L[b] \setminus L[a]))$
7:　　$L'[b] \leftarrow append(L[b], reverse(L[a] \setminus L[b]))$
8: **return** $Kendall\text{-}Distance(L'[a], L'[b])$
9: **end function**
10: **function** $HITS(N)$
Input: : N, An adjacency matrix representing a graph
Output: : An order list of vertices
11:　　$S \leftarrow$ Using HITS calculation, get hub or authority scores for each node
12:　　$L \leftarrow$ sort vertices of N based on S
13: **return** L
14: **end function**

Table 1. Pedersen Benchmark: Comparison of correlations with[8]

	Physician		Coder	
	sct-umls	msh-umls	sct-umls	mesh-umls
path	0.35	0.49	0.5	0.58
Leacock & Chodorow	0.35	0.49	0.5	0.58
Wu & Palmer	-	0.45	-	0.53
Nguyen& Al-Mubaid	-	0.45	-	0.55
	Wikipedia		Wikipedia	
WLM	0.71		0.68	
$co\text{-}citation$	0.68		0.65	
$coupling$	0.66		0.63	
$amsler$	0.72		0.68	
$SimRank$	0.6		0.64	
$rvs\text{-}SimRank$	0.20		0.08	
$P\text{-}Rank$	0.39		0.41	
$HITS\text{-}sim_{aut}$	0.69		0.7	
$HITS\text{-}sim_{hub}$	0.69		0.63	
$HITS\text{-}sim$	**0.74**		**0.7**	

Table 2. Pedersen Benchmark: Comparison of correlations with[4]

	Physicians		Coders	
Ontology Based	Sct-umls	umls	Sct-uml	umls
Intrinsic IC*-Lin	0.41	0.72	0.49	0.70
Intrinsc IC-Path	0.35	0.69	0.45	0.69
Intrinsic IC-lch	0.35	0.69	0.45	0.69
PPR-Taxonomy relations	0.56	0.67	0.70	0.76
PPR-ALL relations	0.19	0.63	0.26	0.73
Wikipedia based	Physicians		Coders	
$HIT - SIM$	0.75		0.70	

*Information Content

1. Wikipedia based methods clearly outperform the most well known ontology based methods using different experiments. Only the combination of both SNOMED-CT, MeSH and other 60 terminologies can in some cases outperform Wikipedia based methods.
2. $HITS$-Sim with Wikipedia gives the best results among all methods in most of the cases.
3. Our proposed HITS based methods and especially $HITS$-Sim are performing well and give the best results in all cases.

4 Future Work

4.1 Theoretic Improvements

Although the idea of ranking neighbors is logical, the amount of improvement is less than our expectations. We believe that following modifications should shed more light on the problem:

- Decreasing the size of the neighborhood graph. Getting rid of even less important nodes by graph sampling algorithms has shown to improve the results of HITS and other ranking algorithms.
- Other ranking methods instead of HITS can be used straightforwardly, such as PageRank[1].
- Ranking neighbors to decrease the clutter around the nodes is not an alternative to the idea of SimRank which is propagation of the similarity. Hence we can use our idea to improve SimRank.

4.2 Further Evaluations

In this direction, we are interested in evaluating our similarity method on more datasets or in real applications.

– Automatic mapping of concepts to Wikipedia articles and disambiguation is a necessary module in case of dealing with large datasets. This task is more challenging in domain specific applications as the disambiguated sense should remain in the same domain.
– Experimenting with new datasets; although not many of such datasets exist, but recent studies[4] have used less reliable but bigger datasets (such as [9] and [10]).
– Applying our relatedness measure in other tasks, specifically *tag-generalization* and *Automatic indexing* of biomedical papers.

References

1. Brin, S., Page, L.: The anatomy of a large-scale hypertextual web search engine. Comput. Netw. ISDN Syst. 30(1-7), 107–117 (1998)
2. Budanitsky, A.: Lexical Semantic Relatedness and its Application in Natural Language Processing. Ph.D. thesis, University of Toronto, Toronto, Ontario (1999)
3. Fagin, R., Kumar, R., Sivakumar, D.: Comparing top k lists. In: Proceedings of the Fourteenth Annual ACM-SIAM Symposium on Discrete Algorithms, SODA 2003, pp. 28–36. Society for Industrial and Applied Mathematics, Philadelphia (2003)
4. Garla, V., Brandt, C.: Semantic similarity in the biomedical domain: an evaluation across knowledge sources. BMC Bioinformatics 13(1), 1–13 (2012)
5. Jeh, G., Widom, J.: Simrank: a measure of structural-context similarity. In: Proceedings of the Eighth ACM SIGKDD International Conference on Knowledge Discovery and Data Mining, KDD 2002, pp. 538–543. ACM, New York (2002)
6. Kleinberg, J.M.: Authoritative sources in a hyperlinked environment. J. ACM 46(5), 604–632 (1999)
7. Landauer, T.K., Foltz, P.W., Laham, D.: An introduction to latent semantic analysis. Discourse Processes 25(2-3), 259–284 (1998)
8. McInnes, B.T., Pedersen, T., Pakhomov, S.V.: UMLS-Interface and UMLS-Similarity: open source software for measuring paths and semantic similarity. In: AMIA Annu. Symp. Proc. 2009, pp. 431–435 (2009)
9. Pakhomov, S., McInnes, B., Adam, T., Liu, Y., Pedersen, T., Melton, G.B.: Semantic Similarity and Relatedness between Clinical Terms: An Experimental Study. In: AMIA Annu. Symp. Proc. 2010, pp. 572–576 (2010)
10. Pakhomov, S.V.S., Pedersen, T., McInnes, B., Melton, G.B., Ruggieri, A., Chute, C.G.: Towards a framework for developing semantic relatedness reference standards. J. of Biomedical Informatics 44(2), 251–265 (2011)
11. Pedersen, T., Pakhomov, S.V., Patwardhan, S., Chute, C.G.: Measures of semantic similarity and relatedness in the biomedical domain. Journal of Biomedical Informatics 40(3), 288–299 (2007)
12. Yeh, E., Ramage, D., Manning, C.D., Agirre, E., Soroa, A.: Wikiwalk: random walks on wikipedia for semantic relatedness. In: Proceedings of the 2009 Workshop on Graph-based Methods for Natural Language Processing, TextGraphs-4, pp. 41–49. Association for Computational Linguistics, Stroudsburg (2009)
13. Zhao, P., Han, J., Sun, Y.: P-rank: a comprehensive structural similarity measure over information networks. In: Proceedings of the 18th ACM Conference on Information and Knowledge Management, CIKM 2009, pp. 553–562. ACM, New York (2009)

Semantic Management of Scholarly Literature: A Wiki-Based Approach

Bahar Sateli

Semantic Software Lab.
Department of Computer Science and Software Engineering
Concordia University, Montréal, Canada

1 Introduction

The abundance of available literature in online repositories poses more challenges rather than expediting the task of retrieving content pertaining to a knowledge worker's information need. The rapid growth of the number of scientific publications has encouraged researchers from various domains to look for automatic approaches that can extract knowledge from the vast amount of available literature. Recently, a number of desktop and web-based applications have been developed to aid researchers in retrieving documents or enhancing them with semantic annotations [1,2]; yet, an integrated, collaborative environment that can encompass various activities of a researcher from assessing the writing quality of a paper to finding complementary work of a subject is not readily available. The hypothesis behind the proposed research work is that knowledge-intensive literature analysis tasks can be improved with semantic technologies. In this paper, we present the Zeeva system, as an empirical evaluation platform with integrated intelligent *assistants* that collaboratively work with humans on textual documents and use various techniques from the Natural Language Processing (NLP) and Semantic Web domains to manage and analyze scholarly publications.

2 Research Question

The motivation behind our research is to improve the management and analysis of scholarly literature and aid knowledge workers in tasks like literature surveys and peer reviews, where a deep semantic processing of literature on a large scale is needed. To this end, we aim to develop an extensible platform to support the mentioned processes based on customizable literature analysis workflows. Provided with an intuitive user interface, various user groups, such as graduate students, peer reviewers or business analysts can create custom analysis workflows by combining sequences of system services (e.g., document clustering, entity extraction) to help them with their task at hand. In addition to generic bibliographic management services, the Zeeva system will provide its users with state-of-the-art NLP services that enrich the literature with semantic metadata suitable for both human and machine processing techniques.

M. Sokolova and P. van Beek (Eds.): Canadian AI 2014, LNAI 8436, pp. 387–392, 2014.

We hypothesize that our tool can improve knowledge-intensive literature analysis tasks like reviewing papers with automatic detection of semantic entities in scholarly publications through a novel collaboration pattern between humans and AI *assistants*. To corroborate our hypothesis, the proposed research will be performed in four phases in an iterative and evolutionary fashion:

During the first phase, an extensive requirements elicitation will be performed in order to detect distinct and overlapping requirements of various user groups. The gathered requirements will help us to accommodate as many task-specific needs as possible in the design of our system with future extensibility and adaptation in mind. So far, more than 50 web-based and desktop bibliographic management systems, online digital libraries, open access repositories, scientific indexing engines and academic social networking websites have been studied to extract researchers' patterns and analysis workflows, as well as popular system features. An initial set of requirements have been gathered and a functional prototype has been built targeting researchers as the user group under study. The second phase encompasses the tasks of developing an extensible platform where various user groups can define custom analysis tasks by combining reusable processing components. For example, a graduate student can combine domain-specific *Information Retrieval* and *Extraction* services to obtain an overview of contributions from a set of papers, a journal reviewer can semantically compare the text of a submission against existing literature for plagiarism, and a business analyst can analyze trending topics in research by extracting affiliations and their corresponding contributions from literature of a specific domain. Although some generic NLP services such as *Named Entity Recognition* can be readily offered to Zeeva users through existing third-party libraries, other value-added, research-specific services identified from the elicited user requirements will be designed, developed and intrinsically evaluated for integration in the Zeeva system during the third phase. Finally, several user studies will be conducted with representative samples of each target user group to measure both the time needed to analyze a paper with and without the Zeeva text mining capabilities, as well as the quality of the generated results.

3   Background

In this section we provide a brief introduction of the tools and techniques used in the Zeeva prototype system, namely, the natural language processing domain and a framework for remote NLP service execution.

3.1   Natural Language Processing

Natural Language Processing (NLP) is an active domain of research that uses various techniques from the Artificial Intelligence and Computational Linguistics areas to process text written in a natural language. One popular technique from the NLP domain is *text mining* which aims at extracting high-quality structured information from free form text and representing them in a (semi-)structured format based on specific heuristics. As an essential part of the literature curation process, text mining techniques prove to be effective in terms of the time needed to extract and formalize the knowledge contained within a document [3]. Motivated by similar goals, the Zeeva system aims to provide its users with a unified access point to a variety of text mining techniques that can help them with reading and analyzing scholarly publications.

3.2   Semantic Assistants

As the use of NLP techniques in software applications is being gradually adopted by developers, various tools have emerged to allow developers to include NLP capabilities in their applications through using third-party tools, such as OpenCalais,[1] without the need for a concrete knowledge of the underlying language processing techniques. In addition, a number of NLP frameworks, such as GATE [4], have also emerged to allow linguists to easily develop both generic and sophisticated language processing pipelines without a profound knowledge in software development. However, a seamless integration of these NLP techniques within external applications is still a major hindering issue that is addressed by the Semantic Assistants project [5]. The core idea behind the Semantic Assistants framework is to create a wrapper around NLP pipelines and publish them as W3C[2] standard Web services, thereby allowing a vast range of software clients to consume them within their context. Since different tasks in semantic literature support will need diverse analysis services, the service-oriented architecture of the Semantic Assistants framework facilitates our experiments by allowing us to easily add or remove services without the need to modify our system's concrete implementation.

4   Zeeva Wiki

One major user group of the Zeeva system are knowledge workers aiming at extracting recent trends, advancements, claims and contributions from available publications in a domain of interest. Researchers and curators usually deal with a large amount of information available on multiple online repositories and are in need of a centralized approach to manage and analyze them. One of the prominent characteristics of Zeeva's system design is to not only provide a collaborative environment for scholars to store scientific literature, but also to aid them in the literature analysis process using state-of-the-art techniques from the NLP and Semantic Computing domains.

Previously, in the Wiki-NLP [6] project, we investigated the feasibility and impact of integrating NLP capabilities within a wiki environment. The Zeeva prototype has been developed based on a wiki platform and makes extensive use of the Wiki-NLP integration in its fundamental core. The Zeeva wiki uses Medi-

---

[1] OpenCalais, http://www.opencalais.com
[2] World Wide Web Consortium, http://www.w3.org

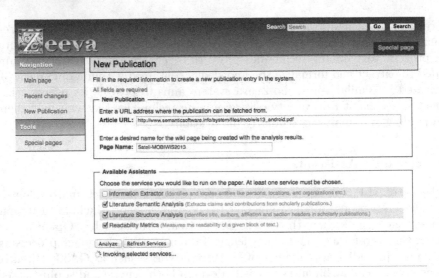

Fig. 1. Analyzing a sample publications with multiple assistants

aWiki[3] as its front-end and expands its core functionality through the installation of the *Semantic MediaWiki*[4] (SMW) extension. The underlying MediaWiki engine provides the basic wiki functionalities, such as user management and content revisioning, while the SMW extension allows us to augment the wiki content with so-called *semantic markup* and later query them directly within wiki pages.

The three aforementioned components, namely, the Zeeva Wiki, the Wiki-NLP integration and the Semantic Assistants server communicate with each other over the HTTP protocol. Users interact directly with the Zeeva wiki interface to create publication entries and verify the NLP results extracted from each document. In order for the NLP pipelines to have access to the text of articles, Zeeva provides users with a *special* page in the wiki, shown in Fig. 1, through which users can provide a URL and a desired name for the paper to be analyzed, as well as selecting one or multiple NLP *assistants* for the analysis task. Provided that the Wiki-NLP integration has adequate permissions to retrieve the article (e.g., from an open access repository or through an institutional license), the article is then passed on to all of the NLP pipelines chosen by the user. Each successful NLP service execution on the wiki generates metadata in form of SMW markup that is made persistent in the wiki database as semantic triples, hence, enriching the paper with a formal representation of the automatically extracted knowledge. Once all the pipelines are executed, the user is automatically redirected to the newly created page with the analysis results, transformed into user-friendly wiki templates like lists or graphs. Fig. 2 shows a list of claims and contributions extracted by Zeeva text mining pipelines from a sample paper. As the Zeeva's underlying wiki engine revisions all the changes to the wiki pages, users can review the services' output and modify the content in case of erroneous results.

---

[3] MediaWiki, http://www.mediawiki.org
[4] Semantic MediaWiki, http://semantic-mediawiki.org

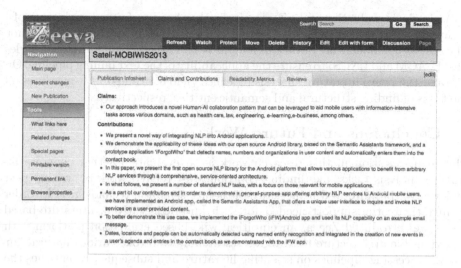

Fig. 2. Extracted semantic metadata from a sample publication

4.1   Literature Metadata Extraction

The integrated text mining capabilities of the Zeeva system is what distinguishes it from other literature management systems. In particular, Zeeva's literature analysis pipelines can extract two types of entities:

- *Structural Entities* refer to parts of the text that uniquely identify a paper, e.g., title or authors, as well as the parts that represent the structure of the paper, such as the abstract, section headers and references. Although, identification of such entities is a relatively easy task for human curators that can be achieved with a fair amount of effort, they stand essential as a leverage for extraction and disambiguation of the semantic entities.
- *Semantic Entities* refer to parts of the text that describe the contributions, claims, findings and conclusions postulated by the paper's authors. This is the most time-consuming task during the analysis process, as it requires the curators to manually read through the full text of each article.

Zeeva's literature structural and semantic analysis pipelines are developed based on the General Architecture for Text Engineering (GATE) [4] and uses the ANNIE plug-in to pre-process the text and detect named entities, such as persons and organizations. Zeeva transducers then look for specific sequences of person, location and organization named entities to generate structural entities, e.g., affiliations. Subsequently, the Zeeva pipelines try to extract semantic entities, in particular, claims and contributions from the paper by finding specific sequences of word tokens based on three gazetteer lists: *(i)* a segmentation trigger list that looks for variation of tokens such as "*Results*" and "*Discussion*", *(ii)* a list of tokens to extract authors' contributions such as "*our framework*" or "*the proposed approach*", and *(iii)* a list of claim triggers to extract sentences with comparative voice, e.g., "Our approach is faster than X", or claims of novelty.

An intrinsic evaluation of the Zeeva pipeline has been performed on a corpus of 5 manually-annotated open access Computer Science papers written by the author. The corpus documents are of various lengths (4 to 12 pages) and formatting styles (IEEE, ACM and LNCS) to ensure the applicability of the text mining pipeline. On average, the Zeeva literature analysis pipelines yield 0.85 and 0.77 F-measures on the tasks of finding structural and semantic entities, respectively.

5   Conclusions and Future Work

The main question in this research work is to evaluate whether concrete literature analysis tasks, like finding contributions, related work or writing peer reviews can be improved using state-of-the-art semantic technologies. In order to investigate the support needed for such knowledge-driven, literature-based tasks, we introduced Zeeva, an empirical wiki-based evaluation platform with an extensible architecture that allows researchers to invoke various natural language processing pipelines on scientific literature and subsequently, enriches the documents with semantic metadata for further human and machine processing. As an example, we developed two literature analysis pipelines that can automatically extract structural and semantic entities from a given full-text paper and generate semantic metadata from the results. Throughout our research, we are planning to identify literature analysis tasks that can be improved with semantic technologies and develop more NLP services relevant to the context of literature analysis. We are also planning to perform an extrinsic evaluation of our hypothesis using the Zeeva platform to assess the usability and efficiency of the proposed approach.

References

1. Zeni, N., Kiyavitskaya, N., Mich, L., Mylopoulos, J., Cordy, J.R.: A lightweight approach to semantic annotation of research papers. In: Kedad, Z., Lammari, N., Métais, E., Meziane, F., Rezgui, Y. (eds.) NLDB 2007. LNCS, vol. 4592, pp. 61–72. Springer, Heidelberg (2007)
2. Yang, Y., Akers, L., Klose, T., Yang, C.B.: Text mining and visualization tools Impressions of emerging capabilities. World Patent Information 30(4), 280–293 (2008)
3. Sateli, B., Meurs, M.J., Butler, G., Powlowski, J., Tsang, A., Witte, R.: IntelliGenWiki: An Intelligent Semantic Wiki for Life Sciences. NETTAB 2012 8 (Suppl. B)., Como, Italy, 50–52. EMBnet.journal (2012)
4. Cunningham, H., Maynard, D., Bontcheva, K., Tablan, V., Aswani, N., Roberts, I., Gorrell, G., Funk, A., Roberts, A., Damljanovic, D., Heitz, T., Greenwood, M.A., Saggion, H., Petrak, J., Li, Y., Peters, W.: Text Processing with GATE (Version 6). University of Sheffield, Department of Computer Science (2011)
5. Witte, R., Gitzinger, T.: Semantic Assistants – User-Centric Natural Language Processing Services for Desktop Clients. In: Domingue, J., Anutariya, C. (eds.) ASWC 2008. LNCS, vol. 5367, pp. 360–374. Springer, Heidelberg (2008)
6. Sateli, B., Witte, R.: Natural Language Processing for MediaWiki: The Semantic Assistants Approach. In: The 8th International Symposium on Wikis and Open Collaboration (WikiSym 2012), Linz, Austria. ACM (2012)

Author Index